Topics in Heterocyclic Chemistry

Volume 63

The series Topics in Heterocyclic Chemistry presents critical reviews on present and future trends in the research of heterocyclic compounds. Overall the scope is to cover topics dealing with all areas within heterocyclic chemistry, both experimental and theoretical, of interest to the general heterocyclic chemistry community. The series consists of topic related volumes edited by renowned editors with contributions of experts in the field. All chapters from Topics in Heterocyclic Chemistry are published OnlineFirst with an individual DOI. In references, Topics in Heterocyclic Chemistry is abbreviated as Top Heterocycl Chem and cited as a journal.

Ross David Jansen-van Vuuren • Janez Košmrlj
Editors

Modern Advances in Click Chemistry

Synthesis of Compounds

With contributions by

S. M. Arteta · L. D. Perez · S. Payamifar · A. Poursattar Marjani · P. Milbeo · C. Zanato · S. Rajkhowa · M. S. Yadav · V. K. Tiwari · A. Sharma · F. Campana · G. Brufani · L. Vaccaro · D. Lanari · F. C. Zacconi · R. D. Jansen-van Vuuren

Springer

Editors
Ross David Jansen-van Vuuren
Faculty of Chemistry and Chemical Technology
University of Ljubljana
Ljubljana, Slovenia

Janez Košmrlj
Faculty of Chemistry and Chemical Technology
University of Ljubljana
Ljubljana, Slovenia

ISSN 1861-9282 ISSN 1861-9290 (electronic)
Topics in Heterocyclic Chemistry
ISBN 978-3-032-24579-3 ISBN 978-3-032-24580-9 (eBook)
https://doi.org/10.1007/978-3-032-24580-9

This Springer imprint is published by the registered company Springer Nature Switzerland AG
The registered company address is: Gewerbestrasse 11, 6330 Cham, Switzerland

Preface

The copper-catalyzed azide–alkyne cycloaddition (CuAAC) reaction developed by Meldal and Sharpless stimulated a surge of innovative protocols, catalyst systems, and applications within click chemistry. In this volume, the first of a two-volume set, with the second dedicated to biological applications of click chemistry, we present an updated and comprehensive overview of recent advances in the field. This edition revisits key topics covered in the earlier volume (Košmrlj, J., Ed., 2012. Click Triazoles. Topics in Heterocyclic Chemistry, Vol. 28. Springer Berlin Heidelberg) while highlighting new methodological developments and emerging synthetic applications of click chemistry.

In the Chapter "Click Chemistry in the Synthesis of Functional Polymers," Professor Leon Perez reviews recent advances in the application of click chemistry to the design and synthesis of functional and smart polymers for contemporary technological applications. In the Chapter "β-Cyclodextrin-Based Catalyst Systems in Click Reactions," Professor Ahmad Marjani provides an overview of newly developed green and environmentally benign β-cyclodextrin-based catalytic systems for click transformations. The chapter "[2+2] Cycloaddition-Retroelectrocyclization (CA- RE)Click Chemistry," authored by Professor Tsuyoshi Michinobu, focuses on a branch of click chemistry developed in the 1980s, but experiencing renewed interest due to shifting research priorities toward new functional materials, sustainability, and modular synthetic strategies – areas to which this methodology is particularly well suited.

In the chapter "The Thiol-ene "Click" Reaction for Synthesis of Biomolecular Thioether and Thioester Derivatives," Associate Professors Pierre Milbeo and Chiara Zanato discuss the thiol-ene "click" reaction, which has likewise seen resurgence owing to its mild and frequently photoinitiated conditions, orthogonality to a broad range of functional groups, and expanding utility in advanced materials, polymer modification, bioconjugation, and additive manufacturing. The chapter "Growing Impact of "MCR/Tandem Click Chemistry": A Vital Tool to Access Biologically Privileged Functionalized 1,2,3-Triazole Analogues" features Professor Vinod K. Tiwari and co-authors, who introduce multicomponent click reactions that enable the efficient synthesis of diverse libraries of biologically privileged functionalized analogues.

In the Chapter "Bio-click Chemistry," Professor Flavia Zacconi examines the development and challenges associated with integrating biocatalysis into click chemistry transformations. This is followed by the Chapter "The Sustainability Aspects of Click-Chemistry Transformations," in which Professor Luigi Vaccaro outlines greener and more sustainable strategies for click chemistry, approaches that are critical in addressing global sustainability initiatives such as the United Nations Sustainable Development Goals, the Stockholm Declaration, and the European Green Deal.

Finally, in the Chapter "Click Immobilization of(Bio)Catalysts in Batch and Continuous Flow Chemistry Systems" Dr. Ross Jansen-van Vuuren and co-authors explore the application of click chemistry in the immobilization of (bio)catalysts onto surfaces, polymers, and other materials for both batch and continuous-flow processes. These strategies enable systems in which catalyst recovery, recyclability, and process sustainability are central considerations.

Ljubljana, Slovenia — Ross David Jansen-van Vuuren
Ljubljana, Slovenia — Janez Košmrlj

Contents

Click Chemistry in the Synthesis of Functional Polymers 1
Sandra M. Arteta and Leon D. Perez

β-Cyclodextrin-Based Catalytic Systems in Click Reactions 37
Sara Payamifar and Ahmad Poursattar Marjani

[2 + 2] Cycloaddition-Retroelectrocyclization (CA-RE) Click Chemistry 69
Xu Chen and Tsuyoshi Michinobu

The Thiol-ene "Click" Reaction for Synthesis of Biomolecular Thioether and Thioester Derivatives 123
Pierre Milbeo and Chiara Zanato

"MCR/Tandem Click Chemistry": A Vital Tool to Access Biologically Privileged Functionalized 1,2,3-Triazole Analogues 161
Sanchayita Rajkhowa, Mangal S. Yadav, Anindra Sharma, and Vinod K. Tiwari

Bio-click Chemistry 205
Diego F. Rodríguez and Flavia C. Zacconi

The Sustainability Aspects of Click-Chemistry Transformations 225
Filippo Campana, Giulia Brufani, Daniela Lanari, and Luigi Vaccaro

Click Immobilization of (Bio)Catalysts in Batch and Continuous Flow Chemistry Systems 257
Ross D. Jansen-van Vuuren, L. Jedlovčnik, and J. Cerkovnik

Top Heterocycl Chem (2026) 63: 1–36
https://doi.org/10.1007/7081_2025_99

Published online: 30 December 2025

Click Chemistry in the Synthesis of Functional Polymers

Sandra M. Arteta and Leon D. Perez

Contents

1 Introduction 3
2 Azide-Alkyne Cycloadditions in Polymer Science 4
  2.1 Introduction to Azide-Alkyne Cycloaddition 4
  2.2 Strain-Promoted Azide-Alkyne Cycloaddition 6
3 Approaches to Polymer Synthesis 7
  3.1 Step-Growth Polymerization via Azide-Alkyne Cycloaddition 8
  3.2 Block Copolymerization 10
4 Applications of Click Reactions in the Production of Functional Polymers 14
  4.1 Biomedical Polymers 14
  4.2 Proton- and Electron-Conducting Polymers 24
5 Conclusion 29
References 29

Abstract Click chemistry has revolutionized polymer science by providing efficient, orthogonal, and highly selective strategies for the synthesis and functionalization of macromolecules. Among these, copper(I)-catalyzed azide-alkyne cycloaddition (CuAAC) and strain-promoted azide-alkyne cycloaddition (SPAAC) stand out as powerful tools for constructing complex polymer architectures under mild conditions. These transformations have enabled unprecedented control over polymer composition, topology, and functionality, supporting the development of advanced materials for biomedical, electronic, and energy-related applications. This chapter offers a comprehensive overview of azide-alkyne cycloadditions in polymer science, emphasizing their applications in step-growth polymerization, block copolymer synthesis, and grafting techniques. Furthermore, we explore how CuAAC and SPAAC have facilitated the creation of smart materials,

S. M. Arteta and L. D. Perez (✉)
Department of Chemistry, Universidad Nacional de Colombia – Sede Bogotá, Bogotá, DC, Colombia
e-mail: ldperezp@unal.edu.co

including polymeric prodrugs, gene delivery vectors, hydrogels, and proton- and electron-conductive membranes.

Abbreviations

6MP	6-Mercaptopurine
AAC	Azide-alkyne cycloaddition
ABMA	4-Azidobenzyl methacrylate
ATRP	Atom transfer radical polymerization
BCN	Bicyclo[6.1.0]nonyne
BPY	2,2′-Bipyridine
β-CD	Beta-cyclodextrin
CL	ε-Caprolactone
CuAAC	Copper(I)-catalyzed azide-alkyne cycloaddition
DBCO	Dibenzocyclooctyne
DEAEMA	2-(Diethylamino)ethyl methacrylate
DIBOD	sym-Dibenzo-1,5-cyclooctadiene-3,7-diyne – a diyne
DIPBA	2,6-Diisopropylphenyl bisazide
DIPEA	*N*,*N*-diisopropylethylamine
DMF	Dimethylformamide
DOX	Doxorubicin
DSPAAC	Double-strain-promoted azide-alkyne cycloaddition
IC_{50}	Half-maximal inhibitory concentration
MCP	5-Methyl-5-propargyloxycarbonyl-1,3-dioxane-2-one
mPEG	Methoxy poly(ethylene glycol)
NMP	Nitroxide-mediated polymerization
NMR	Nuclear magnetic resonance
OEGMA	Oligo(ethylene glycol) methyl ether methacrylate
PCL	Poly(ε-caprolactone)
PDMAEMA	Poly(2-(dimethylamino)ethyl methacrylate)
PDPAO	Poly(DEAEMA)-b-poly(ABMA-co-OEGMA)
PEG	Poly(ethylene glycol)
PEI	Polyethylenimine
PGMA	Poly(glycidyl methacrylate)
PI	Polyisoprene
PLA	Poly(L-lactide)
PLL	Poly(L-lysine)
PMDETA	Pentamethyldiethylenetriamine
Pox	Poly(2-oxazoline)
PP	Polypropylene
PPLG	Poly(γ-propargyl-L-glutamate)
PS	Polystyrene
PVDF	Poly(vinylidene fluoride)

RAFT	Reversible addition–fragmentation chain transfer
ROP	Ring-opening polymerization
RuAAC	Ruthenium-catalyzed azide-alkyne cycloaddition
SEC	Size-exclusion chromatography
SPAAC	Strain-promoted azide-alkyne cycloaddition
THC	Δ^9-Tetrahydrocannabinol
THF	Tetrahydrofuran

1 Introduction

A functional polymer is a macromolecule engineered to exhibit targeted physicochemical or biological activities beyond simply structural support. These materials respond to external stimuli, interact selectively with biological systems, or engage in energy transfer, enabling precise control over their performance [1–3]. The integration of active molecular components into polymer chains has enabled the development of versatile systems for applications ranging from drug delivery and regenerative medicine to optoelectronics and smart coatings [4–8].

Click chemistry has become a foundational tool in modern polymer science, offering orthogonal, chemoselective, and high-yielding reactions for precise modification of macromolecular architectures [9–12]. Among these, copper(I)-catalyzed azide-alkyne cycloaddition (CuAAC) is a benchmark reaction for constructing well-defined functional polymers, enabling efficient conjugation with minimal byproducts and a broad substrate scope [9, 13]. Complementing CuAAC, strain-promoted azide-alkyne cycloaddition (SPAAC) allows catalyst-free click reactions under physiological conditions, significantly advancing the design of bioorthogonal systems for applications in drug delivery, tissue engineering, and nanomedicine [14]. Together, these methodologies have revolutionized the synthesis of functional polymeric materials across various disciplines.

This chapter explores some of the most relevant applications of azide-alkyne cycloadditions in the synthesis of novel functional polymer structures, with a particular focus on the CuAAC and SPAAC click reactions. It begins with an overview of click chemistry, including its mechanisms and key characteristics. Subsequently, the chapter discusses its role in step-growth polymerization and copolymerization for constructing polymer backbones and examines its application in grafting backbones with polymer chains and small molecules to create branched architectures. Finally, representative examples are presented to illustrate the versatility of click chemistry in generating functional polymers in various fields.

2 Azide-Alkyne Cycloadditions in Polymer Science

2.1 Introduction to Azide-Alkyne Cycloaddition

Azide-alkyne cycloaddition (AAC) is one of the most versatile and widely utilized reactions in click chemistry, providing an efficient, selective, and modular strategy for molecular construction. Click chemistry highlights high-yielding, stereospecific, and broad scope reactions that proceed under mild conditions. These reactions are characterized by high reliability, strong thermodynamic driving forces, and minimal byproduct formation [9]. Among the AAC reactions, CuAAC has emerged as the gold standard because of its excellent regioselectivity for 1,4-disubstituted 1,2,3-triazoles, compatibility with aqueous media, and rapid reactions [15]. Conversely, ruthenium-catalyzed azide-alkyne cycloaddition (RuAAC) enables the regioselective formation of 1,5-disubstituted triazoles involving both terminal and internal alkynes, thus broadening the scope of accessible triazole architectures (Scheme 1) [16, 17]. In comparison, the original Huisgen thermal azide-alkyne cycloaddition proceeds without a metal catalyst but requires elevated temperatures and extended reaction times, yielding a mixture of 1,4- and 1,5-triazole regioisomers [18, 19]. This chapter emphasizes CuAAC and SPAAC owing to their broader applicability, lower cost, and greater relevance in functional polymer synthesis and bioconjugation.

2.1.1 CuAAC

CuAAC is a highly efficient and regioselective method for synthesizing 1,4-disubstituted 1,2,3-triazoles. Catalytically active Cu(I) species can be generated by reducing Cu(II) with reducing agents such as sodium ascorbate [13]. The CuAAC catalytic cycle (Scheme 2) is initiated by the formation of a copper(I) acetylide complex (**I**) via coordination between Cu(I) and the terminal alkyne, facilitating its activation and determining the regioselectivity of the reaction [21]. This step is followed by the association of the azide, yielding a copper-azide-acetylide complex (**II**), wherein cooperative stabilization by the copper centers enhances the reactivity.

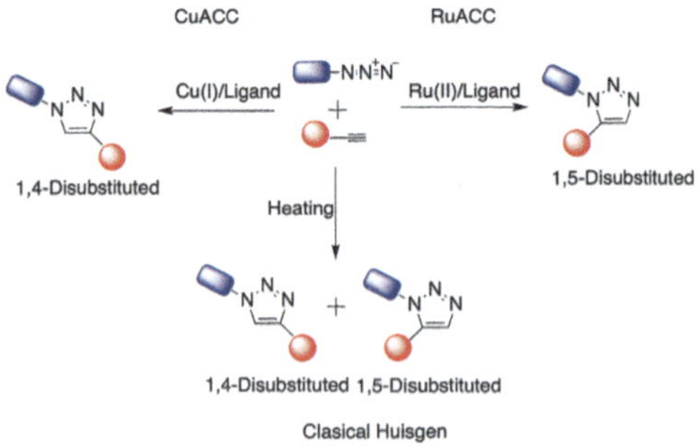

Scheme 1 Comparison of azide-alkyne cycloadditions: copper-catalyzed (CuAAC), ruthenium-catalyzed (RuAAC), and thermal (Huisgen) [20]

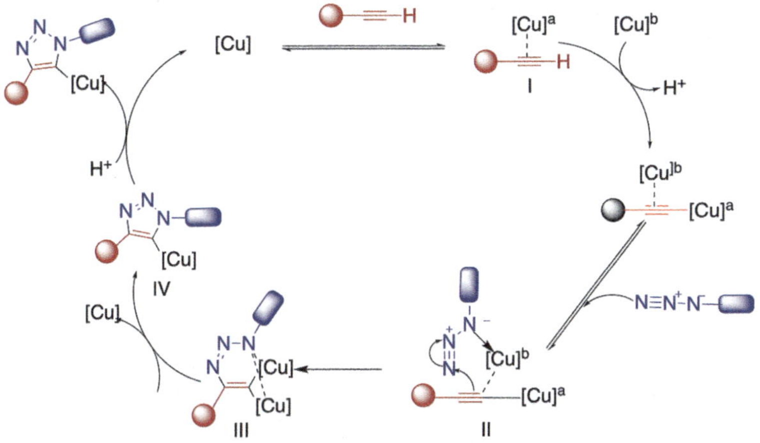

Scheme 2 Proposed CuAAC catalytic cycle featuring a dinuclear copper intermediate involved in 1,4-disubstituted triazole formation [20, 22, 23]. Superscripts **a** and **b** label two Cu centers that interact differently with the acetylene moiety

Subsequently, a cycloaddition event occurs, leading to the formation of a metallacycle intermediate (**III**), which is characterized by the concerted interaction of both copper centers in a bimetallic environment. This intermediate undergoes rearrangement and ring contraction, culminating in the formation of the 1,4-disubstituted triazole product (**IV**). In the final stage of the catalytic cycle, the triazole-Cu (**IV**) complex undergoes protonation, mediated by either a base or a solvent, ultimately affording the desired 1,4-disubstituted 1,2,3-triazole. Concurrently, the copper complex dissociates, restoring the active Cu(I) catalyst and perpetuating the catalytic cycle [20, 22, 23].

In CuAAC reactions, the ligand complexing the Cu(I) catalyst plays a crucial role in modulating its reactivity and stability. Ligands prevent the oxidation of Cu(I) to Cu(II) by forming stable complexes, also improving the solubility and efficiency of the catalyst [24]. Nitrogen-containing ligands, ranging from small amines to polydentate macromolecules, significantly influence the reaction rates, some of which are commercially available, as shown in Scheme 3. It has been found that aliphatic amines accelerate CuAAC reactions more effectively than pyridine-based ligands. This enhancement is attributed to their superior ability to stabilize the Cu-acetylide intermediates, thereby facilitating the cycloaddition step. Moreover, tridentate amine ligands have been found to outperform tetradentate ligands because excessive coordination can hinder substrate binding and reduces the catalytic turnover [25]. Similarly, imidazole derivatives with flexible alkyl chains have been reported to enhance reaction rates, even for sterically hindered alkynes, by optimizing steric effects and ligand flexibility [26].

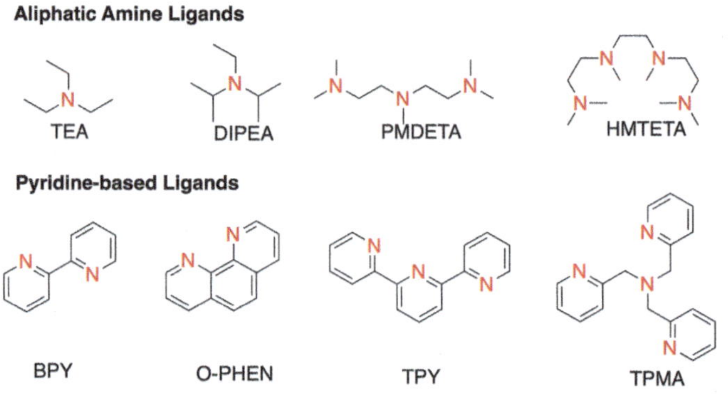

Scheme 3 Structures of representative commercial ligands for CuAAC, including amines and polypyridyl ligands. *TEA* triethylamine, *DIPEA* *N,N*-diisopropylethylamine, *PMDETA* pentamethyldiethylenetriamine, *HMTETA* hexamethyltriethylenetetramine, *TPY* 2,2′:6′,2′-′-terpyridine, *BPY* 2,2′-bipyridine, *O-PHEN* phenanthroline, *TPMA* tris(2-pyridylmethyl)amine

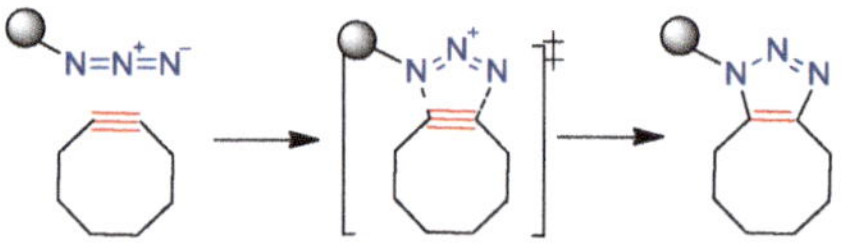

Scheme 4 General mechanism of SPAAC, illustrating strain-induced activation and triazole formation [27]

2.2 Strain-Promoted Azide-Alkyne Cycloaddition

SPAAC is a bioorthogonal reaction widely employed in bioconjugation, live-cell labeling, and in vivo imaging. Introduced by Bertozzi, SPAAC serves as a metal-free alternative to CuAAC, eliminating cytotoxic Cu(I) catalysts, making it particularly advantageous for biomedical applications [14].

The reaction mechanism relies on ring strain activation, where cyclic alkynes such as dibenzocyclooctyne (DBCO), azacyclooctynes, and bicyclononyne exhibit significant strain due to deviations from the ideal *sp*-hybridized linear geometry of alkynes. This strain results in higher-energy π-orbitals, making the alkyne bond more reactive toward azides in a [3+2] cycloaddition reaction (Scheme 4). The SPAAC reaction can follow a concerted mechanism that is either synchronous or asynchronous, depending on the nature of the reagents. The final step yields a stable 1,2,3-triazole product, effectively releasing ring strain and stabilizing the conjugate. This reaction occurs without the need for heat, metal catalysts, or additional reagents, making it highly favorable for bioconjugation under physiological conditions [10, 27, 28]. Figure 1 shows some reported cyclic structures.

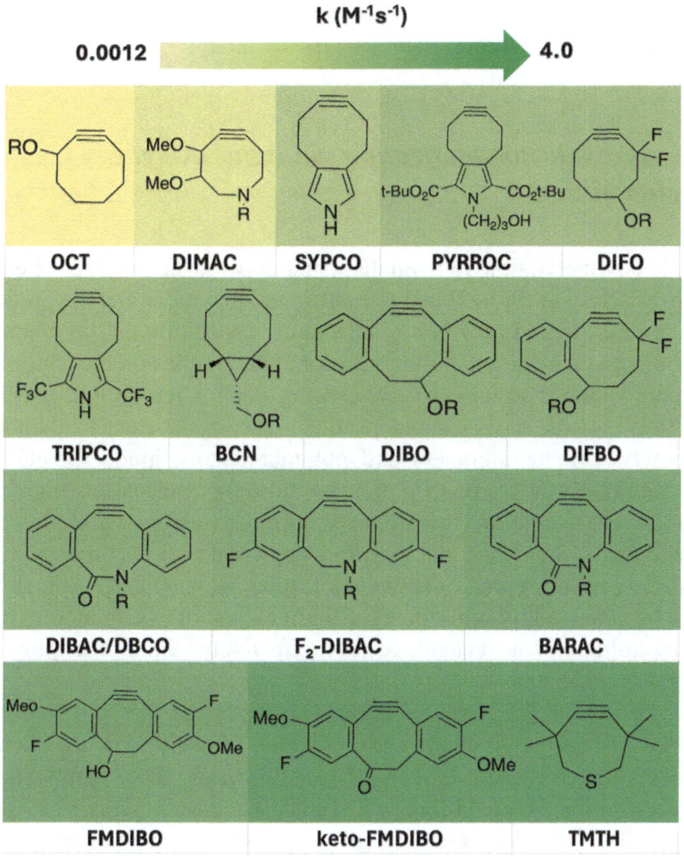

Fig. 1 Chemical structures of representative cyclooctyne derivatives, listed in order of reactivity toward azides (modified from Li et al. [28]). The ordering is based on second-order rate constants (k_2). *OCT* cyclooctyne, *DIMAC* aza-dimethoxycyclooctyne, *SYPCO* symmetrical pyrrolocyclooctyne, *PYRROC* pyrrolocyclooctyne, *DIFO* difluorocyclooctyne, *TRIPCO* α,α-bis (trifluoromethyl)pyrrolocyclooctyne, *BCN* bicyclo[6.1.0]nonyne, *DIBO* dibenzocyclooctyne, *DIFBO* difluorobenzocyclooctyne, *DIBAC (DBCO)* dibenzoazacyclooctyne, *F_2-DIBAC* difluoro-aza-dibenzocyclooctyne, *BARAC* biaryl-azacyclooctynone, *FMDIBO* difluoro-dimethoxy-dibenzocyclooctynol, *keto-FMDIBO* difluoro-dimethoxy-dibenzocyclooctynone, *TMTH* 3,3,6,6-tetramethylthiacycloheptyne. Further details on these compounds can be found in Refs. [29–41]

3 Approaches to Polymer Synthesis

AAC reactions serve as versatile tools for the synthesis of functional polymers using multiple strategies. These include the polymerization of monomers bearing azide and alkyne functionalities, post-functionalization of preformed polymers, and design of copolymers and cross-linked networks. Leveraging the efficiency, regioselectivity, and mild reaction conditions of CuAAC and SPAAC, advanced polymeric materials with tailored properties can be fabricated. The following section explores the diverse applications of click chemistry in polymer synthesis, modification, and structural engineering.

3.1 Step-Growth Polymerization via Azide-Alkyne Cycloaddition

Step-growth polymerization is a fundamental process in polymer chemistry that involves the reaction of bifunctional or multifunctional monomers to produce linear, branched, or cross-linked polymer architectures, depending on the monomer functionality. Unlike chain-growth polymerization, which relies on the continuous addition of monomer units to an actively growing chain, step-growth polymerization proceeds through successive condensation reactions between monomers and between newly formed oligomers and polymer chains, gradually increasing the molecular weight. In step-growth polymerization, the molecular weight is directly dependent on the extent of functional group conversion, with high molecular weights achieved only when the reaction approaches full conversion (~100%). Additionally, precise control over the molecular weight and terminal functional groups can be achieved by introducing a slight excess of one monomer or incorporating a monofunctional reagent, which effectively limits the polymer chain growth [42].

The incorporation of AAC reactions into step-growth polymerization has emerged as a highly efficient and versatile strategy for the synthesis of functionalized, tunable, and biocompatible polymers. The significance of this approach lies in its compatibility with click chemistry principles, particularly CuAAC and SPAAC, which enable high-yield and selective polymerization under mild conditions [43, 44]. In addition, these reactions are compatible with aqueous and organic media, and the resulting polymer contains reactive azide or alkyne groups, allowing their posterior functionalization [9].

In step-growth polymerization based on AAC, polymer formation can occur through the reaction of two bifunctional monomers (AA and BB types), such as diazide and dialkyne, or via AB-type monomers, where a single molecule contains both azide and alkyne functional groups (e.g., α-azide-ω-alkyne compounds), as shown in Scheme 5. However, the AA + BB approach presents significant challenges, primarily because of the stringent requirements of high monomer purity and precise stoichiometric balance. Failure to achieve these conditions can result in low-molecular-weight polymers, making the AB monomer strategy a more efficient and controlled alternative for synthesizing well-defined polytriazole-based materials.

This approach was exemplified by Binauld et al. (2008), who successfully employed CuAAC click chemistry to copolymerize α-azide-ω-alkyne monomers (Scheme 6), demonstrating the ability of click chemistry to construct well-defined polytriazoles with tunable architectures and properties [45]. Polymerization was conducted in chloroform (0.2 M monomer concentration) using $Cu(PPh_3)_3Br$ as the catalyst and DIPEA the base. ^1H NMR spectroscopy confirmed successful

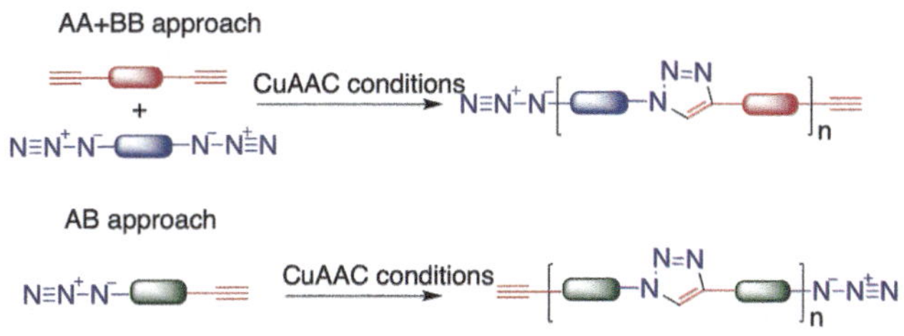

Scheme 5 Step-growth polymerization via CuAAC using AA + BB and AB-type monomers

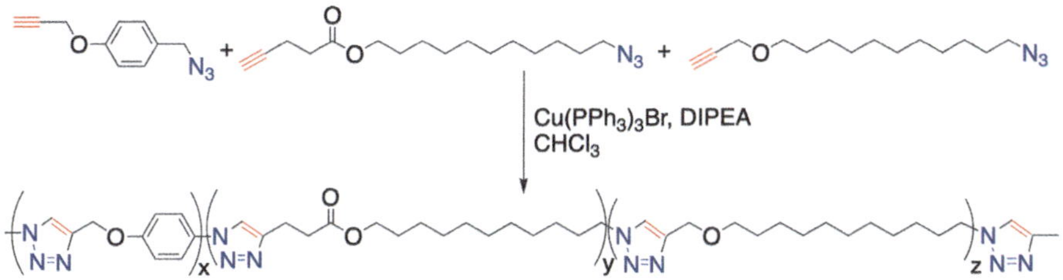

Scheme 6 CuAAC step-growth copolymerization of α-azide-ω-alkyne monomers, as reported in Ref. [45]

polymerization through the disappearance of the alkyne signal and the appearance of the triazole-specific signal. Additionally, size-exclusion chromatography (SEC) analysis showed the absence of residual monomers, indicating a high conversion rate and demonstrating the effectiveness of CuAAC-mediated step-growth polymerization for the synthesis of highly stable polytriazoles.

Owing to the versatility of AAC reactions, SPAAC has been effectively utilized for polymer synthesis via step-growth mechanisms. McNelles et al. (2019) reported the efficient fabrication of high-molecular-weight polytriazoles by reacting a bifunctional strained dibenzoazacyclooctyne (DBCO) monomer with various bifunctional azide-containing comonomers (Scheme 7) [43]. This catalyst-free polymerization demonstrated remarkable efficiency, reaching high conversion within minutes and yielding polymers with molecular weights ranging from 27 to 52 kDa, depending on the diazide compound, and dispersity higher than 1.5, as confirmed by SEC and ^{1}H NMR spectroscopy.

To expand the scope of AAC reactions, double-strain-promoted azide-alkyne cycloaddition (DSPAAC) has emerged as a powerful strategy for synthesizing ultrahigh-molecular-weight polymers through step-growth polymerization. Zhang et al. (2019) developed an efficient DSPAAC-based polymerization method utilizing sym-dibenzo-1,5-cyclooctadiene-3,7-diyne (DIBOD) as the bifunctional alkyne monomer and 2,6-diisopropylphenyl bisazide (DIPBA) as the comonomer (Scheme 8). This catalyst-free polymerization proceeded rapidly under ambient conditions, yielding polymers with molecular weights exceeding 10^6 g/mol [46]. The process leverages a reactive intermediate mechanism, in which the initial azide-alkyne

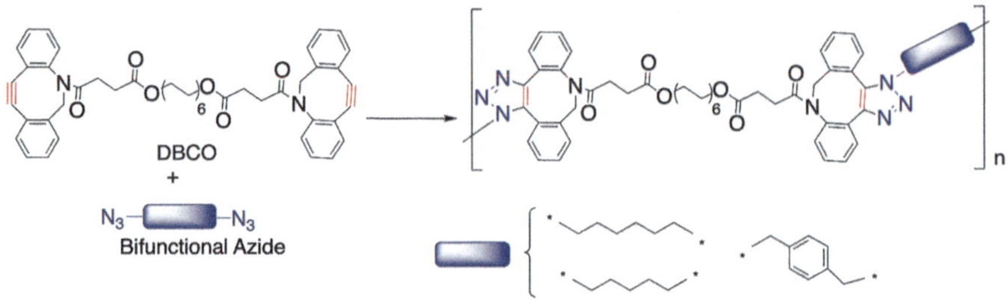

Scheme 7 SPAAC step-growth polymerization reported in Ref. [43] for an AA + BB monomer system

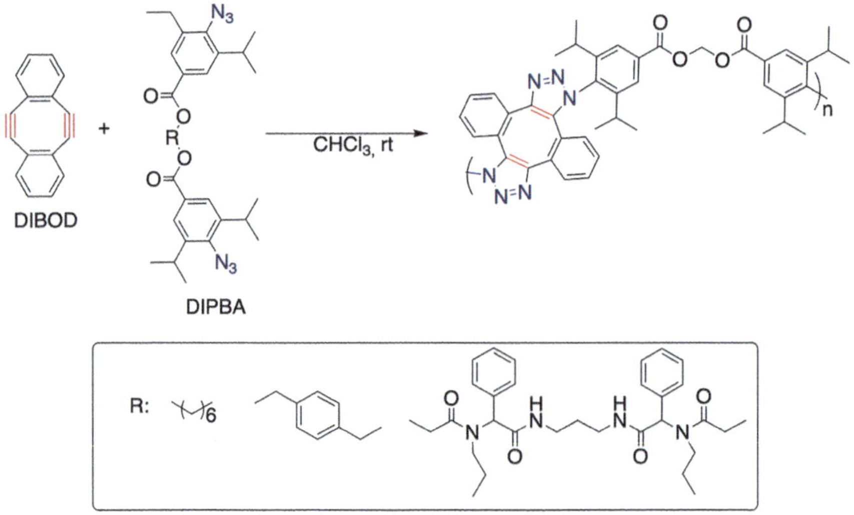

Scheme 8 Reaction model for DSPAAC between DIBOD and 2,6-diisopropylphenyl azide group, as reported in [46]

reaction increases the ring strain in the DIBOD unit, accelerating the second cycloaddition step and facilitating high-conversion polymerization.

3.2 Block Copolymerization

Block copolymers have become materials of paramount importance across various fields because of their ability to integrate two or more distinct polymer blocks connected via covalent bonds. These macromolecules can adopt a diverse range of molecular architectures, from simple linear arrangements to more complex comb-like, starlike, and multi-arm structures. Their structural versatility has enabled their application in nanomedicine, smart materials, and high-performance coatings, making them an essential class of functional polymers [47–49].

Traditionally, block copolymers have been synthesized by combining different polymerization techniques, including anionic polymerization, controlled free radical polymerization, and ring-opening polymerization (ROP), depending on the nature of the monomers involved [50, 51]. Despite the high degree of molecular weight control achievable with these techniques, fine-tuning of the copolymer architecture remains challenging, requiring precise adjustment of the reaction conditions to obtain well-defined structures. Achieving block connectivity with high selectivity and efficiency remains a bottleneck in polymer chemistry.

Click chemistry has emerged as a powerful tool for tailoring block copolymers, offering modularity, orthogonality, and high efficiency for polymer conjugation. This approach allows for the post-polymerization connection of individual pre-synthesized and purified polymer blocks, each carrying terminal azide or alkyne functionalities. This strategy was reported by Vukicevic et al. (2012) for the synthesis of polystyrene (PS)-poly(vinylidene fluoride) (PVDF) copolymers. In their study, styrene was first polymerized via nitroxide-mediated polymerization (NMP) using an initiator containing an alkyne functional group [52]. The resulting alkyne-terminated PS was then reacted with azide-terminated PVDF via CuAAC to form a PS-PVDF block copolymer (Scheme 9a). This reaction, conducted over 7 days, highlights the effectiveness of click chemistry in enabling the synthesis of block copolymers, particularly when dealing with monomers with vastly different reactivities that are challenging to copolymerize directly.

Touris and Hadjichristidis (2011) demonstrated the successful application of CuAAC in the synthesis of multiblock and cyclic structures composed of polystyrene (PS) and polyisoprene (PI) [53]. Their approach involved sequential anionic polymerization to synthesize a PS-b-PI block copolymer incorporating terminal alkyne and azide functional groups at the chain ends. These prefunctionalized polymers were subsequently coupled via CuAAC, leading to the formation of well-defined cyclic and multiblock PS-b-PI copolymers. The resulting polymer architecture was highly influenced by the reaction parameters, including the solvent selection and bifunctional copolymer concentration, as illustrated in Scheme 9b.

3.2.1 Synthesis of Comblike Copolymers

Comblike polymers, distinguished by their densely grafted side chains along a linear backbone, have attracted significant attention because of their potential as biomaterials and nanostructured materials and drug delivery and self-assembled systems [54–56]. Comblike copolymers can be obtained using three different grafting methods. The grafting-onto approach involves attaching preformed polymer side chains to a functionalized backbone via selective reactions, such as AAC click chemistry, offering excellent molecular weight control but facing steric hindrance at high grafting densities [57]. In contrast, the grafting-from method relies on in situ polymerization from an initiator-functionalized backbone, enabling higher grafting densities and more uniform structures. The grafting-through strategy employs the copolymerization of macromonomers with polymerizable groups, allowing the

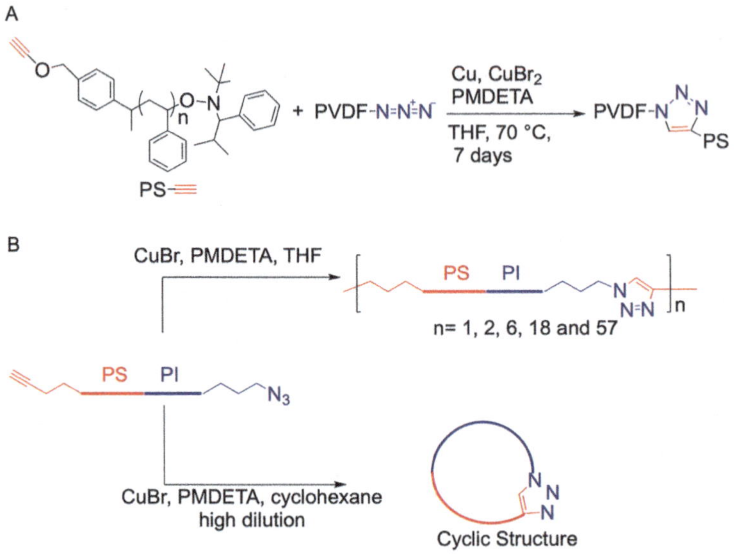

Scheme 9 Examples of block copolymer synthesis via CuAAC. (**a**) Synthesis of PS-b-PVDF via CuAAC, starting from a PS synthesized via NMP employing 2,2,5-trimethyl-3-(10-(p-propargyloxymethyl)phenylethoxy)-4-phenyl-3-azahexane-3-oxyl as initiator and PVDF azido terminated described in [52]. (**b**) Synthesis of multiblock copolymers and cyclic structures starting from PS-b-PI copolymers obtained by anionic polymerization reported in [53]

incorporation of side chains during polymerization and providing fine control over the molecular weight and composition, while maintaining a well-defined polymer architecture [58]. In this chapter, the grafting-onto methodology is reviewed more extensively, as it represents the polymer modification strategy in which click chemistry has had the most profound impact.

Acik et al. (2018) employed CuAAC to synthesize polypropylene-graft-poly(L-lactide) (PP-g-PLA) copolymers by modifying preformed polypropylene [59]. The process involved a two-step functionalization, first converting chlorinated polypropylene into azide-functionalized polypropylene (PP-N_3) via nucleophilic substitution with trimethylsilyl azide in tetrahydrofuran (THF). Separately, alkyne-terminated PLA (PLA-alkyne) was synthesized via the ring-opening polymerization (ROP) of L-lactide using propargyl alcohol as an initiator and $Sn(Oct)_2$ as a catalyst in toluene. The final CuAAC coupling reaction was conducted in toluene at room temperature for 24 h using Cu(I)Br and PMDETA as the ligands. Despite using an excess of PLA-alkyne, steric hindrance and reaction kinetics limited the grafting efficiency to 74%, highlighting the challenges in achieving complete functionalization of the polymer. Similarly, Chu et al. (2017) utilized CuAAC click chemistry to synthesize poly(glycidyl methacrylate)-graft-poly(2-oxazoline) (PGMA-g-POx) polymer brushes, which encountered incomplete grafting due to steric constraints [60]. The grafting densities ranged from 47% to 85%, depending on the molecular weight and composition of the grafted chains.

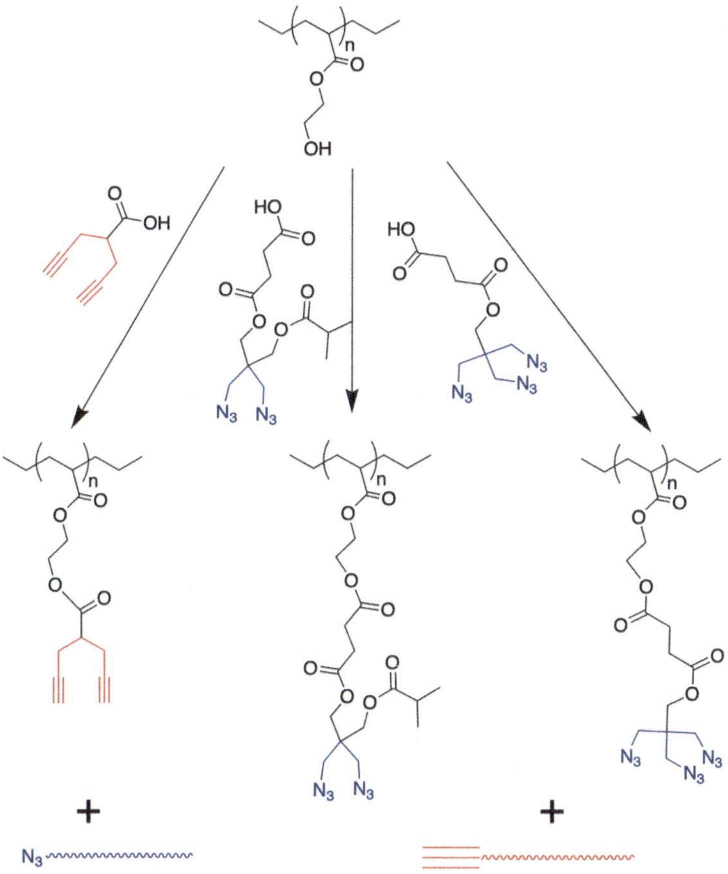

Scheme 10 Synthesis of molecular brushes with ultrahigh side-chain density using accelerated CuAAC [61]

As an alternative to achieve higher grafting efficiency, Gan et al. (2017) used an accelerated CuAAC grafting-onto methodology to obtain ultrahigh grafting density molecular brushes, leveraging the Cu(I)/triazole complexation effect to enhance the reaction kinetics and grafting efficiency (Scheme 10) [61]. A polymethacrylate backbone with multiple azido or alkyne groups served as the reactive sites for attaching alkyne- or azide-terminated polymer side chains, including poly(ethylene oxide), poly(methyl acrylate), and polydimethylacrylamide. The CuAAC reaction, conducted in dimethylformamide (DMF) at room temperature with $CuSO_4{\cdot}5H_2O$, PMDETA, and ascorbic acid, yielded grafting densities of up to 1.34 side chains per backbone carbon, surpassing previous molecular brush synthesis efforts.

Engler et al. (2009) demonstrated an effective strategy to enhance grafting efficiency in the grafting-onto approach by synthesizing poly(γ-propargyl-L-glutamate)-graft-poly(ethylene glycol) (PPLG-g-PEG) copolymers using CuAAC click

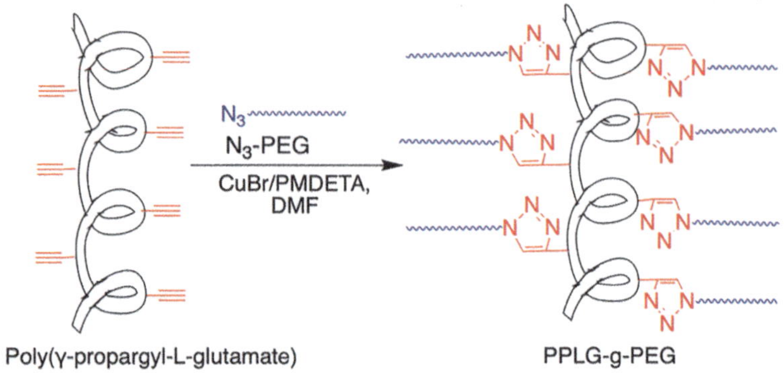

Scheme 11 High-efficiency synthesis of PPLG-g-PEG via CuAAC using a helical PPLG backbone to reduce steric hindrance [62]

chemistry (Scheme 11) [62]. The key to achieving exceptionally high grafting efficiencies (96.3–98.9%) was the α-helical conformation of the PPLG backbone, which minimized steric hindrance by exposing reactive alkyne groups outward, facilitating efficient coupling with azide-terminated PEG chains. The CuAAC reaction, conducted in DMF with CuBr/PMDETA as the catalyst, employed a 1:2 alkyne/azide ratio, ensuring near-complete functionalization. This strategy represents a promising alternative for overcoming steric limitations in grafting-onto methods, significantly improving polymer functionalization.

Hoogenboom et al. (2006) demonstrated the efficacy of CuAAC click chemistry for the synthesis of star-shaped poly(ε-caprolactone) (PCL) copolymers using microwave irradiation (Scheme 12) [63]. The synthesis began with the ROP of ε-caprolactone (CL), using 5-hexyn-1-ol as an initiator and $Sn(Oct)_2$ as a catalyst, ensuring a controlled molecular weight distribution ($Đ < 1.20$). The second step involved CuAAC click chemistry, where alkyne-functionalized PCL (PCL-alkyne) was grafted onto heptakis-azido-β-cyclodextrin (β-CD-N_3) under microwave irradiation at 100°C, which significantly accelerated the reaction while maintaining high selectivity. The Cu(I)-catalyzed cycloaddition facilitated efficient coupling, yielding star-shaped heptakis-PCL-β-cyclodextrin with a 9:1 feed ratio of PCL-alkyne to β-CD-N_3.

4 Applications of Click Reactions in the Production of Functional Polymers

4.1 Biomedical Polymers

The integration of click chemistry into biomedical polymer synthesis has revolutionized the development of materials for tissue engineering, controlled drug delivery, and manufacture of medical devices. Its unparalleled specificity and efficiency

Scheme 12 Synthesis of star-shaped poly (ε-caprolactone) via CuAAC grafting onto azide-functionalized cyclodextrin [63]

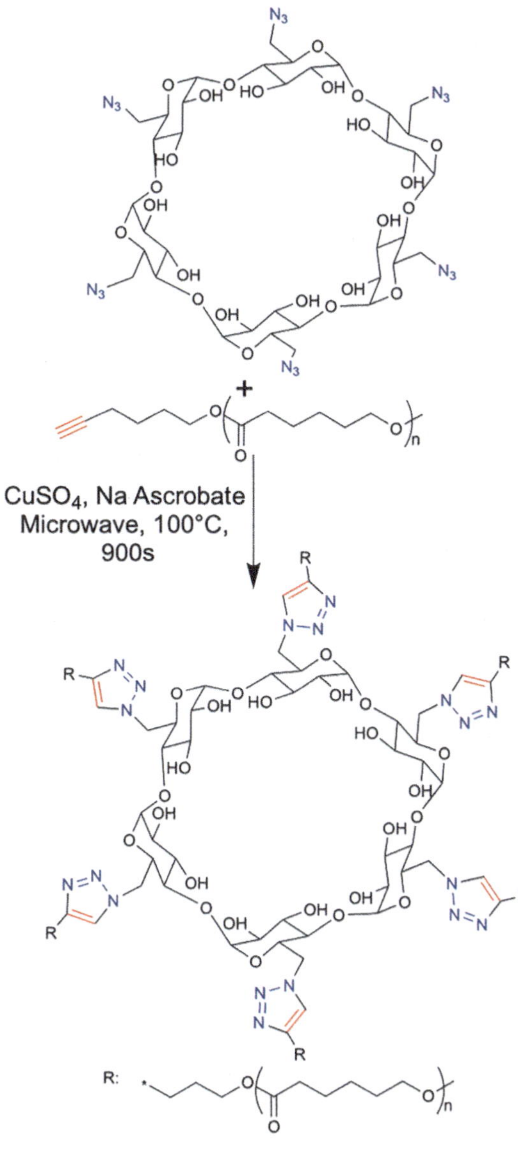

enable the conjugation of polymeric blocks and biomolecules without generating undesirable byproducts, ensuring high material purity, which is an essential factor in biomedical applications [64–66]. Moreover, its compatibility with mild reaction conditions, such as room temperature and aqueous environments, preserves the biological activity of proteins, peptides, and other sensitive macromolecules, facilitating their incorporation into advanced biomaterials. Click chemistry also enables post-synthetic modifications, allowing precise control over polymer properties, including hydrophilicity, surface charge, and responsiveness to stimuli, thereby optimizing their performance in biological systems [65–67]. Although the applications of AAC in biomedicine are numerous, some of the most relevant examples are listed below.

4.1.1 Drug Delivery Systems

The AAC reaction has emerged as a key strategy for engineering polymeric drug delivery systems, enabling the precise functionalization of polymer nanoparticles, including micelles, polymersomes, and hydrogels for controlled drug release and targeted delivery [68, 69].

Porras et al. synthesized PEGylated amphiphilic block copolymers by combining ROP with CuAAC reaction [70, 71]. First, CL and 5-methyl-5-propargyloxycarbonyl-1,3-dioxane-2-one (MCP) were copolymerized via ROP using methoxy poly(ethylene glycol) (mPEG) as an initiator to obtain PEG-b-P(CL-co-MCP) copolymers with a target number of alkyne functionalities per chain. These alkyne-functionalized copolymers were subsequently grafted with azide-bearing lauric, oleic, and linoleic acid derivatives via CuAAC (Scheme 13).

To reduce copper-associated cytotoxicity, the reaction was optimized using an azide:Cu(I) molar ratio of 1:0.05, which ensured a high conversion efficiency and simplified purification. The resulting amphiphilic block copolymers self-assembled into stable micelles, efficiently encapsulating Δ^9-tetrahydrocannabinol (THC)-rich cannabis extracts, and exhibited enhanced colloidal stability and prolonged

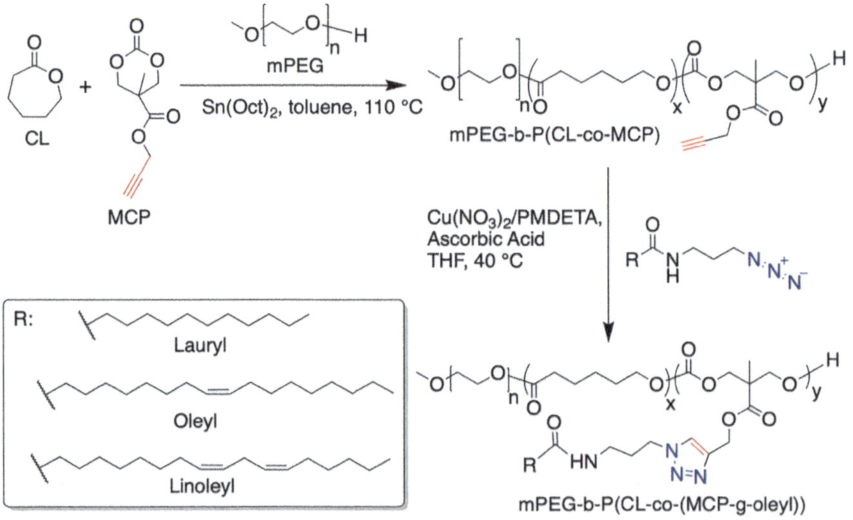

Scheme 13 Synthesis of PEGylated amphiphilic block copolymers comprising fatty acid moieties by combining ROP and CuAAC click chemistry

analgesic effects in murine models. Compared to PEG-b-PCL micelles, these CuAAC-derived nanocarriers exhibited superior drug encapsulation efficiency and controlled release properties, highlighting their potential for the oral delivery of lipophilic therapeutics.

Deng et al. developed a pH-responsive polymeric micelle system based on copper-free click chemistry, designed to enhance tumor retention, reduce systemic clearance, and improve immunochemotherapy efficacy [72]. The system leveraged the strain-promoted azide-alkyne cycloaddition (SPAAC) reaction between DBCO-modified poly(ε-caprolactone)-b-poly(ethylene glycol) (PCL-PEG-DBCO) micelles and azide-functionalized PCL-PEG micelles (PCL-PEG-N_3), which remained well dispersed under physiological conditions (pH 7.4) but rapidly aggregated in the acidic tumor microenvironment (pH 6.5–6.8), as shown in Scheme 14. This in situ micellar cross-linking leads to increased tumor accumulation, reduced systemic clearance, and improved intracellular drug retention.

The micelles were engineered for the co-delivery of doxorubicin (DOX) and the Smad3 selective inhibitor SIS3, two agents with synergistic anticancer and immunomodulatory properties. Compared to conventional micellar systems, this SPAAC-driven self-assembling nanoplatform significantly improved tumor residence time, enhanced natural killer cell activation, and effectively suppressed melanoma growth.

4.1.2 Polymer Prodrugs

Polymeric prodrugs offer an effective alternative for improving drug delivery, ensuring targeted organ distribution, reduced toxicity, and enhanced therapeutic efficacy. These pharmacologically inactive drug derivatives undergo enzymatic or chemical activation within the body, enhancing their solubility, bioavailability, and stability while enabling controlled release. Compared to conventional prodrugs, polymeric prodrugs provide greater structural versatility, prolonged circulation, and tunable degradation profiles, making them particularly valuable in oncology, antiviral therapy, and CNS disorders [73]. Additionally, spatially and temporally controlled activation triggered by tumor-specific enzymes, redox conditions, or external stimuli minimizes systemic toxicity and improves pharmacokinetics, thereby addressing the solubility and permeability challenges for efficient drug delivery [74, 75]. Given the versatility of AAC reactions, they serve as powerful tools for synthesizing diverse polymeric prodrug architectures, enabling the seamless integration of functional polymer backbones with covalently bound therapeutic agents. This approach facilitates the exploration of synergistic interactions between the polymer carriers and conjugated drugs. Polymeric prodrugs are synthesized by conjugating a functionalized polymer backbone bearing azide or alkyne groups with a drug derivative carrying complementary functionality.

A notable example of polymeric prodrug synthesis is the dual-drug pH/reduction-sensitive micellar delivery system developed by Liao et al. (2020), who used CuAAC click chemistry to integrate 6-mercaptopurine (6MP) and DOX for cancer therapy [76]. They synthesized an amphiphilic copolymer poly(DEAEMA)-b-poly

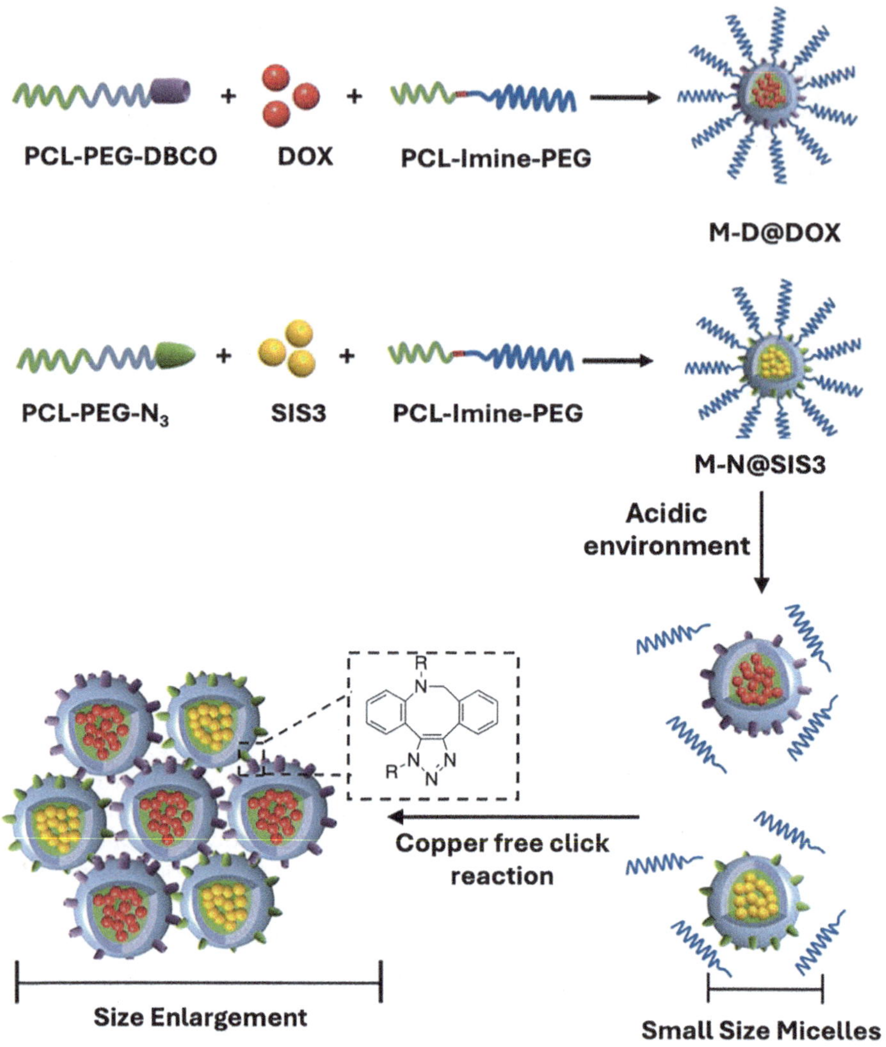

Scheme 14 Schematic illustration of pH-triggered assembly and dual-drug delivery of DOX and SIS3 via SPAAC. The DBCO-functionalized micelles (M-D@DOX) and azide-functionalized micelles (M-N@SIS3) remain dispersed at physiological pH. In acidic tumor environments, PEG detachment exposes reactive groups, enabling click cross-linking and micellar aggregate formation. (Reprinted with permission from [72]. Copyright 2025 American Chemical Society)

(ABMA-co-OEGMA) (PDPAO) via reversible addition–fragmentation chain transfer (RAFT) polymerization of 2-(diethylamino)ethyl methacrylate (DEAEMA), 4-azidobenzyl methacrylate (ABMA), and oligo(ethylene glycol) methyl ether methacrylate (OEGMA). The click reaction was performed in DMF at room temperature using CuBr/PMDETA as the catalyst, ensuring the efficient conjugation of DOX and 6MP (Scheme 15). Cytotoxicity studies on HeLa and HL-60 leukemia cells indicated

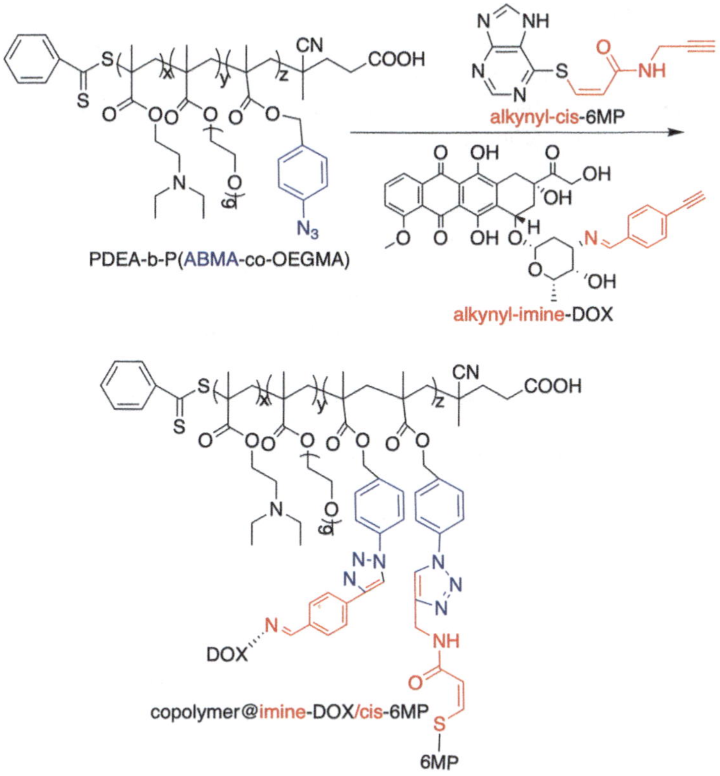

Scheme 15 Schematic illustration of the synthesis of PDPAO@imine-DOX/cis-6MP prodrugs containing pH- and glutathione-sensitive linkers. The reaction was carried out employing CuBr/PMDETA as catalyst at room temperature during 48 h. PDEAEMA-b-P(ABMA-co-OEGMA) was obtained via RAFT polymerization [76]

a probable synergistic therapeutic effect, as deduced from the comparison of IC_{50} values for the prodrug formulations and free drugs.

4.1.3 Gene Delivery Vectors

Gene delivery vectors are engineered systems designed to transport genetic material into target cells, thereby facilitating gene expression or silencing for therapeutic applications. These vectors play a crucial role in gene therapy, RNA-based treatments, and genome-editing technologies such as CRISPR-Cas9. Although highly efficient, traditional viral vectors pose challenges related to immunogenicity, limited payload capacity, and potential oncogenic risk. As an alternative, nonviral gene delivery systems based on functional polymers have gained significant attention because of their biocompatibility, tunability, and lower immunogenicity [77–81].

Polymeric gene delivery vectors offer significant advantages in nucleic acid therapeutics because they do not elicit an immunological response. They can encapsulate and protect nucleic acids from enzymatic degradation, facilitate targeted cellular uptake, and enable controlled intracellular release [82, 83]. Additionally, their synthesis can be efficiently scaled up, making gene therapies more cost-effective.

The most widely used polymers in gene delivery are polyethylenimine (PEI), poly(L-lysine) (PLL), poly(2-(dimethylamino)ethyl methacrylate) (PDMAEMA), and polysaccharide-based carriers such as chitosan and hyaluronic acid derivatives. These polymers exhibit high transfection efficiency; however, their cytotoxicity increases with molecular weight, as larger polymer chains tend to be more effective but also more damaging to cells. To address this limitation, chemical modifications are often required to enhance biocompatibility while maintaining or improving the gene delivery performance [84–87].

Maze et al. developed a hybrid gene delivery system utilizing SPAAC to minimize polymer use and reduce DNA/polymer polyplex toxicity [88]. Plasmid DNA encoding the luciferase gene was functionalized with 3,6,9-trioxaundecan-1-{4-[(2-chloroethyl)ethylamino]-benzylamino},11-azide (CEBA) as seen in Scheme 16a, introducing an azide moiety for subsequent SPAAC with DBCO-modified l-PEI (Scheme 16b). In addition, DBCO-SS-l-PEI incorporated disulfide bridges for redox-responsive modifications. Polyplex stability was assessed using 20 μM dextran sulfate (DS), a polyplex dissociation agent. Agarose gel electrophoresis showed complete dissociation of the pDNA complexed with unmodified l-PEI, whereas DBCO-l-PEI and DBCO-SS-l-PEI polyplexes remained intact, confirming that click chemistry-mediated conjugation effectively stabilized DNA condensation and prevented premature release in the presence of DS.

Díaz et al. (2021) synthesized amphiphilic graft copolymers by integrating mPEG, PCL, and PDMAEMA to improve gene delivery while mitigating cytotoxicity [89]. The synthesis began with the ROP of CL and MCP initiated by mPEG. In a one-pot procedure, 2-(2-azidoethoxy)ethyl-bromoisobutyrate (N3EBBr) was grafted onto the polymer backbone via CuAAC using CuBr/BPY as the catalyst. Subsequently, DMAEMA polymerization was conducted via atom transfer radical polymerization (ATRP) using the same catalytic system, as shown in Scheme 17. This sequential click-ATRP strategy ensured a precise polymer architecture, balancing hydrophilic, hydrophobic, and cationic segments for enhanced DNA complexation, transfection efficiency, and biocompatibility. Compared to conventional linear polyethylenimine (l-PEI), click-functionalized copolymers exhibited higher transfection efficiency in both HEK-293 and Jurkat cells, highlighting the potential of CuAAC-driven polymer architectures for gene delivery applications.

4.1.4 Development of Hydrogels

Hydrogels are highly hydrated, biocompatible materials that have attracted significant interest in controlled drug release, tissue engineering, and regenerative

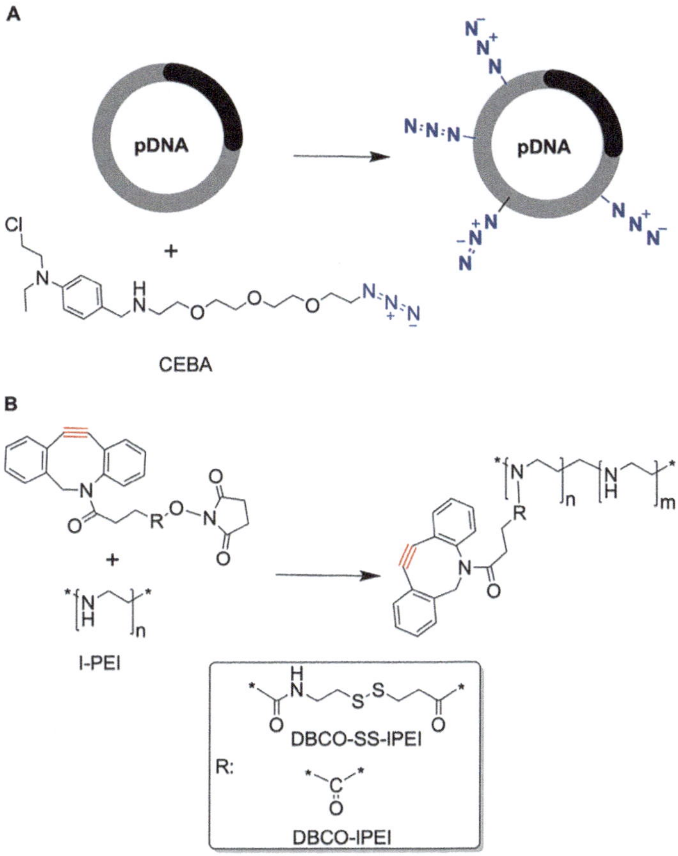

Scheme 16 Synthesis of azide and alkyne precursors in l-PEI plasmid conjugation. (**a**) Functionalization of the plasmid with CEBA and (**b**) modification of l-PEI with dibenzocyclooctyl groups reported in Ref. [88]

medicine because of their ability to mimic the extracellular matrix (ECM). The integration of azide-alkyne click chemistry into hydrogel design has enabled the precise functionalization and cross-linking of hydrogel networks, offering advantages such as tailored mechanical properties, stimuli-responsive behavior, and bioactive modifications [65]. Particularly, SPAAC is increasingly favored for the synthesis of hydrogels intended for biomedical applications, ensuring high biocompatibility, selective functionalization, and minimal cytotoxicity.

Fu et al. (2017) developed injectable hydrogels based on hyaluronic acid (HA) and PEG cross-linked via SPAAC [90]. HA was first cyclooctyne-functionalized by coupling 2-(aminoethoxy)cyclooctyne to its carboxylates (cyclooctyne-HA), and a tetra-azide PEG (azide-PEG) served as the cross-linker. This catalyst-free, bioorthogonal reaction enables rapid in situ gelation under

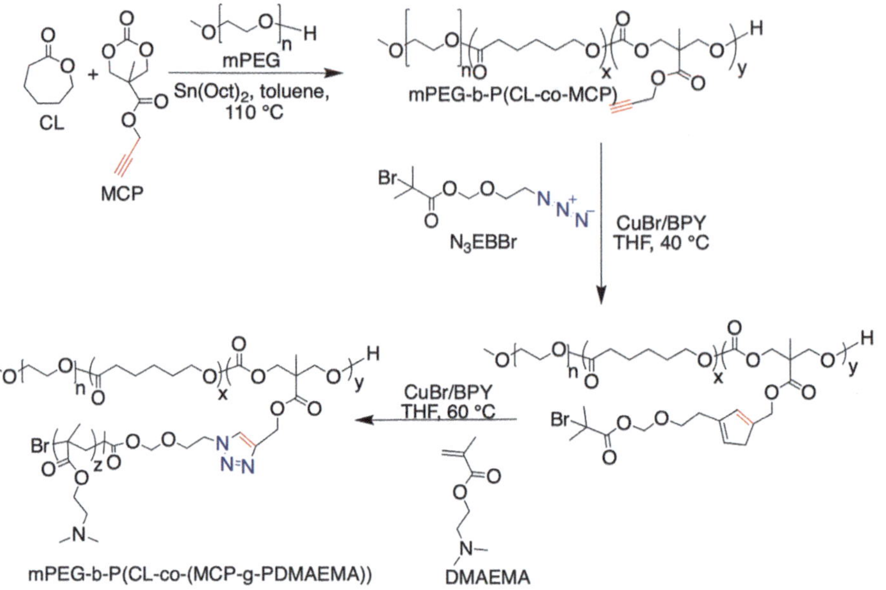

Scheme 17 Synthesis of mPEG-b-P(CL-co-(MCP-g-PDMAEMA)) via sequential ROP, CuAAC, and ATRP [89]

physiological conditions, yielding a cytocompatible, stable HA–PEG network. The building blocks for SPAAC are shown in Fig. 2a, while Fig. 2b–f displays freeze-dried microstructures with an interconnected porous architecture whose pore size can be tuned by precursor concentration and azide-alkyne ratio.

Jain et al. (2021) developed a novel strategy for modifying alginate hydrogels through SPAAC to enhance their bioactivity and cell adhesion [91]. Alginate, widely used as an ECM mimic in tissue engineering, is typically modified using carbodiimide chemistry; however, this approach often results in nonspecific conjugation and reduced peptide bioactivity. To overcome these limitations, the authors functionalized alginate with an azide-reactive strained cyclooctyne (BCN-amine), followed by click conjugation with an azide-modified cyclic RGD peptide, as depicted in Scheme 18. Compared to conventional carbodiimide-based conjugation, SPAAC-mediated peptide grafting preserved ligand bioactivity and significantly enhanced cell adhesion, spreading, and proliferation in osteoblast and bone marrow stromal cell cultures.

4.1.5 Antimicrobial Polymers

The growing threat of antibiotic resistance has intensified the need for alternative antimicrobial strategies, particularly for medical devices, protective coatings, packaging materials, and biomaterials. A promising approach involves the integration of

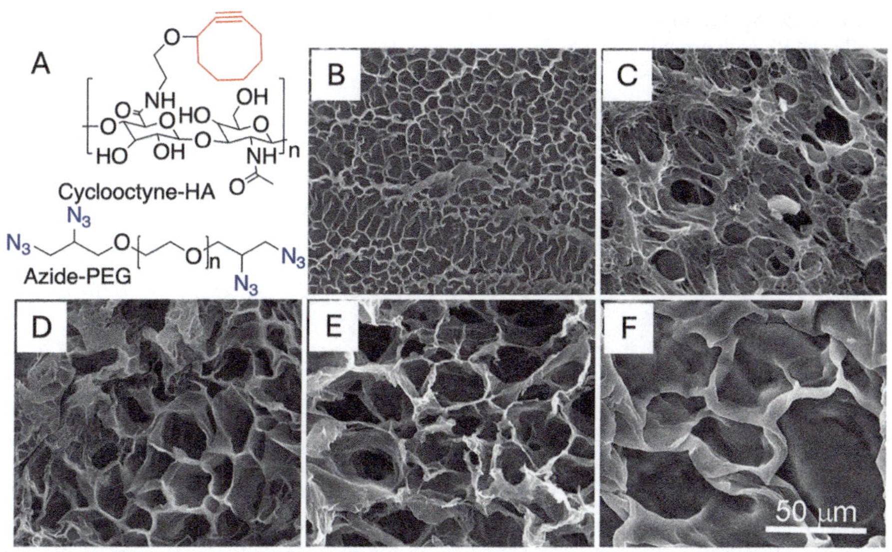

Fig. 2 (**a**) Schematic of the SPAAC between cyclooctyne-HA and azide-PEG, forming triazole-cross-linked HA–PEG hydrogels. (**b–f**) SEM micrographs of freeze-dried hydrogels showing an interconnected porous network; pore size increases as the overall precursor concentration decreases and/or the azide-alkyne ratio is lowered. (Reprinted with permission from [90]. Copyright 2025 Elsevier)

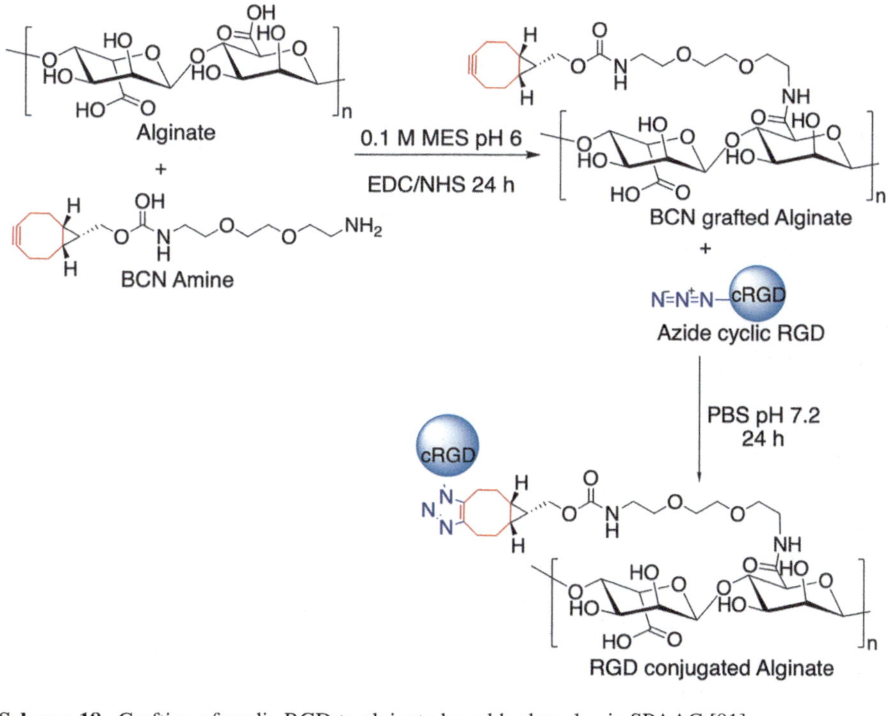

Scheme 18 Grafting of cyclic RGD to alginate-based hydrogels via SPAAC [91]

Scheme 19 Modification of PP and PE with penicillin to obtain antibacterial polymeric materials [93]

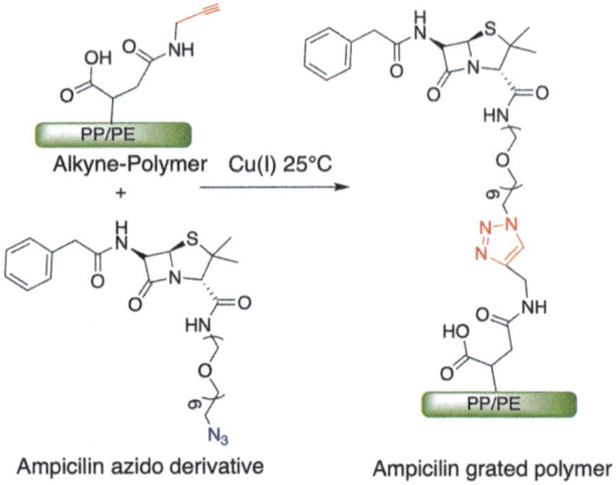

azide-alkyne click chemistry into polymer synthesis, enabling precise functionalization with antimicrobial agents such as quaternary ammonium salts (QAS), metal nanoparticles, antimicrobial peptides, and bioactive macromolecules [92]. The modification of conventional polymers with antimicrobial functionalities has emerged as a versatile and effective strategy for designing microbial-killing surfaces with enhanced stability, bioactivity, and long-term efficacy.

A notable example is Pearson and Urban (2014), who developed a solvent-free, microwave-assisted modification strategy for polyethylene (PE) and polypropylene (PP) surfaces. Their approach involved maleic anhydride plasma treatment to introduce carboxyl (–COOH) functionalities, followed by alkyne functionalization with propargylamine [93]. These alkyne-modified surfaces were covalently grafted with azide-functionalized ampicillin via CuAAC click chemistry (Scheme 19), achieving a 97–99.8% reduction in *Staphylococcus aureus* biofilm formation, demonstrating a highly effective antibacterial coating strategy.

Similarly, Acik et al. (2018) employed CuAAC click chemistry to functionalize polypropylene (PP) with quaternary ammonium salts (QAS), an established antimicrobial agent [94]. The process began with the azidation of chlorinated polypropylene (PP-Cl) to obtain azide-functionalized PP (PP-N_3). An alkyne-functionalized QAS was synthesized by quaternizing benzyl bromide with 3-dimethylamino-1-propyne (Scheme 20). The final CuAAC reaction between PP-N_3 and QAS-alkyne yielded PP-QAS films with broad-spectrum antimicrobial properties, effectively inhibiting both Gram-positive (*S. aureus*) and Gram-negative (*E. coli*) bacterial growth.

4.2 *Proton- and Electron-Conducting Polymers*

4.2.1 Electron-Conductive Polymers

The CuAAC reaction has emerged as a highly efficient tool for the step-growth synthesis of π-conjugated polymers, providing a valuable alternative to conventional

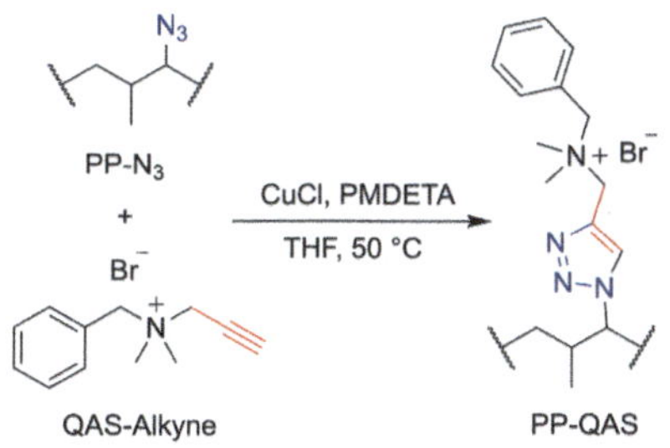

Scheme 20 Synthesis of PP-QAS starting from QAS-alkyne and PP-N3 via CuAAC reported in [94]

transition metal-catalyzed coupling reactions such as Stille, Suzuki, and Heck. Although these traditional methods have enabled the development of numerous organic semiconductors, they suffer from inherent drawbacks, including the use of toxic organometallic reagents, poor atom economy, and residual metal impurities, which can interfere with device performance and necessitate extensive purification.

By leveraging CuAAC chemistry, triazole-linked π-conjugated polymers have been successfully synthesized, in which the triazole moiety acts as an electron-accepting unit, modulating the charge transport properties, optical absorption, and nonlinear optical behavior. These materials have demonstrated strong potential for applications in organic light-emitting diodes, organic field-effect transistors, and organic photovoltaics [95].

A significant example of this approach was demonstrated by Van Steenis et al. (2005), who employed CuAAC click chemistry to synthesize fluorene-based π-conjugated polymers for light-emitting applications [96]. Polymerization was conducted using 2,7-diazidofluorene monomers (C6 and C12 side chains) and various diethynyl monomers in THF/MeCN at room temperature and −10°C with a Cu(II)/Cu(0) catalyst system, as depicted in Scheme 21. The resulting polymers (M_w 2,600–48,000 g/mol) exhibited enhanced fluorescence quantum yields, with those polymerized at −10°C showing twice the quantum efficiency of their room-temperature counterparts. The long alkyl chains improved the solubility, making the materials suitable for thin-film processing.

In addition to direct polymerization, CuAAC has been widely used for post-functionalization, in which pre-synthesized conjugated polymers bearing azide or alkyne groups are modified to introduce electron-donating or electron-accepting functionalities. This strategy enables the fine-tuning of optical absorption, solubility, and electronic properties, which are essential for optimizing the device performance [97].

Bu et al. (2011) demonstrated that the post-polymerization functionalization of conducting poly(3,4-ethylenedioxythiophene) (PEDOT) via CuAAC is a versatile method for modifying polymeric surfaces with various functional moieties [98]. The authors synthesized azide-functionalized PEDOT (PEDOT-N_3) via electropolymerization of azidomethyl-substituted EDOT monomers, forming

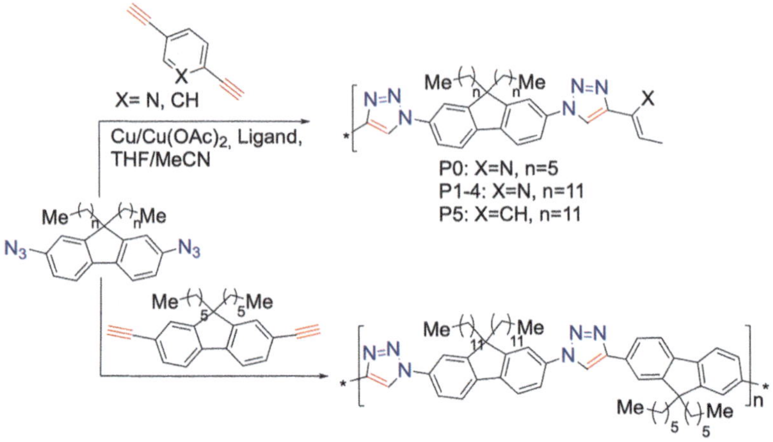

Scheme 21 Synthesis of fluorene-based polymers via CuAAC growth polymerization [96]

reactive coatings that were subsequently modified under heterogeneous CuAAC conditions (Scheme 22). The reaction efficiently conjugated diverse electron-accepting, electron-donating, and neutral alkynes, affording surface-grafted PEDOT derivatives with tunable electronic and optical properties. This approach overcomes the steric and electronic limitations often encountered when introducing functional groups during pre-polymerization.

4.2.2 Proton-Conductive Polymers

The AAC reaction has become a pivotal tool in the design of proton-conductive polymers, offering a precise and efficient strategy for developing materials tailored for energy storage and fuel cell applications. AAC enhances ionic conductivity, chemical stability, and water retention, all of which are critical parameters for high-performance proton-exchange membranes (PEMs), by enabling the controlled incorporation of proton-conducting units into the polymer backbone. In this sense, Norris et al. (2010) successfully employed CuAAC click chemistry to functionalize polysulfones with pendant azide moieties, introducing sodium 3-(prop-2-ynyloxy) propane-1-sulfonate to endow proton conductivity for being used as fuel cell membranes (Scheme 23a) [99]. Additionally, to enhance mechanical stability and reduce methanol permeability, which are essential factors in direct methanol fuel cells (DMFCs), the polymer membranes were cross-linked using 1,7-octadiyne, leading to improved structural integrity (Scheme 23b). By systematically tuning the degree of sulfonation and cross-linking, the resulting membranes exhibited high selectivity, superior proton conductivity, and significantly lower methanol crossover compared to conventional Nafion-based membranes.

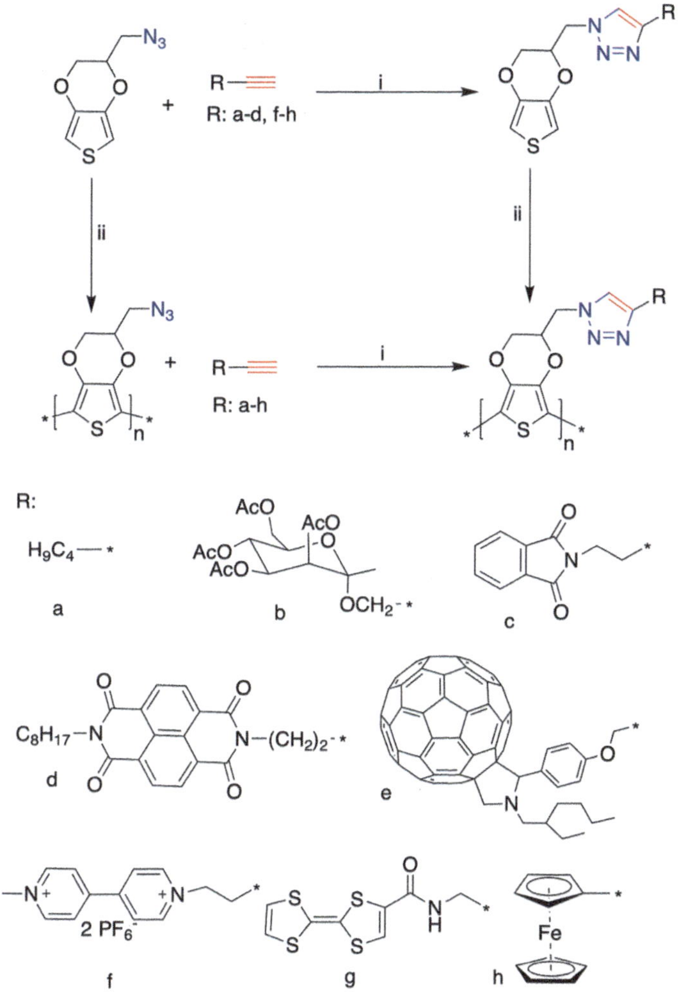

Scheme 22 Pre- and post-polymerization functionalization of conducting PEDOT via CuAAC with alkyne-containing moieties. The synthesis conditions were as follows: (i) Cu/Cu $(CH_3CN)_4PF_6/CH_3CN$ or THF or PhCN, r.t., and (ii) CH_3CN or $PhCN/Bu_3NPF_6$, $ne^{-}/-nH^{+}$ [98]

CuAAC step-growth polymerization has also been employed to integrate sulfonated proton-conducting units into polymer backbones. In this framework, Ghorai et al. (2022) utilized CuAAC click chemistry to synthesize trifluoromethyl-functionalized sulfonated polytriazoles (PTPPS and PTPS) using a diphenylphosphine oxide-based dialkyne monomer (POALK) polymerized with sulfonated diazide (DSAZ) and phosphine oxide-containing diazide (PFAZ) as depicted in Scheme 24 [100]. This strategy led to high-molecular-weight polytriazoles (59,600–75,800 g/mol) with superior oxidative and thermal stabilities,

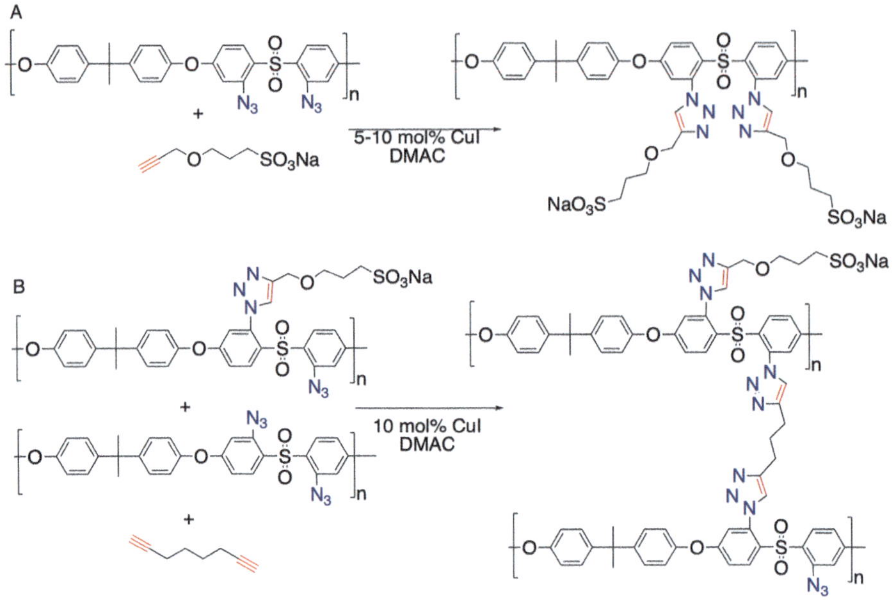

Scheme 23 (**a**) Functionalization of a poly(sulfone) via CuAAC starting from an alkyne sulfonate with 3-(prop-2-ynyloxy)propane-1-sulfonate. (**b**) Cross-linking of functionalized sulfone employing 1,7-octadiyne as the cross-linking agent

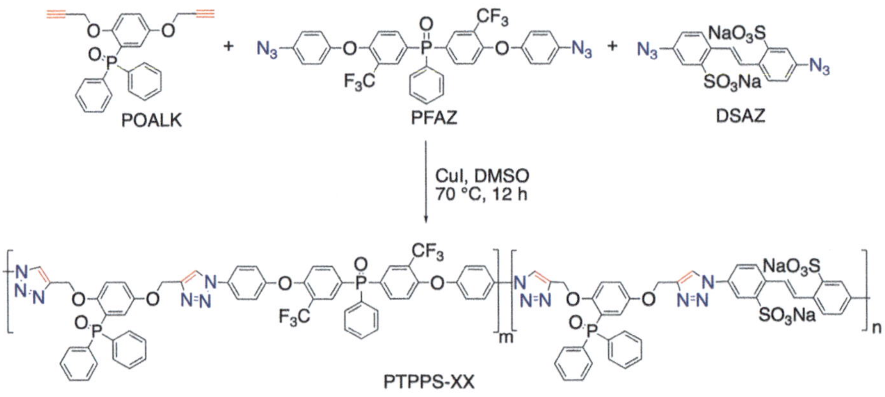

Scheme 24 Step-growth polymerization of POALK, PFAZ, and DSAZ yielded different n values ranging from 0.7 to 1. For all the compositions, $n + m = 1$ [100]

which are key attributes for PEM applications. Notably, the incorporation of phosphine oxide groups facilitated the formation of ionic nanochannels, optimizing proton transport and ensuring high proton conductivity (118 mS/cm at 80°C). These membranes significantly outperformed conventional PEM materials, demonstrating their potential as next-generation fuel cell membranes by balancing high conductivity, durability, and reduced methanol crossover, which is crucial for DMFCs.

5 Conclusion

The integration of AAC reactions, particularly CuAAC and SPAAC variants, has profoundly transformed the landscape of functional polymer synthesis. These click reactions offer unparalleled control, efficiency, and modularity in the construction of complex macromolecular architectures for biomedical, electronic, and energy-related applications. CuAAC has demonstrated broad utility in step-growth polymerization, surface modification, drug delivery, gene vectors, and antimicrobial coatings, enabling precise spatial and functional control. Meanwhile, SPAAC has extended the applicability of click chemistry to highly sensitive biological systems, allowing for metal-free reactions under physiological conditions and advancing hydrogel design, prodrugs, and in vivo bioconjugation.

Using carefully selected examples, this chapter highlights how AAC reactions promote the synthesis of linear and grafted polymers, block copolymers, conjugated semiconductors, polymeric prodrugs, and polyelectrolyte membranes. Notably, the ability to fine-tune grafting efficiency, charge distribution, hydrophilicity, and degradation kinetics has expanded the frontier of polymer design for drug delivery, tissue engineering, and energy devices. The synergy between click chemistry and polymer science continues to drive innovation, offering new opportunities to create multifunctional materials with precise molecular architectures and high performance.

Future developments will likely focus on expanding the chemical diversity of clickable monomers, improving the reaction kinetics under green and aqueous conditions, and enabling orthogonal multi-click strategies to further increase the complexity and functionality of polymeric systems. As demonstrated throughout this chapter, CuAAC and SPAAC remain pivotal tools in the polymer chemist's toolbox, bridging synthetic precision with application-driven versatility in next-generation material design.

Acknowledgments The authors are grateful to Atenea for the support provided through grant number 290-2024, awarded to Leon Perez.

References

1. Elsayed R, Teow YH (2025) Advanced functional polymer materials for biomedical applications. J Appl Polym Sci 142:e56391
2. Farinha JP, Baleizão C, Martinho JMG (2024) Functional polymer-based nanostructured materials. In: synthesis and applications in chemistry and materials. Biomass and waste

valorisation, functional materials, energy conversion and supercritical systems, vol 14. World Scientific, pp 317–371

3. Zhang H (2023) Functional polymeric systems for advanced industrial applications. Polymers (Basel) 15:1277
4. El-Husseiny HM, Mady EA, El-Dakroury WA, Zewail MB, Noshy M, Abdelfatah AM, Doghish AS (2022) Smart/stimuli-responsive hydrogels: state-of-the-art platforms for bone tissue engineering. Appl Mater Today 29. https://doi.org/10.1016/j.apmt.2022.101560
5. Municoy S, Álvarez Echazú MI, Antezana PE, Galdopórpora JM, Olivetti C, Mebert AM, Foglia ML, Tuttolomondo MV, Alvarez GS, Hardy JG, Desimone MF (2020) Stimuli-responsive materials for tissue engineering and drug delivery. Int J Mol Sci 21:1–39. https://doi.org/10.3390/ijms21134724
6. Pathan N, Shende P (2021) Strategic conceptualization and potential of self-healing polymers in biomedical field. Mater Sci Eng C 125. https://doi.org/10.1016/j.msec.2021.112099
7. Urban MW (2009) Stratification, stimuli-responsiveness, self-healing, and signaling in polymer networks. Prog Polym Sci (Oxford) 34:679–687. https://doi.org/10.1016/j.progpolymsci.2009.03.004
8. Ghosal K, Bhattacharyya SK, Mishra V, Zuilhof H (2024) Click chemistry for biofunctional polymers: from observing to steering cell behavior. Chem Rev. https://doi.org/10.1021/acs.chemrev.4c00251
9. Kolb HC, Finn MG, Sharpless KB (2001) Click chemistry: diverse chemical function from a few good reactions. Angew Chem Int Ed 40:2004–2021. https://doi.org/10.1002/1521-3773(20010601)40:11<2004::AID-ANIE2004>3.0.CO;2-5
10. Li K, Fong D, Meichsner E, Adronov A (2021) A survey of strain-promoted azide–alkyne cycloaddition in polymer chemistry. Chem Eur J 27:5057–5073. https://doi.org/10.1002/chem.202003386
11. Van Dijk M, Rijkers DTS, Liskamp RMJ, Van Nostrum CF, Hennink WE (2009) Synthesis and applications of biomedical and pharmaceutical polymers via click chemistry methodologies. Bioconjug Chem 20:2001–2016. https://doi.org/10.1021/bc900087a
12. Tiwari VK, Jaiswal MK, Rajkhowa S, Singh SK (2024) Click chemistry in polymer science. In: Tiwari VK, Jaiswal MK, Rajkhowa S, Singh SK (eds) Click chemistry. Springer, Singapore, pp 335–362. https://doi.org/10.1007/978-981-97-4596-8_11
13. Presolski SI, Hong VP, Finn MG (2011) Copper-catalyzed azide–alkyne click chemistry for bioconjugation. Curr Protoc Chem Biol 3:153–162. https://doi.org/10.1002/9780470559277.ch110148
14. Sletten EM, Bertozzi CR (2011) From mechanism to mouse: a tale of two bioorthogonal reactions. Acc Chem Res 44:666–676. https://doi.org/10.1021/ar200148z
15. Rostovtsev VV, Green LG, Fokin VV, Sharpless KB (2002) A stepwise Huisgen cycloaddition process: copper(I)-catalyzed regioselective "ligation" of azides and terminal alkynes. Angew Chem Int Ed 41:2596–2599
16. Boren BC, Narayan S, Rasmussen LK, Zhang L, Zhao H, Lin Z, Jia G, Fokin VV (2008) Ruthenium-catalyzed azide-alkyne cycloaddition: scope and mechanism. J Am Chem Soc 130:8923–8930. https://doi.org/10.1021/ja0749993
17. Ferrara F, Beke-Somfai T, Kann N (2024) Recent developments in the ruthenium-catalyzed azide alkyne cycloaddition (RuAAC) reaction. Eur J Org Chem 27. https://doi.org/10.1002/ejoc.202400113
18. Kasana S, Nigam V, Singh S, Das Kurmi B, Patel P (2024) A new insight into the Huisgen reaction: heterogeneous copper catalyzed azide-alkyne cycloaddition for the synthesis of 1,4-disubstituted triazole (from 2018–2023). Chem Biodivers 21. https://doi.org/10.1002/cbdv.202400109
19. Cheng G, Zeng X, Shen J, Wang X, Cui X (2013) A metal-free multicomponent cascade reaction for the regiospecific synthesis of 1,5-disubstituted 1,2,3-triazoles. Angew Chem Int Ed 52:13265–13268. https://doi.org/10.1002/anie.201307499

20. Tiwari VK, Jaiswal MK, Rajkhowa S, Singh SK (2025) Materials horizons: from nature to nanomaterials. Springer
21. Zhu L, Brassard CJ, Zhang X, Guha PM, Clark RJ (2016) On the mechanism of copper(I)-catalyzed azide-alkyne cycloaddition. Chem Rec 16:1501–1517. https://doi.org/10.1002/tcr.201600002
22. Ahlquist M, Fokin VV (2007) Enhanced reactivity of dinuclear copper(I) acetylides in dipolar cycloadditions. Organometallics 26:4389–4391. https://doi.org/10.1021/om700669v
23. Li L, Zhang Z (2016) Development and applications of the copper-catalyzed azide-alkyne cycloaddition (CuAAC) as a bioorthogonal reaction. Molecules 21. https://doi.org/10.3390/molecules21101393
24. Hein JE, Fokin VV (2010) Copper-catalyzed azide-alkyne cycloaddition (CuAAC) and beyond: new reactivity of copper(I) acetylides. Chem Soc Rev 39:1302–1315. https://doi.org/10.1039/b904091a
25. Golas PL, Tsarevsky NV, Sumerlin BS, Matyjaszewski K (2006) Catalyst performance in "click" coupling reactions of polymers prepared by ATRP: ligand and metal effects. Macromolecules 39:6451–6457. https://doi.org/10.1021/ma061592u
26. Asano K, Matsubara S (2010) Effects of a flexible alkyl chain on a ligand for CuAAC reaction. Org Lett 12:4988–4991. https://doi.org/10.1021/ol101990d
27. Svatunek D, Houszka N, Hamlin TA, Bickelhaupt FM, Mikula H (2019) Chemoselectivity of tertiary azides in strain-promoted alkyne-azide cycloadditions. Chem Eur J 25:754–758. https://doi.org/10.1002/chem.201805215
28. Debets MF, Van Berkel SS, Dommerholt J, Dirks AJ, Rutjes FPJT, Van Delft FL (2011) Bioconjugation with strained alkenes and alkynes. Acc Chem Res 44:805–815. https://doi.org/10.1021/ar200059z
29. Agard NJ, Prescher JA, Bertozzi CR (2004) A strain-promoted [3 + 2] azide−alkyne cycloaddition for covalent modification of biomolecules in living systems. J Am Chem Soc 126:15046–15047. https://doi.org/10.1021/ja044996f
30. Sletten EM, Bertozzi CR (2008) A hydrophilic azacyclooctyne for Cu-free click chemistry. Org Lett 10:3097–3099. https://doi.org/10.1021/ol801141k
31. Lis C, Rubner S, Gröst C, Hoffmann R, Knappe D, Berg T (2018) iSPAAC: isomer-free generation of a Bcl-xL-inhibitor in living cells. Chem Eur J 24:13762–13766. https://doi.org/10.1002/chem.201803032
32. Codelli JA, Baskin JM, Agard NJ, Bertozzi CR (2008) Second-generation difluorinated cyclooctynes for copper-free click chemistry. J Am Chem Soc 130:11486–11493. https://doi.org/10.1021/ja803086r
33. Lis C, Berg T (2019) Synthesis of TRIPCO: a new cyclooctyne for iSPAAC. Synlett 30:939–942
34. Dommerholt J, Schmidt S, Temming R, Hendriks LJA, Rutjes FPJT, van Hest JCM, Lefeber DJ, Friedl P, van Delft FL (2010) Readily accessible bicyclononynes for bioorthogonal labeling and three-dimensional imaging of living cells. Angew Chem Int Ed 49:9422–9425. https://doi.org/10.1002/anie.201003761
35. Ning X, Guo J, Wolfert MA, Boons G-J (2008) Visualizing metabolically labeled glycoconjugates of living cells by copper-free and fast Huisgen cycloadditions. Angew Chem Int Ed 47:2253–2255. https://doi.org/10.1002/anie.200705456
36. Sletten EM, Nakamura H, Jewett JC, Bertozzi CR (2010) Difluorobenzocyclooctyne: synthesis, reactivity, and stabilization by β-cyclodextrin. J Am Chem Soc 132:11799–11805. https://doi.org/10.1021/ja105005t
37. Debets MF, van Berkel SS, Schoffelen S, Rutjes FPJT, van Hest JCM, van Delft FL (2010) Aza-dibenzocyclooctynes for fast and efficient enzyme PEGylation via copper-free (3+2) cycloaddition. Chem Commun 46:97–99. https://doi.org/10.1039/B917797C
38. Chadwick RC, Van Gyzen S, Liogier S, Adronov A (2014) Scalable synthesis of strained cyclooctyne derivatives. Synthesis (Stuttgart) 46:669–677. https://doi.org/10.1055/s-0033-1340509

39. Jewett JC, Sletten EM, Bertozzi CR (2010) Rapid Cu-free click chemistry with readily synthesized biarylazacyclooctynones. J Am Chem Soc 132:3688–3690. https://doi.org/10.1021/ja100014q
40. Terzic V, Pousse G, Méallet-Renault R, Grellier P, Dubois J (2019) Dibenzocyclooctynes: effect of aryl substitution on their reactivity toward strain-promoted alkyne–azide cycloaddition. J Org Chem 84:8542–8551. https://doi.org/10.1021/acs.joc.9b00895
41. de Almeida G, Sletten EM, Nakamura H, Palaniappan KK, Bertozzi CR (2012) Thiacycloalkynes for copper-free click chemistry. Angew Chem Int Ed 51:2443–2447. https://doi.org/10.1002/anie.201106325
42. Odian G (2004) Principles of polymerization. John Wiley & Sons
43. McNelles SA, Pantaleo JL, Meichsner E, Adronov A (2019) Strain-promoted azide-alkyne cycloaddition-mediated step-growth polymerization. Macromolecules 52:7183–7187. https://doi.org/10.1021/acs.macromol.9b01609
44. Sheng X, Mauldin TC, Kessler MR (2010) Kinetics of bulk azide/alkyne "click" polymerization. J Polym Sci A Polym Chem 48:4093–4102. https://doi.org/10.1002/pola.24196
45. Binauld S, Damiron D, Hamaide T, Pascault JP, Fleury E, Drockenmuller E (2008) Click chemistry step growth polymerization of novel α-azide-ω-alkyne monomers. Chem Commun:4138–4140. https://doi.org/10.1039/b805164j
46. Zhang L, Ren X, Zhang Y, Zhang K (2019) Step-growth polymerization method for ultrahigh molecular weight polymers. ACS Macro Lett 8:948–954. https://doi.org/10.1021/acsmacrolett.9b00475
47. Singh AN, Thakre RD, More JC, Sharma PK, Agrawal YK (2015) Block copolymer nanostructures and their applications: a review. Polym Plast Technol Eng 54:1077–1095. https://doi.org/10.1080/03602559.2014.986811
48. Samaddar P, Deep A, Kim KH (2018) An engineering insight into block copolymer self-assembly: contemporary application from biomedical research to nanotechnology. Chem Eng J 342:71–89. https://doi.org/10.1016/j.cej.2018.01.062
49. Rodríguez-Hernández J, Chécot F, Gnanou Y, Lecommandoux S (2005) Toward "smart" nano-objects by self-assembly of block copolymers in solution. Prog Polym Sci (Oxford) 30: 691–724. https://doi.org/10.1016/j.progpolymsci.2005.04.002
50. Guo X, Choi B, Feng A, Thang SH (2018) Polymer synthesis with more than one form of living polymerization method. Macromol Rapid Commun 39. https://doi.org/10.1002/marc.201800479
51. Dey A, Haldar U, De P (2021) Block copolymer synthesis by the combination of living cationic polymerization and other polymerization methods. Front Chem 9. https://doi.org/10.3389/fchem.2021.644547
52. Vukićević R, Schwadtke U, Schmücker S, Schäfer P, Kuckling D, Beuermann S (2012) Alkyne-azide coupling of tailored poly(vinylidene fluoride) and polystyrene for the synthesis of block copolymers. Polym Chem 3:409–414. https://doi.org/10.1039/c1py00427a
53. Touris A, Hadjichristidis N (2011) Cyclic and multiblock polystyrene-block-polyisoprene copolymers by combining anionic polymerization and azide/alkyne "click" chemistry. Macromolecules 44:1969–1976. https://doi.org/10.1021/ma102900d
54. Wang C, He W, Wang F, Yong H, Bo T, Yao D, Zhao Y, Pan C, Cao Q, Zhang S, Li M (2024) Recent progress of non-linear topological structure polymers: synthesis, and gene delivery. J Nanobiotechnol 22. https://doi.org/10.1186/s12951-024-02299-6
55. Deng H, Dai F, Ma G, Zhang X (2015) Theranostic gold nanomicelles made from biocompatible comb-like polymers for thermochemotherapy and multifunctional imaging with rapid clearance. Adv Mater 27:3645–3653. https://doi.org/10.1002/adma.201501420
56. Sharma A, Pande PP (2019) Recent advancement in comb-like polymers: a review. J Polym Mater 36:175–194. https://doi.org/10.32381/JPM.2019.36.02.6
57. Diaz IL, Jérôme V, Freitag R, Perez LD (2021) Development of poly (ethyleneimine) grafted amphiphilic copolymers: evaluation of their cytotoxicity and ability to complex DNA. J Bioact Compat Polym 36:447–463

58. Shi H, Zhao Y, Dong X, Zhou Y, Wang D (2013) Frustrated crystallisation and hierarchical self-assembly behaviour of comb-like polymers. Chem Soc Rev 42:2075–2099. https://doi.org/10.1039/c2cs35350d
59. Acik G, Altinkok C, Tasdelen MA (2018) Synthesis and characterization of polypropylene-graft-poly(l-lactide) copolymers by CuAAC click chemistry. J Polym Sci A Polym Chem 56: 2595–2601. https://doi.org/10.1002/pola.29241
60. Chu Y, Li H, Huang H, Zhou H, Chen Y, Andreas B, Liu L, Chen Y (2018) Uni-molecular nanoparticles of poly(2-oxazoline) showing tunable thermoresponsive behaviors. J Polym Sci A Polym Chem 56:174–183. https://doi.org/10.1002/pola.28889
61. Gan W, Shi Y, Jing B, Cao X, Zhu Y, Gao H (2017) Produce molecular brushes with ultrahigh grafting density using accelerated CuAAC grafting-onto strategy. Macromolecules 50:215–222. https://doi.org/10.1021/acs.macromol.6b02388
62. Engler AC, Lee H, Hammond PT (2009) Highly efficient "grafting onto" a polypeptide backbone using click chemistry. Angew Chem 121:9498–9502. https://doi.org/10.1002/ange.200904070
63. Hoogenboom R, Moore BC, Schubert US (2006) Synthesis of star-shaped poly(-ε-caprolactone) via "click" chemistry and "supramolecular click" chemistry. Chem Commun:4010–4012. https://doi.org/10.1039/b608313g
64. Paats JWD, Hamelmann NM, Paulusse JMJ (2024) Dual-reactive single-chain polymer nanoparticles for orthogonal functionalization through active ester and click chemistry. J Control Release 373:117–127. https://doi.org/10.1016/j.jconrel.2024.07.003
65. Degirmenci A, Sanyal R, Sanyal A (2024) Metal-free click-chemistry: a powerful tool for fabricating hydrogels for biomedical applications. Bioconjug Chem 35:433–452. https://doi.org/10.1021/acs.bioconjchem.4c00003
66. Kaur J, Saxena M, Rishi N (2021) An overview of recent advances in biomedical applications of click chemistry. Bioconjug Chem 32:1455–1471. https://doi.org/10.1021/acs.bioconjchem.1c00247
67. Arnold RM, Huddleston NE, Locklin J (2012) Utilizing click chemistry to design functional interfaces through post-polymerization modification. J Mater Chem 22:19357–19365. https://doi.org/10.1039/c2jm31708g
68. Lallana E, Sousa-Herves A, Fernandez-Trillo F, Riguera R, Fernandez-Megia E (2012) Click chemistry for drug delivery nanosystems. Pharm Res 29:1–34. https://doi.org/10.1007/s11095-011-0568-5
69. Saraiva NM, Alves A, Costa PC, Correia-da-Silva M (2024) Click chemistry in polymersome technology. Pharmaceuticals 17. https://doi.org/10.3390/ph17060747
70. Porras JD, Román Y, Palacio J, Blandón-Naranjo L, Benjumea D, Pérez LD (2024) Amphiphilic block copolymers bearing fatty acid derivatives as vehicles for THC in the development of analgesic oral formulations. React Funct Polym 195. https://doi.org/10.1016/j.reactfunctpolym.2023.105811
71. Porras JD, Diaz IL, Perez LD (2024) Synthesis of PEGylated amphiphilic block copolymers with pendant linoleic moieties by combining ring-opening polymerization and click chemistry. Biopolymers 115. https://doi.org/10.1002/bip.23582
72. Deng M, Guo R, Zang S, Rao J, Li M, Tang X, Xia C, Li M, Zhang Z, He Q (2021) PH-triggered copper-free click reaction-mediated micelle aggregation for enhanced tumor retention and elevated immuno-chemotherapy against melanoma. ACS Appl Mater Interfaces 13:18033–18046. https://doi.org/10.1021/acsami.1c02567
73. Rautio J, Kumpulainen H, Heimbach T, Oliyai R, Oh D, Järvinen T, Savolainen J (2008) Prodrugs: design and clinical applications. Nat Rev Drug Discov 7:255–270. https://doi.org/10.1038/nrd2468
74. Markovic M, Ben-Shabat S, Dahan A (2020) Prodrugs for improved drug delivery: lessons learned from recently developed and marketed products. Pharmaceutics 12:1–12. https://doi.org/10.3390/pharmaceutics12111031

75. Ioele G, Chieffallo M, Occhiuzzi MA, De Luca M, Garofalo A, Ragno G, Grande F (2022) Anticancer drugs: recent strategies to improve stability profile, pharmacokinetic and pharmacodynamic properties. Molecules 27. https://doi.org/10.3390/molecules27175436
76. Liao J, Peng H, Wei X, Song Y, Liu C, Li D, Yin Y, Xiong X, Zheng H, Wang Q (2020) A bio-responsive 6-mercaptopurine/doxorubicin based "Click Chemistry" polymeric prodrug for cancer therapy. Mater Sci Eng C 108. https://doi.org/10.1016/j.msec.2019.110461
77. Arabi F, Mansouri V, Ahmadbeigi N (2022) Gene therapy clinical trials, where do we go? An overview. Biomed Pharmacother 153. https://doi.org/10.1016/j.biopha.2022.113324
78. Dunbar CE, High KA, Joung JK, Kohn DB, Ozawa K, Sadelain M (2018) Gene therapy comes of age. Science (1979) 359. https://doi.org/10.1126/science.aan4672
79. Liu WW, Zheng SQ, Li T, Fei YF, Wang C, Zhang S, Wang F, Jiang GM, Wang H (2024) RNA modifications in cellular metabolism: implications for metabolism-targeted therapy and immunotherapy. Signal Transduct Target Ther 9. https://doi.org/10.1038/s41392-024-01777-5
80. Alzahrani MS, Almutairy B, Althobaiti YS, Alsaab HO (2024) Recent advances in RNA interference-based therapy for hepatocellular carcinoma: emphasis on siRNA. Cell Biochem Biophys. https://doi.org/10.1007/s12013-024-01395-6
81. Feng X, Li Z, Liu Y, Chen D, Zhou Z (2024) CRISPR/Cas9 technology for advancements in cancer immunotherapy: from uncovering regulatory mechanisms to therapeutic applications. Exp Hematol Oncol 13. https://doi.org/10.1186/s40164-024-00570-y
82. Sarvari R, Nouri M, Agbolaghi S, Roshangar L, Sadrhaghighi A, Seifalian AM, Keyhanvar P (2022) A summary on non-viral systems for gene delivery based on natural and synthetic polymers. Int J Polym Mater Polym Biomater 71:246–265. https://doi.org/10.1080/00914037.2020.1825081
83. Yue Y, Wu C (2013) Progress and perspectives in developing polymeric vectors for in vitro gene delivery. Biomater Sci 1:152–170. https://doi.org/10.1039/c2bm00030j
84. Sanchez-Martos M, Martinez-Navarrete G, Bernabeu-Zornoza A, Humphreys L, Fernandez E (2021) Evaluation and optimization of poly-d-lysine as a non-natural cationic polypeptide for gene transfer in neuroblastoma cells. Nanomaterials 11. https://doi.org/10.3390/nano11071756
85. Casper J, Schenk SH, Parhizkar E, Detampel P, Dehshahri A, Huwyler J (2023) Polyethylenimine (PEI) in gene therapy: current status and clinical applications. J Control Release 362:667–691. https://doi.org/10.1016/j.jconrel.2023.09.001
86. Dong L, Li Y, Cong H, Yu B, Shen Y (2024) A review of chitosan in gene therapy: developments and challenges. Carbohydr Polym 324. https://doi.org/10.1016/j.carbpol.2023.121562
87. Dou J, Yu S, Zhang Y (2024) A facile and scalable method to synthesize PEGylated PDMAEMA for gene delivery. Biopolymers 115. https://doi.org/10.1002/bip.23584
88. Maze D, Pichon C, Midoux P (2023) Reversible stabilization of DNA/PEI complexes by reducible click-linkage between DNA and polymer. A new polyplex concept for lowering polymer quantity. Gene Ther 30:783–791. https://doi.org/10.1038/s41434-023-00386-1
89. Diaz Ariza IL, Jérôme V, Pérez Pérez LD, Freitag R (2021) Amphiphilic graft copolymers capable of mixed-mode interaction as alternative nonviral transfection agents. ACS Appl Bio Mater 4:1268–1282
90. Fu S, Dong H, Deng X, Zhuo R, Zhong Z (2017) Injectable hyaluronic acid/poly(ethylene glycol) hydrogels crosslinked via strain-promoted azide-alkyne cycloaddition click reaction. Carbohydr Polym 169:332–340. https://doi.org/10.1016/j.carbpol.2017.04.028
91. Jain E, Neal S, Graf H, Tan X, Balasubramaniam R, Huebsch N (2021) Copper-free azide-alkyne cycloaddition for peptide modification of alginate hydrogels. ACS Appl Bio Mater 4: 1229–1237. https://doi.org/10.1021/acsabm.0c00976
92. Luo H, Yin XQ, Tan PF, Gu ZP, Liu ZM, Tan L (2021) Polymeric antibacterial materials: design, platforms and applications. J Mater Chem B 9:2802–2815. https://doi.org/10.1039/d1tb00109d

93. Pearson HA, Urban MW (2014) Simple click reactions on polymer surfaces leading to antimicrobial behavior. J Mater Chem B 2:2084–2087. https://doi.org/10.1039/c3tb21865a
94. Acik G, Altinkok C, Olmez H, Tasdelen MA (2018) Antibacterial film from chlorinated polypropylene via CuAAC click chemistry. Prog Org Coat 125:73–78. https://doi.org/10.1016/j.porgcoat.2018.08.029
95. Marrocchi A, Facchetti A, Lanari D, Santoro S, Vaccaro L (2016) Click-chemistry approaches to π-conjugated polymers for organic electronics applications. Chem Sci 7:6298–6308. https://doi.org/10.1039/c6sc01832g
96. Van Steenis DJVC, David ORP, Van Strijdonck GPF, Van Maarseveen JH, Reek JNH (2005) Click-chemistry as an efficient synthetic tool for the preparation of novel conjugated polymers. Chem Commun:4333–4335. https://doi.org/10.1039/b507776a
97. Terayama K, Michinobu T (2024) Organic semiconducting polymers and their post-functionalization for optoelectronics. In: Wong W-Y, Ma Y (eds) Optical and optoelectronic polymers. Royal Society of Chemistry, pp 435–457. https://doi.org/10.1039/BK9781837673551-00435
98. Bu HB, Götz G, Reinold E, Vogt A, Schmid S, Segura JL, Blanco R, Gómez R, Bäuerle P (2011) Efficient post-polymerization functionalization of conducting poly (3,4-ethylenedioxythiophene) (PEDOT) via 'click'-reaction. Tetrahedron 67:1114–1125. https://doi.org/10.1016/j.tet.2010.12.022
99. Norris BC, Li W, Lee E, Manthiram A, Bielawski CW (2010) 'Click'-functionalization of poly (sulfone)s and a study of their utilities as proton conductive membranes in direct methanol fuel cells. Polymer (Guildf) 51:5352–5358. https://doi.org/10.1016/j.polymer.2010.09.041
100. Ghorai A, Kamble R, Banerjee S (2022) Trifluoromethyl functionalized sulfonated polytriazoles from diphenylphosphine oxide-based dialkyne via click polymerization: effect of high content of phosphorus on the proton exchange membrane properties. Eur Polym J 162. https://doi.org/10.1016/j.eurpolymj.2021.110869

Top Heterocycl Chem (2026) 63: 37–68
https://doi.org/10.1007/7081_2025_97

Published online: 24 December 2025

β-Cyclodextrin-Based Catalytic Systems in Click Reactions

Sara Payamifar and Ahmad Poursattar Marjani

Contents

1 Introduction 38
  1.1 Click Chemistry 39
  1.2 Some Notable Features of CuAAC in Click Chemistry 41
  1.3 Importance of Click Chemistry for 1,2,3-Triazoles 42
2 β-CD Chemistry 42
  2.1 Properties and Advantages of β-CD 44
  2.2 Application of β-CD as the Catalyst in Organic Reactions 46
  2.3 Use the β-CD-Based Catalyst System in Click Reactions 49
3 Conclusion 60
References 62

Abstract This chapter highlights the efficiency of the click reaction, particularly for synthesizing 1,2,3-triazole, which is widely applied in organic and medicinal chemistry due to its high yields and broad utility. The reaction typically involves aryl/alkyl halides, sodium azide, and alkynes under mild conditions. Special focus is placed on β-cyclodextrin (β-CD)-based copper catalysts, which are highly selective, efficient, and reusable. These biodegradable and eco-friendly catalysts enhance sustainability by forming supramolecular inclusion complexes, making the process green, safe, and effective. The chapter emphasizes the role of CD-Cu systems in advancing environmentally friendly click chemistry.

Keywords 1,2,3-Triazole · Azide-alkyne cycloaddition · Click chemistry · Cu(I)-catalyzed click reaction · Green chemistry · β-Cyclodextrin (β-CD)

The original version of the chapter has been revised. The Figure 2 has been replaced with a revised version.

S. Payamifar and A. Poursattar Marjani (✉)
Department of Organic Chemistry, Faculty of Chemistry, Urmia University, Urmia, Iran
e-mail: a.poursattar@urmia.ac.ir

Abbreviations

Ar-$B(OH)_2$	Boronic acids
CD	Cyclodextrin
CTAB	Cetyltrimethylammonium bromide
Cu@β-CD@SPIONs	Copper supported β-cyclodextrin-grafted superparamagnetic iron oxide nanoparticles
CuAAC	Cu(I) azide/alkyne cycloaddition
FT-IR	Fourier transform infrared
GO	Graphene oxide
h-BN	Hexagonal boron nitride
LMMs	Low melting mixtures
MesoGO	Mesoporous silica nanoparticles-graphene oxide
MNPs	Metal nanoparticles
NaN_3	Sodium azide
NPs	Nanoparticles
PEG	Polyethylene glycol
SPION	Superparamagnetic iron oxide nanoparticle
TEOS	Tetraethoxysilane
TESPIC	3-(Triethoxysilyl) propyl isocyanate
TOF	Turnover frequency
β-CD	β-Cyclodextrin

1 Introduction

Developing safe and green routes for the construction of chemical generation is a key goal in modern organic chemistry. The noticeable strategy in organic synthesis is to expand procedures by operating on green materials without causing chemical contamination. One of the fast-growth areas of study is associated with developing green catalysts for chemical transformations [1]. β-Cyclodextrin (β-CD) is a green catalyst recently attracting notable attention. They have green biomimetic compounds and a unique nature, showing excellent catalytic performance in various organic transformations [2–4]. The easy accessibility, greenness, low price, renewability, nontoxicity, and fantastic activity for organic conversion are some advantages of β-CD that have drawn the attention of the research team universally. β-CD is well known as an essential building block in the organic chemistry world. Relating to green chemistry's aims, β-CD can open a new path to designing and developing hopeful catalysts for diverse organic reactions [5–7]. The click reactions are prevalent and best recognized for green chemistry's extreme productivity and selectivity for yielding 1,2,3-triazoles by striking medical and biological functions. This famous reaction comprises alkyl/aryl halides, alkynes, and sodium azide to provide crucial platforms named 1,2,3-triazoles in environmental states. Thus, discovering eco-friendly methods for preparing 1,2,3-triazoles is particularly interesting [8]. Recently, noteworthy advances have been performed in designing modified β-CD compounds as green catalysts for click reactions to prepare 1,2,3-triazole compounds. Employing renewable β-CD has created a way that is green, environment-friendly, and sustainable. The click reaction has furnished gain with diverse 1,2,3-triazoles vital for medical products and displays various pharmaceutical activities [9].

1.1 Click Chemistry

Click chemistry is the strongest 1,3-dipolar cycloaddition reaction for organic conversion. It is sufficient that the hoped molecule is required to be functionalized with the alkyne or azide group and then is prepared to bind to the aimed compound or substrate via a click reaction. The four kinds of click chemistry reported are shown in Scheme 1. The Huisgen 1,3-dipolar cycloaddition with Cu(I)-catalyzed (CuAAC) is the most typical class, and it is nowadays usually cited as the "click chemistry reaction" [10]. Scheme 2 displays a mechanism of azide-alkyne cycloaddition. The CuAAC mechanism involves the following key steps to form 1,4-disubstituted 1,2,3-triazoles: (a) Alkyne activation: A terminal alkyne reacts with a Cu (I) species, forming a π-complex or σ-alkynyl-copper intermediate, which increases the electrophilicity of the alkyne. (b) Azide coordination: The organic azide (R-N_3) coordinates with the activated alkyne-Cu complex. (c) Cycloaddition (triazole ring formation): A [3 + 2] cycloaddition occurs between the azide and the Cu-acetylide, leading to a metallacyclic intermediate. (d) Protonation and triazole formation: The intermediate undergoes protonation and ring closure, releasing the 1,4-disubstituted 1,2,3-triazole and regenerating the copper catalyst.

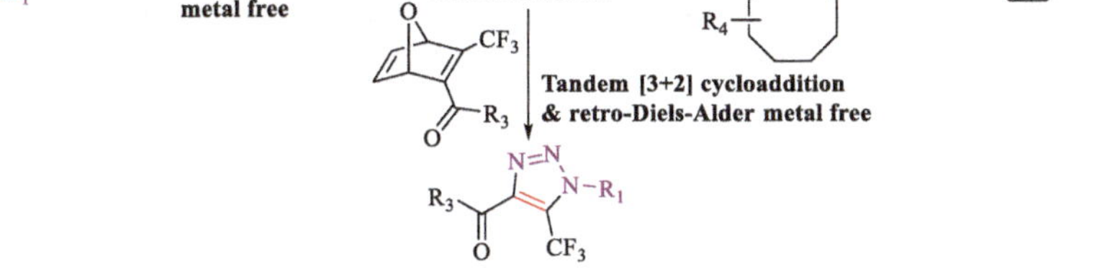

Scheme 1 Various kinds of routes for click chemistry

Scheme 2 The mechanism pathway of CuAAC

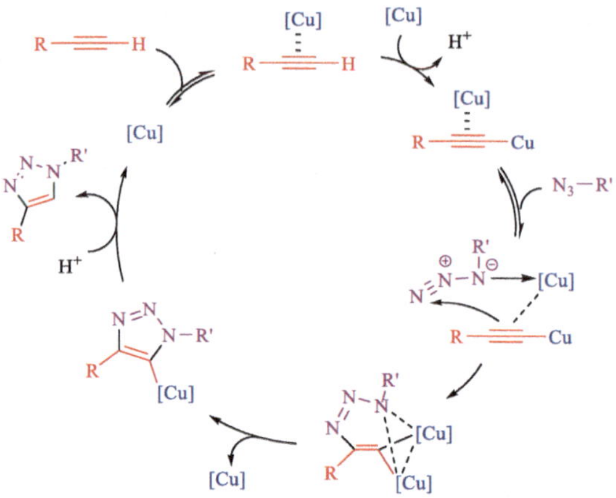

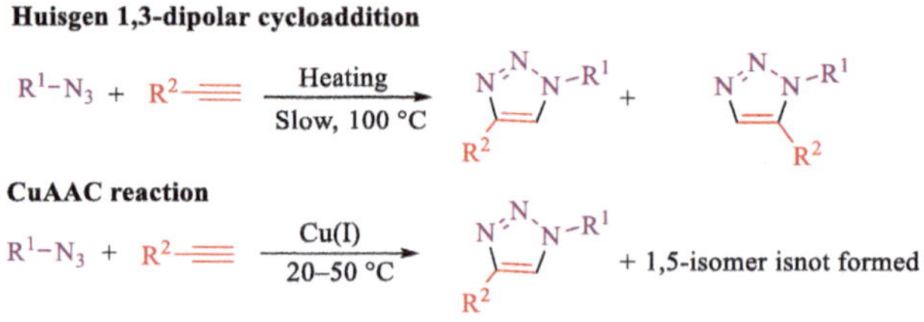

Scheme 3 Reaction for preparing triazoles in the existence and absence of copper(I) catalyst

The term "click" presented reactions that could be performed efficiently by removing solvents and with no chromatographic purification. Click reactions can contain many reactions with various pathways if they follow a shared reaction route [11]. For the first time, the phrase "click chemistry" was presented by Sharpless's team in 1999 at the annual meeting of the 217th American Chemical Society [12]; afterward, 2 years later, Kolb, Finn, and Sharpless, in their landmark review, explained it as an idea for executing organic conversion, which was according to the hypothesis that organic synthesis must concentrate notably on incredibly selective, easy orthogonal reactions that provide heteroatom-linkage systems with excellent efficacy in moderate conditions [13].

Click chemistry can also be regarded as a guide to discovering how to fulfill the needs of the modern chemistry community. Reactions need to be modular, with a broader scope, high yield, and capability to produce stable products under physiological conditions. They should be executed with easily accessible reagents, be stereospecific (not necessarily enantioselective), and generate only non-dangerous by-products. Among all the reactions that attain "click status," the Huisgen 1,3-dipolar cycloaddition to produce triazoles is the top-ranking instance of the click reaction [14–18]. The click reaction was not famous until Cu was used as a catalyst, which fixed the problem. The discovery of the CuAAC in 2002 (Scheme 3) transformed C–C bond formation from an innovative concept to a recognized fact. This reaction is commonly performed at moderate states employing various accessible reactants, extensively producing 1,4-regioisomers of 1,2,3-triazoles as the only products instead of regio-random triazole, which conforms entirely to the intention of the theoretical click reaction [19]. The key difference in regioselectivity between RuAAC and CuAAC is that CuAAC selectively forms 1,4-disubstituted 1,2,3-triazoles, whereas RuAAC yields 1,5-disubstituted isomers.

The birth of click chemistry has profoundly impacted nearly every part of the chemical science world. In this regard, in 2022, the Nobel Prize in chemistry was granted to Bertozzi, Meldal, and Sharpless for their efforts and contributions in establishing the fundamental principles of click and bio-orthogonal chemistry [20].

1.2 Some Notable Features of CuAAC in Click Chemistry

CuAAC (Cu(I)-catalyzed reactions) are classic click reactions uncovered and identified by Sharpless and Medal in 2001. An alkyne and an azide are the functional groups that merge to create a five-membered ring triazole compound. The reaction might or might not be catalyzed, depending on the preliminary product needed. Cu (I) catalysis is crucial for enhancing the functional groups' low reactivity and attaining just the 1,4-regioisomer [21]. Cu(I)-catalyzed 1,2,3-triazoles, which create a reaction between azides with terminal alkynes, have evolved into an excellent example of click chemistry owing to their biocompatibility, specificity, and reliability. The CuAAC, or click chemistry, has been recognized as one of the most prosperous, versatile, and modular methods for fast and regioselective building of 1,4-disubstituted 1,2,3-triazoles [22]. Until now, an extensive assortment of copper catalysts has been utilized in the CuAAC reaction. Among the various catalysts tried, Cu(I) in Cu(I)iodide and Cu(I)bromide are the most suitable, specifically Cu(I)-iodide, since it can be readily created in the lab. In addition to this, Cu(II) is also broadly employed in the existence of a reducing reagent like sodium ascorbate, which transforms it from Cu(II) to Cu(I) [23]. Thus, catalyst accessibility is one of the most essential benefits of this approach. Also, separating the catalyst from the reaction mixture is easy; it can be eliminated by rinsing it in water and a dilute hydrochloric acid solution. Most of the time, alkyne and alkyl azides do not react with different functional groups, but for click reactions, they participate in chemoselectivity [24]. The further noteworthy benefit of click chemistry is associated with the monitoring of reactions, which does not require complex methods. The completion of this reaction is followed by the FT-IR method due to alkyne and azide having specific peaks, respectively, in 2260–2100 and 2160–2120 cm^{-1}. Therefore, as soon as these peaks vanished, the click reaction ended [25]. Click reactions have some features, including the following: fast, highly selective, irreversible, thermodynamically driven reactions are commonly exothermic; moderate reaction conditions are needed (ideally nonsensitive to H_2O and O_2), without a solvent, a benign solvent (e.g., H_2O), or a readily eliminated solvent; it is a one-pot procedure and can be regiospecific and stereospecific; and it offers nontoxic by-products, quantitative yields, and easy product isolation, ideally without purification or using non-chromatographic techniques (e.g., distillation and crystallization) [26].

1.3 Importance of Click Chemistry for 1,2,3-Triazoles

The most extensive and one of the classical branches of organic chemistry is the heterocycle area. Heterocyclic compounds have an incredible variety of chemistry and physical and biological attributes that afford them a broad scope of practical uses. Heterocycles are of tremendous significance not only medically but also industrially. Heterocyclic compounds have a crucial role in pharmaceuticals and agrochemicals. Synthetic chemistry furnishes many heterocyclic systems, and more than 90% of drugs comprise heterocycle rings, thereby bridging the chemistry and biology community interface. Triazoles have been extensively studied due to their fascinating biological activities, including anticancer, antibiotic, anti-HIV, antifungal, muscarinic, and hypercholesterolemia (Fig. 1). These remarkable biological characteristics cause 1,2,3-triazole to be a promising nominee in designing drug delivery systems [27–29]. Today, 1,2,3-triazole products are broadly utilized in a vast scope of uses, comprising drug design [30], erosion inhibitors [31], dyes [32], and photographic materials [33]. The construction of five-membered heterocycles (comprising triazole rings) has been the subject of immense attention in *N*-heterocyclic chemistry in the 1960s by Huisgen for the 1,3-dipolar cycloaddition reaction between alkynes with azides [22, 34–38].

2 β-CD Chemistry

Recently, supramolecular chemistry has attracted massive interest in diverse areas of chemistry. For the first time, Cram, Lehn, and Pedersen presented their investigation of host-guest chemistry, and in 1987, they received the Nobel Prize [39–42]. Among the diverse supramolecules, CDs and their derivatives are captivating to research due to their ability to encapsulate smaller guests and create inclusion complexes,

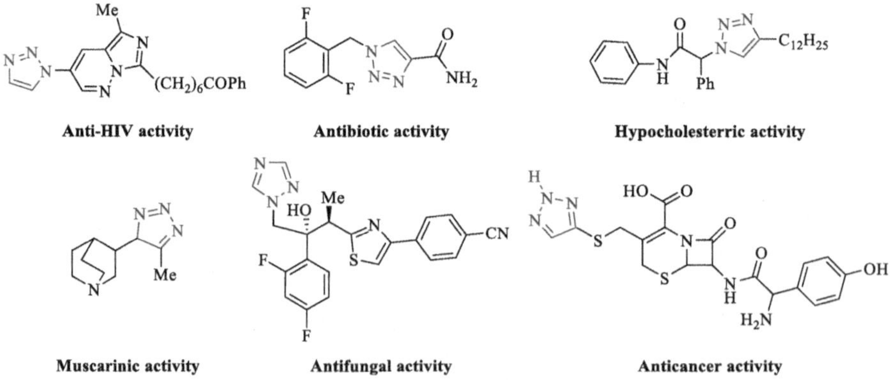

Fig. 1 The chemical structure of some bioactive 1,2,3-triazoles

indicating host-guest associations [43]. These macromolecules are highly practical and beneficial for drug delivery systems due to their nontoxicity in small doses [44, 45] and for clearing and eliminating contaminants from wastewater [46–48]. Cyclodextrin (CD) is a cyclic oligosaccharide, which is made of glucose groups that are connected by an α-1,4-glycosidic. The framework carbons and ethereal oxygens of the glucopyranose remainder cover the center hollow of the CDs. Therefore, it is hydrophobized. It provided a lipophilic microenvironment that could accommodate appropriately sized medicine molecules. A drug molecule creates a complex with a CD compound [49, 50]. CDs have well-known chemical structures with multiple sites for chemical modification, accessibility of various cavity sizes, specific hydrophilicity, low toxicity, and protection impact of the conjugated medicines from biodegradation. CDs and their derivatives could be carriers for therapeutically crucial proteins, peptides, and oligonucleotides among the different organic and inorganic compounds [51, 52]. Extensive investigations into compound recognition using CDs and functionalized CDs as compound receptors have confirmed that the corn-form aquaphobic hollow of CDs can selectively bind various molecules to form host-guest complexes and supramolecular structures [53–56], providing a perfect ideal arrangement that mimics enzymes [57, 58].

For the first time, Villiers isolated them from starch by enzymatic degradation. CD displays a donut form with a hydrophobic hollow and a hydrophilic external surface. Three well-known types of cyclodextrins, including α-, β-, and γ- CD, contain 6–8 glucose sections, respectively (Fig. 2). Although well known for over a century, the first applications of β-CD go back to the pharmaceutical and food fields in the 1980s. They can form a water-soluble inclusion complex with various organic compounds through a host-guest interaction within their cavity. According to this unique property, CDs and their related derivatives have been employed as host compounds to enhance the solubility of medicines, fix volatile aromas, and stabilize sensitive combinations. Inclusion in CDs profoundly influences the physicochemical characteristics of guest compounds as they are provisionally caged in the host hollow, providing advantageous changes for guest compounds, which are not attainable in another way. The possible guests for encapsulation with CDs are wholly varied. It comprises chain aliphatic, organic acids, aromatics, ketones, aldehydes, alcohols, fatty acids, and polar compounds such as halogens, amines, and oxyacids. The connection of guest compounds within CDs is not steady but is a dynamic balance. The ability of CDs to construct an inclusion complex depends on two main factors. The first factor is related to sterics. If the guest is the wrong size, it will not fit appropriately into the CD's hollow. The second crucial case is the thermodynamic interactions between the various pieces, such as CD, guest, and solvent. To create a complex, a beneficial net energetic driving force should draw the guest into the CDs [59].

CDs' developments in chemical and industrial fields have diverged into three steps: discovery time, exploratory course, and application course by József Szejtli. CDs took a long time, about 100 years, to move forward from fascinating chemical properties to promoting medical excipients. They were discovered in 1891 and identified in the twentieth century, but it was not until the past 30 years that they

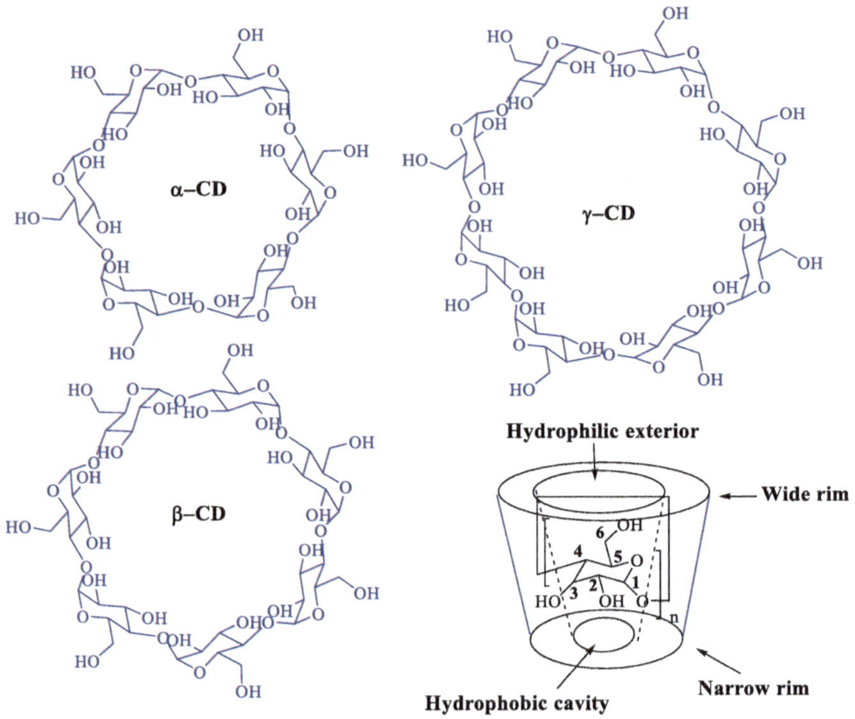

Fig. 2 The α-, β-, and γ-CD structures

became accessible as extremely purified excipients. CDs, including drug formulations, have been confirmed for dealing in Europe, Japan, and, in the previous 10 years, even the USA. At the start, CDs were utilized to increase medicines' aqueous solubility and chemical stability, and these properties were associated with their capacity to create drug/CD inclusion complexes. Therefore, 100 years after the discovery of CD, they are still considered new excipients with unknown opportunities [60].

β-CD should be regarded as being among the natural macrocycles because it is acquired from the hydrolysis of starch. They are fundamentally polymers of glucose, and these macrocycles are well known for their nontoxicity and biodegradability. Thus, they have an expansive spectrum of utility in the pharmaceutical sector [61], nitroaromatic reduction [62], chemical fixation of carbon dioxide [63], drug delivery [64], and catalysis [2–4, 6, 9].

2.1 Properties and Advantages of β-CD

β-CDs is a cyclic heptamer comprised of seven glucose sections joined by α-1,4-bond. The hollow size and the internal hydrophobicity are appropriate for encapsulating different guests, such as aromatic compounds. It is broadly acknowledged that the linkage forces implicated in the inclusion complex creation are van der Waals interactions and hydrophobic interactions between β-CD and guest molecules. Because of such behavior, β-CD catalyzes diverse organic transformations in aqueous media. Per β-CD symbolizes a torus in which the internal surface has a relatively hydrophobic feature, and the exterior one is hydrophilic. The secondary hydroxyl groups encircle the vaster radius of the torus, while the slender ones include all of the primary ones. They are biochemical conversion products from starch decomposition by enzymatic transformation [65]. The main attributes of the mentioned CDs are demonstrated in Table 1 [66].

Owing to the place of the hydroxyl groups on the upper edge (secondary hydroxyls) and lower edge (primary hydroxyls) of the torus, CDs have a hydrophobic cavity, while their external surface is hydrophilic. The principal features, such as low price, solubility, nontoxicity, freely available hydroxyl groups, and ability to create inclusion complexes, make this compound an excellent choice for numerous applications in diverse fields of the chemistry world.

The ability and power of cyclodextrins to make host-guest complexes in hydrophobic holes are remarkable. They offer a significant potential for trapping various guest molecules without forming any covalent bonds. They contain various hollow sizes, and the size of the central hollow plays a crucial role in building inclusion complexes. Owing to the accessibility of several reactive OH parts, the functionality of CDs is significantly enhanced with chemical modification. Via modification, the use of CDs is extended. CDs are modified by replacing diverse functional groups on the primary and secondary face. Modified CDs are helpful due to the substituted functional groups operating for molecular recognition. Also, this feature is utilized in analytical chemistry and aims at drug delivery as modified CDs display enhanced enantioselectivity compared to unfunctionalized CDs [67, 68]. The most vital thing in guest selectivity is the size of the cyclodextrin hollow, which is compatible with that of the guest compound. Once the CD hole accommodates a guest substance, the resulting system can exhibit various attributes that differ significantly from the guest molecule alone. Molecular encapsulation may occur in the solution and/or the solid state. In a solid state, the guest substance could be encapsulated entirely, likely leading to a decreased chance of housing the interior of the CD's cavity [69, 70]. The CDs are well known as invaluable medically, as they can interact with diverse drug

Table 1 The principal attributes of the three CDs

Features	α-CD	β-CD	γ-CD
Glucose section	6	7	8
Molecular mass/g	972	1,135	1,237
Solubility in water (mg/mL)	145	18.5	232
Cavity volume/nm^3	0.174	0.262	0.472
Depth (nm)	0.78	0.78	0.78
pK_a	12.33	12.20	12.08
Interior diameter (nm)	0.57	0.78	0.95
Exterior diameter (nm)	1.37	1.53	1.69

$$\text{Drug}_{\text{free}} + \text{CD}_{\text{free}} \xrightleftharpoons{K} \text{Drug/CD}_{\text{complex}}$$

Scheme 4 Construction of inclusion complexes between CD and drug compounds

compounds to make inclusion complexes; this inclusion complex commonly has a 1: 1 interaction, which is exhibited in Scheme 4 [71].

Different techniques and methods were utilized to prepare inclusion complexes, as indicated in Scheme 5. Spray-drying and freeze-drying approaches were introduced as the most helpful drug/CD inclusion complexation strategies [72].

Over the last few decades, CDs have significantly contributed to the development of novel homogeneous and heterogeneous catalytic strategies. More than straightforward compound scaffolds to design innovative ligands, CDs have produced uncommon reaction mediums like supramolecular hydrogels and low-melting mixtures that can stabilize reactive catalytic species. By employing such replacement and unusual media, increased catalytic performance and selectivities were acquired for diverse transition metal catalytic transformations. The researched catalytic platforms are comfortable to execute and permit the excellent stabilization of organometallic and MNP catalysts. Remarkably, the size of MNPs could be well adjusted via confinement for the CD supramolecular hydrogels. Further, the catalysts could be recycled with easy phase separation. This catalytic phase could be reserved at 25°C under an atmosphere in the solid state and recovered with no notable performance reduction. Finally, such CD catalytic systems significantly enhance procedure safety [73].

2.2 *Application of β-CD as the Catalyst in Organic Reactions*

In recent years, "green chemistry" has been drawing significant attention as an innovative technology for the sustainable development of the world economy with no harm to worldwide ecosystems. In this idea, chemical conversions should be executed with optimal atom efficiency and energy atomic economy without using dangerous solvents and substances. Water is more desirable than organic solvents due to its eco-friendliness, nontoxicity, and safety. Outstanding catalysts are essential for achieving diverse conversions selectively and efficiently under moderate conditions. To prepare complex chemicals from several constituents, the one-pot strategy is preferable to stepwise procedures. These endeavors could reduce the cost and energy required to produce pure products and eliminate unwanted materials [74–78]. Along this line, CDs and their related compound satisfy all these needs and help develop green chemistry goals. Also, their distinctive characteristics can be improved regarding different chemical approaches. CDs, as a part of supramolecular chemistry, nowadays have a growing interest among researchers in using them as catalysts in organic transformations [79–85]. The catalysis-based CD systems are indicated by five features, all of which are vital for green chemistry, including [86–88]:

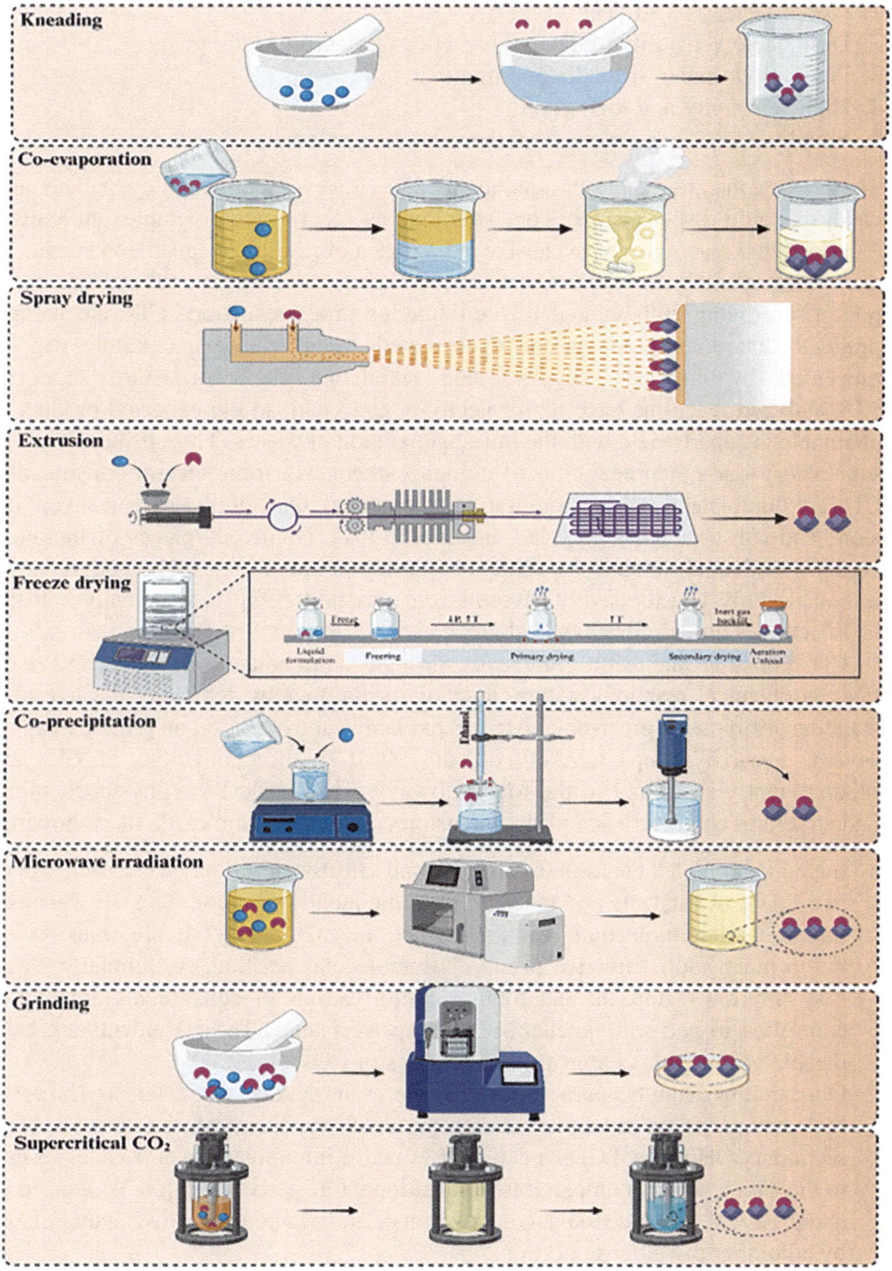

Scheme 5 The usual CD complexation methods [57]

1. Use of water as a solvent
2. Designable versatility of catalytic reactions based on conditions
3. Pretty environment-friendly and nontoxic
4. High selectivity and high yield
5. Simple recovery and reuse after the catalytic process

These reactions are doable in aqueous media. However, aquaphobic reactants and catalysts could disperse in aqueous solutions by incorporating complex structures. CD-catalyzed reactions are expansive at a range and readily designable via covalent and noncovalent modifications of CDs [89–96]. Once essential, several CD compounds are being collaborated for cutting-edge processes. Since CDs are natural products that do not pollute the environment, their release into nature can be prevented by creating polymers and stabilizing them on solid supports. CD-catalyzed reactions have high selectivity and yield, so the expected product is attainable in superb yield with the tiniest generation of wastes. This attribute leads to satisfactory one-pot preparation of complex chemicals from several constituents. CDs are durable and stable enough to be maintained intact during chemical conversions and can be readily recycled and reused [97]. Numerous pieces of literature certify the adequate chemical stability of CDs in traditional states. Their degradation is notable only in excessively severe acidic situations [98]. In conclusion, all the CDs' features are apt for advanced green chemistry.

CDs construct inclusion complexes through intermediates, catalysts, and reactants in chemical reactions within aqueous media, thereby achieving the five outstanding attributes of green chemistry, as explained above. Based on prior studies by several research groups [59, 60, 64, 66, 99–102], these functions of CDs are predominantly according to the four pathways, which reflect the physicochemical and structural characteristics of the inclusion complexes that include the following:

1. Including complex creation for efficient and selective reaction via near and proper placement of catalysts and reactants. Via the inclusion of complex construction, traditional intermolecular reactions (intermolecular catalysis) are transformed into a much more effective pseudo-intramolecular reaction. Additionally, these easy proximity impacts and mutual conformations of constituents are highly controlled to perform excellent reaction performance. These elements are handleable via suitable chemical modifications of CDs.
2. Outstanding catalytic parts are accessible as intramolecular catalysts (inherent hydroxyl parts and functional section presented via chemical modification). The secondary OH parts of CDs function as effective intramolecular acid catalysts due to their high acidity compared to conventional OH parts. Their pK_a is reduced to about 12 with the successive construction of an intramolecular hydrogen linkage by adjoining the glucose portion.
3. Cavities of CDs act as a sterically limited reaction area to control the size and form of the product.
4. Protection of unsteady chemicals from unwanted reactions, decomposition to boost the main reaction, and suppression of undesired transformation.

In multiple CD-mediated reactions, several items of these four mechanisms A–D synergistically work to promote and improve the catalytic performance. Indeed, the solubilization of hydrophobic chemicals with CDs in aqua media is critical for this pathway to operate with satisfactory productivity. Also, several catalytic conversions by CDs include special and general methods.

Hapiot et al. [81] have suggested the present categorization in catalytic uses of CDs:

(A) They enhance the selectivity and speed of reactions catalyzed by water-soluble organometallic complexes.
(B) Developing novel water-soluble ligands in aqueous organometallic catalysis.
(C) The stabilized catalytically active noble metal NPs in aqueous media.
(D) Supported metals catalyze the facilitated reactions in an aqueous medium.

These cyclic oligosaccharides are eco-friendly, nontoxic, and recoverable to perform prominent functions in aqueous medium. Their most noteworthy properties are to create inclusion complexes by intermediates, reactants, and catalysts. Consequently, their hollows act as sterically limited and polar reaction fields to heighten the selectivity and efficiency of reactions. Also, intermediates and unsafe reagents are protected from unwanted side conversions. The range of their utility has been additionally broadened via covalent and noncovalent modifications. Combinations of them by metal catalysis are effective and fruitful. In these influences, diverse chemical transformations are executed with perfect yield and selectivity, so invaluable compounds are prepared in one-pot reactions from multiple components. In addition, CD units are well organized and made with oligomers and polymers to display their participation in cutting-edge features. CD-based metal-organic skeleton and polyoxometalate-CD scaffolding have been created and used for special utility. CDs meet numerous green chemistry needs and will likely contribute considerably to this growing area.

2.3 *Use the β-CD-Based Catalyst System in Click Reactions*

β-CDs have long been recognized as a significant contributor to the growth of catalytic methodologies. They have been utilized in catalytic conversion and generation of innovative catalysts imitating enzymatic performance. This section summarizes the uses of the β-CD catalytic system in click reactions.

2.3.1 Native β-CD

Recent concerns in the CD field have concentrated on the catalytic synthesis of valuable chemical material. The conception of green chemistry has been extremely underscored. In this report, β-CD is used in its native shape with no modification. In 2012, Shin and colleagues presented the preparation of triazoles **3** by utilizing β-CD

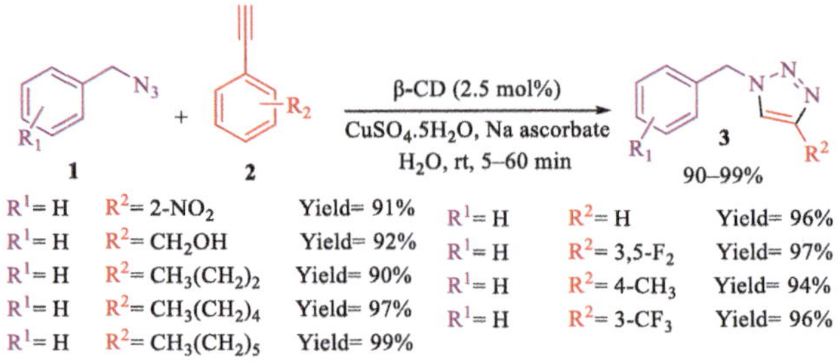

Scheme 6 The CuAAC reaction using β-CD developed by Shin et al. [103]

(2.5 mol%) as a helpful phase transfer catalyst in water media, and the expected triazole derivative **3** was attained with ideal results in a few minutes (Scheme 6) [103]. The authors used the ^{1}H-NMR technique to examine the task of β-CD in the procedure. It contained guest molecules in the β-CD hollow, leading to upfield shifts of the H-3 and H-5, demonstrating their location towards the internal of the β-CD system. Contrarily, the H-1, H-2, and H-4 on the β-CD's external surface remained unchanged. Remarkably, the hydrogens of β-CD are relatively protected at benzyl azide and phenylacetylene, delivering proof of guest-host relationship. A noteworthy benefit of this method is that the reaction could be executed in aqueous media, removing the requirement for harmful solvents, unsafe reagents, and expensive substances. This strategy presents eco-friendly and economic advantages.

2.3.2 Cu(II)-β-CD Complex

In 2012, for the first time, Kaboudin et al. employed Cu(II)-β-CD as a green catalyst in a one-pot preparation of 1,2,3-triazoles from Ar-B$(OH)_2$ via a click cyclization reaction that had not been previously investigated (Scheme 7) [104]. A suitable, fast, and environmentally safe one-pot approach was designed and expanded to create product **5** from **2**, NaN_3, and Ar-B$(OH)_2$ **4** at water at ambient temperature, with the existence of O_2 and with no additives.

In 2017, Gupta et al. utilized this same Cu(II)-β-CD catalyst as a green and water-friendly nanocatalyst in the synthesis of triazole derivatives 7 from **6**, sodium azide, and **2** at 25°C (Scheme 8) [105]. This catalyst operates selectively and efficiently and reacts rapidly (15–70 min). This Cu(II)-β-CD was created to employ a modified prior procedure type and was analyzed using different methods. The triazole compounds were readily separated with a selectivity of up to 99% by easy filter. This catalyst presents benefits compared to earlier studied catalysts, comprising its low price, efficiency, safety, and eco-friendliness. This strategy offers high-yield outcomes with no demand for extra reducing agents. The catalytic activity was assessed five times, and though a little reduction in activity was observed, the catalyst remained recyclable.

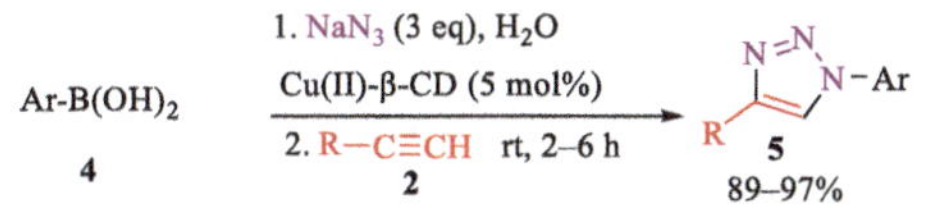

Ar= C_6H_5	R= C_6H_5	Yield= 94%	Ar= β-Naphthyl	R= C_6H_5	Yield= 94%
Ar= 2-FC_6H_4	R= C_6H_5	Yield= 94%	Ar= C_6H_5	R= 4-$H_3CC_6H_4$	Yield= 95%
Ar= 3-$O_2NC_6H_4$	R= C_6H_5	Yield= 95%	Ar= C_6H_5	R= 3-$H_2NC_6H_4$	Yield= 95%
Ar= 3,5-$F_2C_6H_3$	R= C_6H_5	Yield= 96%	Ar= C_6H_5	R= $CH_2CH_2CH_3$	Yield= 93%
Ar= 4-$CH_3C_6H_4$	R= C_6H_5	Yield= 94%	Ar= C_6H_5	R= $CH_2CH_2CH_2CH_3$	Yield= 89%
Ar= 4-$CH_3OC_6H_4$	R= C_6H_5	Yield= 97%	Ar= C_6H_5	R= $CH_2(CH_2)_4CH_3$	Yield= 91%
Ar= 4-ClC_6H_4	R= C_6H_5	Yield= 96%	Ar= C_6H_5	R= CH_2OH	Yield= 92%
Ar= 4-$HOCC_6H_4$	R= C_6H_5	Yield= 94%	Ar= C_6H_5	R= $OHCH(CH_2)_4CH_3$	Yield= 90%
			Ar= C_6H_5	R= CO_2H	Yield= 94%

Scheme 7 The preparation of triazole derivatives **5** from Ar-B$(OH)_2$ by Cu(II)-β-CD developed by Kaboudin et al. [104]

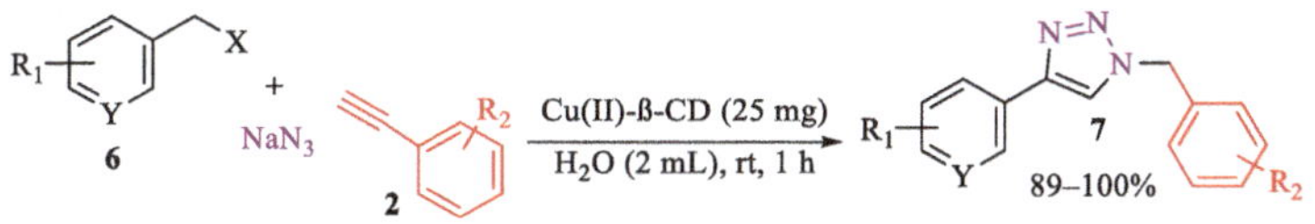

X= Cl, Br R_2= H, CH_3
Y= C, N
R_1= Cl, NO_2, CN, F, CH_3

R_1= 4-F	R_2= H	Yield= 90%
R_1= 4-Cl	R_2= H	Yield= 90%
R_1= 2,4-Cl_2	R_2= H	Yield= 89%
R_1= 2-CN	R_2= H	Yield= 100%
R_1= 4-NO_2	R_2= H	Yield= 100%
R_1= H	R_2= H	Yield= 91%
R_1= H	R_2= CH_3	Yield= 97%
R_1= 2-CH_3 pyridine	R_2= H	Yield= 100%

Scheme 8 The preparation of product **7** using Cu(II)-β-CD catalyst developed by Gupta et al. [105]

$$\text{Cu(II)–β-CD(OH)}_2^{-2} + 4H_2O \longrightarrow 2\,Cu_2^{+} + \text{β-CD} + 6OH^-$$

$$\left.\begin{array}{c} ZrOCl_2 \\ \text{Cu(II)–β-CD(OH)}_2^{-2} \\ NH_4OH \end{array}\right\} \longrightarrow \text{Zr(II)-β-CD-Cu(II)}$$

Scheme 9 Creation of ZrO_2-supported Cu(II)-β-CD developed by Girish et al. [106]

2.3.3 Functionalized β-CD Nanoparticles

In 2014, Girish et al., for the first time, reported an exceedingly proper one-pot procedure in preparing *N*-2-substituted-1,2,3-triazoles using ZrO_2 NP-supported Cu (II)-β-CD [106]. This nanocatalyst was created via a simple one-pot coprecipitation method, including $ZrOCl_2{\cdot}8H_2O$, Cu(II)-β-CD, and ammonia solution (Scheme 9). This process was carried out without extra additives, leading to derivatives **10** from NaN_3, **9**, and **8** (Scheme 10). It is crucial to underscore that prior procedures did not utilize esters as an alkylating reagent, and several had flaws like harsh temperature, dangerous reagents, and prolonged reaction times. In comparison, this approach suggests some benefits over earlier studied techniques. These possess the ease of the method, catalyst recoverability, usage in drug chemistry, regioselectivity, and

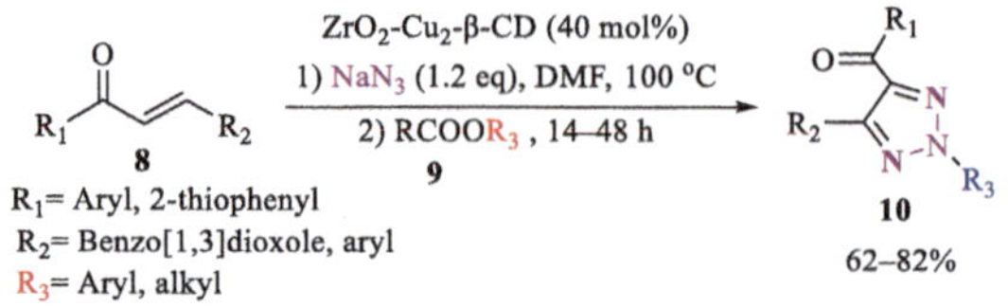

R_1= 2-Thiophenyl	R_2= 4-$CH_3OC_6H_4$	R_3= CH_3	R= 3-ClC_6H_4	Yield= 80%
R_1= 2-Thiophenyl	R_2= 3-4-$(OCH_3)_2C_6H_3$	R_3= CH_3	R= 3-ClC_6H_4	Yield= 82%
R_1= 2-Thiophenyl	R_2= C_6H_5	R_3= CH_3	R= 3-ClC_6H_4	Yield= 76%
R_1= 4-$CH_3OC_6H_4$	R_2= C_6H_5	R_3= CH_3	R= 3-ClC_6H_4	Yield= 81%
R_1= 4-$CH_3OC_6H_4$	R_2= 4-$CH_3C_6H_4$	R_3= CH_3	R= 3-ClC_6H_4	Yield= 82%
R_1= 4-$CH_3OC_6H_4$	R_2= 4-FC_6H_4	R_3= CH_3	R= 3-ClC_6H_4	Yield= 78%
R_1= 2-Thiophenyl	R_2= 4-FC_6H_4	R_3= CH_2CH_3	R= $BrCH_2$	Yield= 72%
R_1= 2-Thiophenyl	R_2= 4-EthylC_6H_4	R_3= CH_3	R= 3-ClC_6H_4	Yield= 79%
R_1= 2-Thiophenyl	R_2= 3,4,5-$(OCH_3)_3$ C_6H_2	R_3= CH_3	R= 3-ClC_6H_4	Yield= 81%
R_1= 2-Thiophenyl	R_2= Benzo[1,3]dioxole	R_3= CH_3	R= 3-ClC_6H_4	Yield= 75%
R_1= 2-Thiophenyl	R_2= 4-EthylC_6H_4	R_3= CH_3	R= 3-ClC_6H_4	Yield= 77%
R_1= 4-$CH_3OC_6H_4$	R_2= 4-BrC_6H_4	R_3= CH_3	R= 3-ClC_6H_4	Yield= 79%
R_1= 4-$CH_3OC_6H_4$	R_2= C_6H_5	R_3= CH_2CH_3	R= 3-ClC_6H_4	Yield= 68%
R_1= 2-Thiophenyl	R_2= 4-FC_6H_4	R_3= CH_3	R= C_6H_5	Yield= 62%

Scheme 10 Synthesizes compounds **10** by employing ZrO_2–Cu_2–β-CD developed by Girish et al. [106]

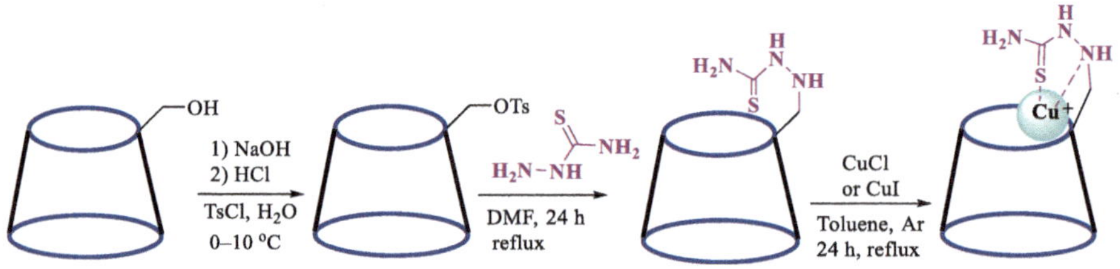

Scheme 11 Stages of catalyst preparation developed by Tajbakhsh and Naimi-Jamal [107]

excellent yields. This nanocatalyst was recycled via filtration and retained its performance for over four cycles with only a minute decline in catalytic activity.

Recently, Tajbakhsh and Naimi-Jamal created and developed a high-stability, homogeneous, and water-soluble catalyst of Cu(I) ions supported on thiosemicarbazide-modified named β-CD (Cu@TSC-β-CD) [107]. The catalyst preparation steps are shown in Scheme 11. This new catalyst presents a one-pot, economical, excellent-performance, and extremely regioselective way of preparing 1,2,3-triazoles via click reaction in aqueous media. Triazole derivatives' yields reached 95% under optimized states (Scheme 12). The presence of thiosemicarbazide in the catalyst structure facilitates strong binding with Cu through coordination, a crucial factor in preventing significant Cu(I) leaching. The nanocatalyst constructed a broad spectrum of 1,2,3-triazoles **3** from **11**, various alkynes **2**, and aryl azides. This catalyst was recycled by filtration and reused for seven runs with no drop in its catalytic performance. The offered route for this click reaction is the same as the known pathways in the CuAAC (Scheme 13).

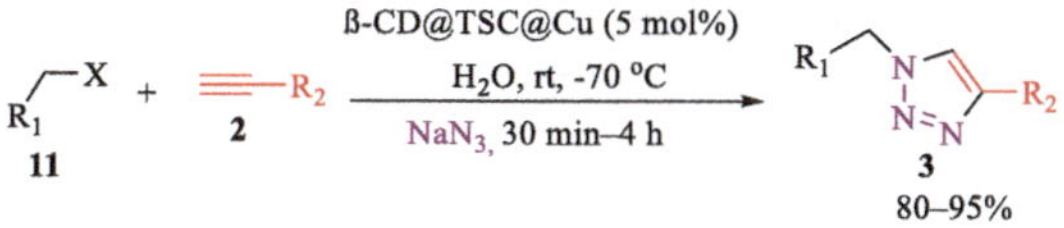

X= Br	R_1= $C_6H_5CH_2$	R_2= Ar	Yield= 95%
X= Cl	R_1= $C_6H_5CH_2$	R_2= Ar	Yield= 92%
X= Br	R_1= 4-$CH_3OC_6H_4$	R_2= Ar	Yield= 90%
X= Br	R_1= $C_6H_5COCH_2$	R_2= Ar	Yield= 85%
X= Br	R_1= 4-$O_2NC_6H_4COCH_2$	R_2= Ar	Yield= 90%
X= Br	R_1= 4-$CH_3C_6H_4COCH_2$	R_2= Ar	Yield= 85%
X= Br	R_1= 4-$BrC_6H_4COCH_2$	R_2= Ar	Yield= 84%

Scheme 12 Cu@TSC-β-CD as a catalyst for the creation of compounds **3** developed by Tajbakhsh and Naimi-Jamal [107]

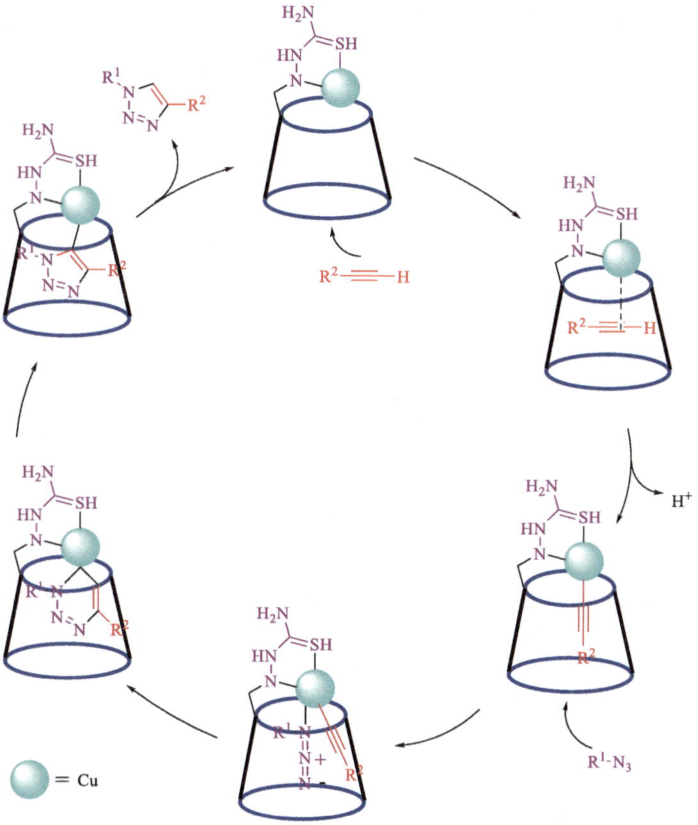

Scheme 13 A suggested pathway in click reaction by using Cu@TSC-β-CD developed by Tajbakhsh and Naimi-Jamal [107]

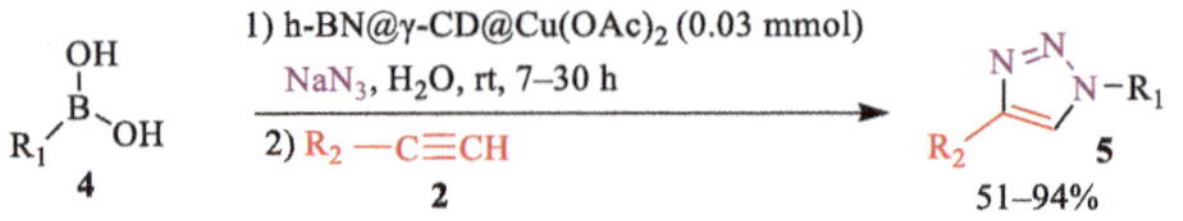

R_1= C_6H_5	R_2= C_6H_5	Yield= 90%	R_1= C_6H_5	R_2= *N*-octyl	Yield= 94%
R_1= 3-$CH_3C_6H_4$	R_2= C_6H_5	Yield= 91%	R_1= C_6H_5	R_2= 2-Hydroxyethyl	Yield= 84%
R_1= 4-$(CH_3)_3CC_6H_4$	R_2= C_6H_5	Yield= 72%	R_1= C_6H_5	R_2= Isobutanol	Yield= 87%
R_1= 4-$CH_3OC_6H_4$	R_2= C_6H_5	Yield= 90%	R_1= C_6H_5	R_2= Aminomethyl	Yield= 80%
R_1= 4-FC_6H_4	R_2= C_6H_5	Yield= 91%	R_1= 3,5-$(CF_3)_2C_6H_3$	R_2= C_6H_5	Yield= 88%
R_1= 3,5-$(Cl)_2C_6H_3$	R_2= C_6H_5	Yield= 71%	R_1= 3-Pyridiniyl	R_2= C_6H_5	Yield= 51%

Scheme 14 This new nanocatalyst is used in the click reaction developed by Nie et al. [108]

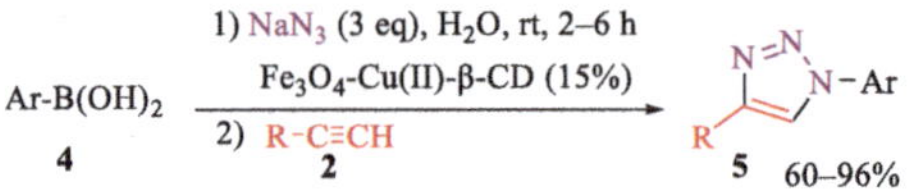

Ar= C_6H_5	R= H	Yield= 95%	Ar= 4-$HOCC_6H_4$	R= H	Yield= 93%
Ar= 2-FC_6H_4	R= H	Yield= 95%	Ar= α-Naphthyl	R= H	Yield= 90%
Ar= 2-$CH_3C_6H_4$	R= H	Yield= 95%	Ar= β-Naphthyl	R= H	Yield= 96%
Ar= 3,5-$(F)_2C_6H_3$	R= H	Yield= 95%	Ar= C_6H_5	R= 4-CH_3	Yield= 94%
Ar= 4-$CH_3C_6H_4$	R= H	Yield= 95%	Ar= C_6H_5	R= 4-OCH_3	Yield= 96%
Ar= 4-$CH_3OC_6H_4$	R= H	Yield= 95%	Ar= C_6H_5	R= 3-NH_2	Yield= 60%
Ar= 4-ClC_6H_4	R= H	Yield= 91%	Ar= C_6H_5	R= $CH_3(CH_2)_3$	Yield= 90%
			Ar= C_6H_5	R= $CH_3(CH_2)_5$	Yield= 93%

Scheme 15 The preparation of triazoles **5** utilizing Cu(II)-β-CD NPs developed by Kaboudin et al. [109]

2.3.4 Heterogeneous Catalysis

In 2016, Nie et al. introduced h-BN@γ-CD@$Cu(OAc)_2$ as a proper and recoverable catalyst in synthesizing triazoles [108]. The hexagonal boron nitride (*h*-BN) was employed as a catalyst support in various chemical reactions because it is chemically and thermally resistant, has excellent mechanical stability, and is structurally similar to graphene. These researchers utilize modified *h*-BN as a stable substance for supporting Cu-γ-CD systems. This heterogeneous catalyst exhibited promising catalytic activity in the reaction of R_1-$B(OH)_2$ **4** with sodium azides and diverse 2 at 25°C in an aqueous medium (Scheme 14). Various techniques identified this heterogeneous catalyst. The expected triazole **5** yielded fine to superior results with minimal Cu loadings (0.03 mmol). The recyclability of this catalyst showed that it is reusable seven times with a slight reduction in performance.

2.3.5 Copper-β-CD-Functionalized Magnetic NPs

Kaboudin et al. developed and explained a new reusable Fe_3O_4-supported Cu (II)-β-CD complex as a nanocatalyst to synthesize triazole **5** from aryl boronic acids **4** (Scheme 15) [109]. The novel catalyst demonstrated excellent activity across a broad range of substrates in aqueous media. The recoverability of this nanocatalyst was checked four times with a tiny reduction in catalytic activity.

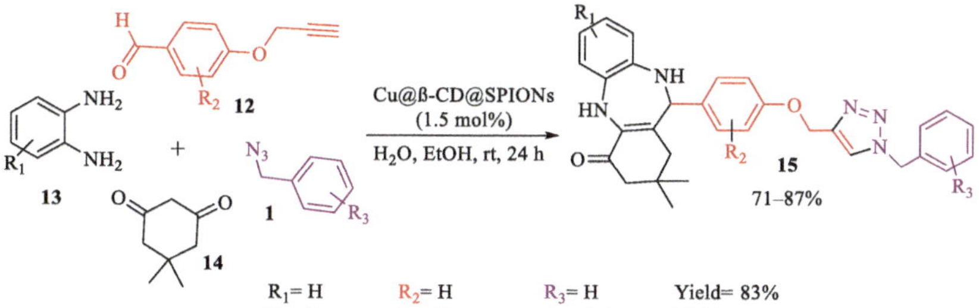

R_1= H	R_2= H	R_3= H	Yield= 83%
R_1= H	R_2= H	R_3= 4-F	Yield= 84%
R_1= H	R_2= H	R_3= 4-NO_2	Yield= 80%
R_1= H	R_2= H	R_3= 2,3-Cl_2	Yield= 82%
R_1= H	R_2= H	R_3= 4-CH_3	Yield= 87%
R_1= H	R_2= H	R_3= 3-Cl	Yield= 81%
R_1= H	R_2= 2-OCH_3	R_3= H	Yield= 79%
R_1= 2-CH_3	R_2= H	R_3= 4-F	Yield= 76%

Scheme 16 Demonstration of the Cu@β-CD@SPION nanocatalyst for the construction of compounds **15** developed by Mahdavi et al. [110]

Mahdavi et al. designed a new magnetic reusable catalyst abbreviated Cu@-β-CD@SPIONs and employed it as a helpful nanocatalyst in preparing novel triazoles **15** through click reaction (Scheme 16) [110]. This novel nanocatalyst was constructed via immobilizing Cu within β-CD covalently linked to Fe_3O_4 NPs (Scheme 17). One of the most notable benefits of this nanocatalyst was that it worked as an excellent nanocatalyst for multicomponent and click reactions. Employing an external magnet, this nanocatalyst was separated and reused over five runs without significantly reducing its activity. A possible route is demonstrated in Scheme 18. At first, an imine is made from the catalytic reaction between **13** and **14**. In this stage, the –NH_2 piece of **13** attacks the carbonyl group of **14**, and the catalyst starts to create **16** by releasing H_2O. A [1,3]-H shift made the enamine. Subsequently, the following –NH_2 section of **13** attacks the C=O group **12**, which is activated with the catalyst for creating **19**. Finally, an intramolecular cyclization was performed on the **20**. A [1,3]-H shift constructs the final compound **15**.

In another report, Bahadorikhalili et al. presented a recoverable nanocatalyst named Cu@βCD-PEG-mesoGO to prepare triazoles **3** and benzimidazoles (Scheme 19) [111]. The synthesis of various stages of the catalyst is shown in Scheme 20. For this nanocatalyst, β-CD works as the ligand in the Cu catalyst, and GO operates as a superior support owing to its extensive surface region. At the same time, PEG passes on hydrophilic attributes to the NPs, making them very water dispersible. Copper plays a crucial role in this nanocatalytic system during the click reaction. The Cu in this catalyst structure is vital because no results were obtained without this nanocatalyst. The catalyst displayed significant recovery and preserved its activity ten successive times with no substantial decline. This nanocatalyst exhibits outstanding characteristics, including high performance, simple workup, moderate conditions, use of H_2O as a green solvent, and recoverability. The writers presented a likely pathway for this nanocatalyst in the click reaction, as in Scheme 21.

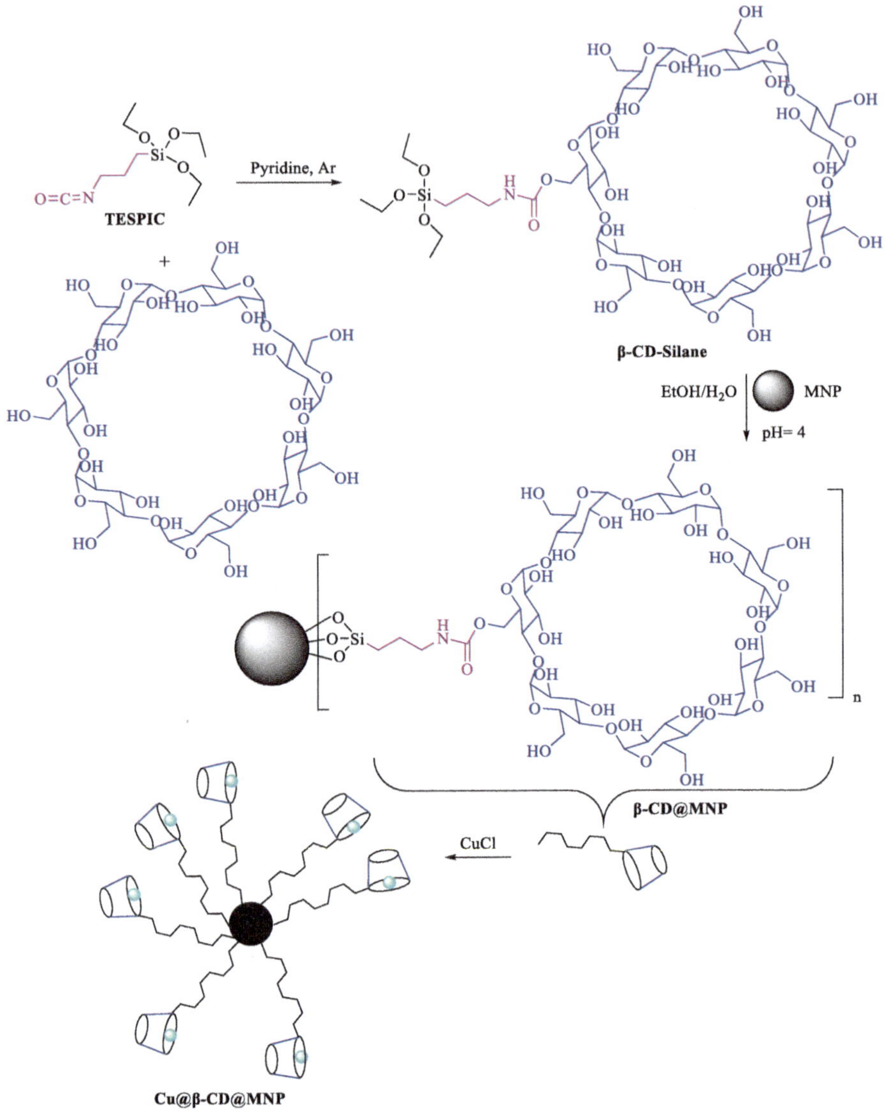

Scheme 17 The synthesis stages of the Cu@β-CD@SPION nanocatalyst developed by Mahdavi et al. [110]

Bahadorikhalili et al. presented a new Cu multifunctional nanocatalyst (Scheme 22) [112]. This nanocatalytic system comprised the usage of Cu-supported β-CD and then was functionalized by $SiO_2@Fe_3O_4$ NPs. This iron oxide-supported copper catalyst demonstrated great activity in click reactions, multicomponent reactions, and C–N linkage oxidation. The β-CD has two duties; both act as a ligand in copper and a phase transfer catalyst. The reaction of **23** with **22** leads to the preparation of

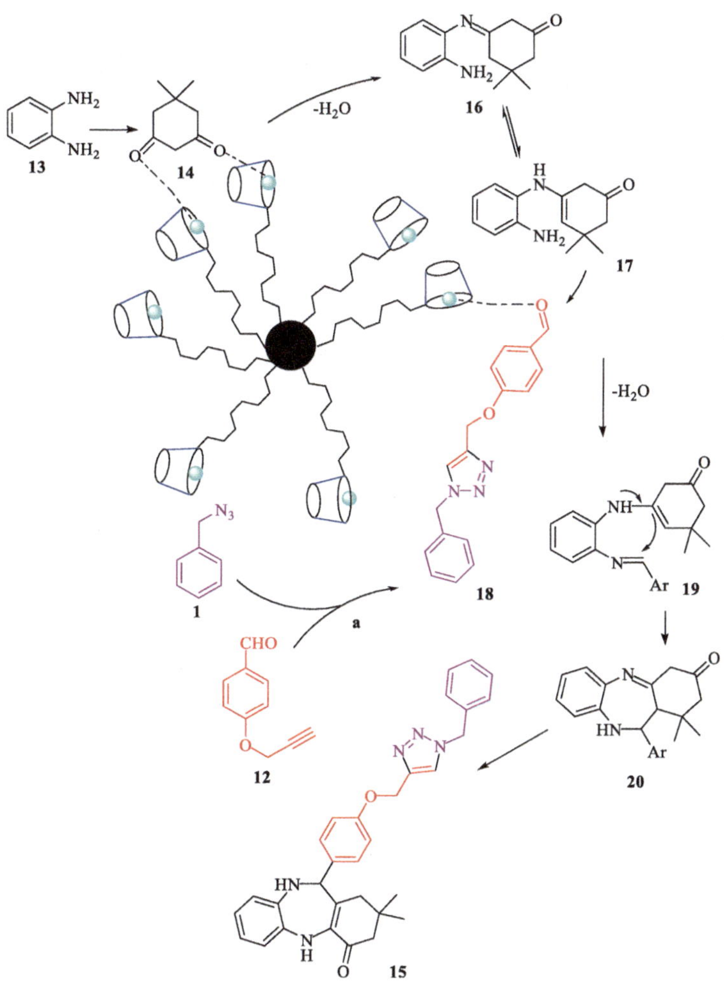

Scheme 18 The route offered for the preparation of products **15** developed by Mahdavi et al. [110]

various scopes of 1,2,3-triazole compounds **24** (Scheme 23). The reactions were effectively conducted in aqueous media operating as a safe solvent, which provided the expected products with excellent results. This nanocatalyst was readily separated utilizing a magnet and recovered with no noteworthy activity drop.

An overview of diverse β-CD catalysts for click reactions is shown in Table 2.

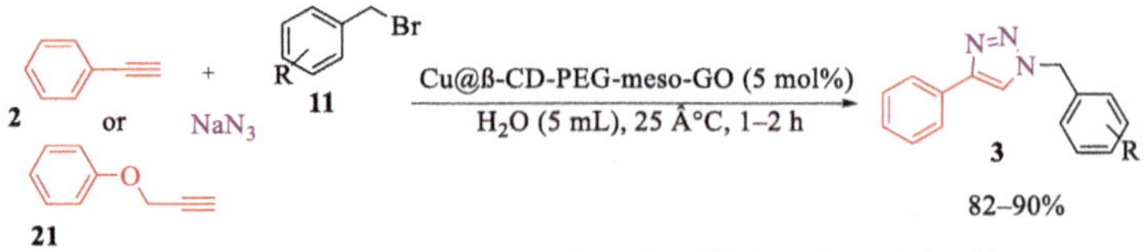

Reaction with phenylacetylene		Reaction with (prop-2-yn-1-yloxy)benzene	
R= H	Yield= 90%	R= H	Yield= 87%
R= 4-NO_2	Yield= 89%	R= 4-NO_2	Yield= 82%
R= 4-CH_3	Yield= 91%	R= 4-CH_3	Yield= 88%
R= 2-Cl	Yield= 90%	R= 2-Cl	Yield= 86%
R= 3-CH_3	Yield= 89%	R= 3-CH_3	Yield= 83%

Scheme 19 Synthesis of triazole derivatives using Cu@βCD-PEG-mesoGO developed by Bahadorikhalili et al. [111]

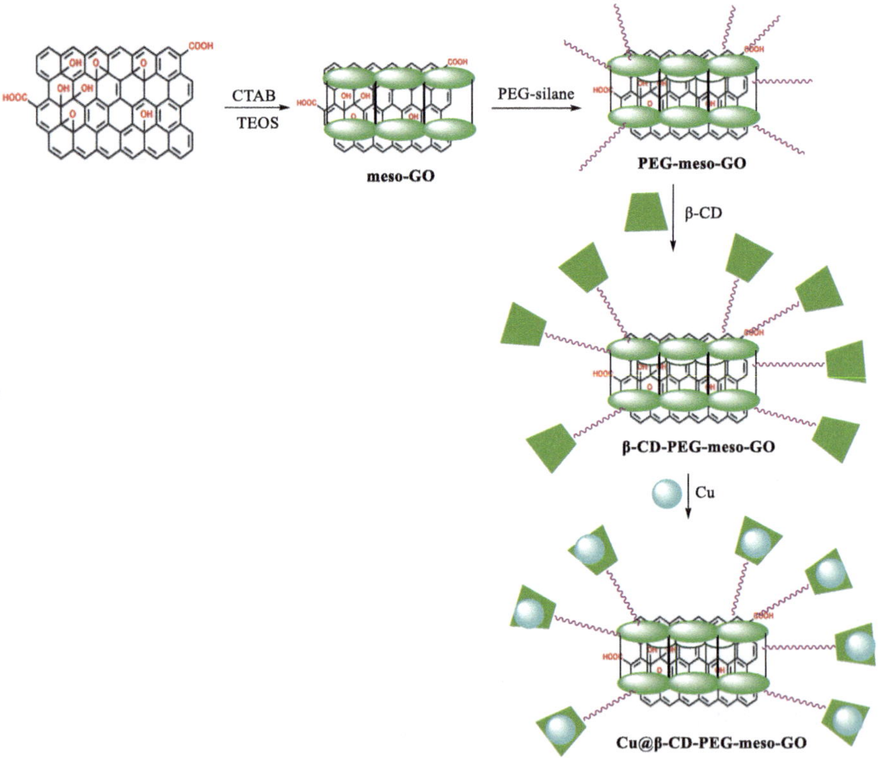

Scheme 20 The diverse steps of production of the Cu@βCD-PEG-mesoGO nanocatalyst developed by Bahadorikhalili et al. [111]

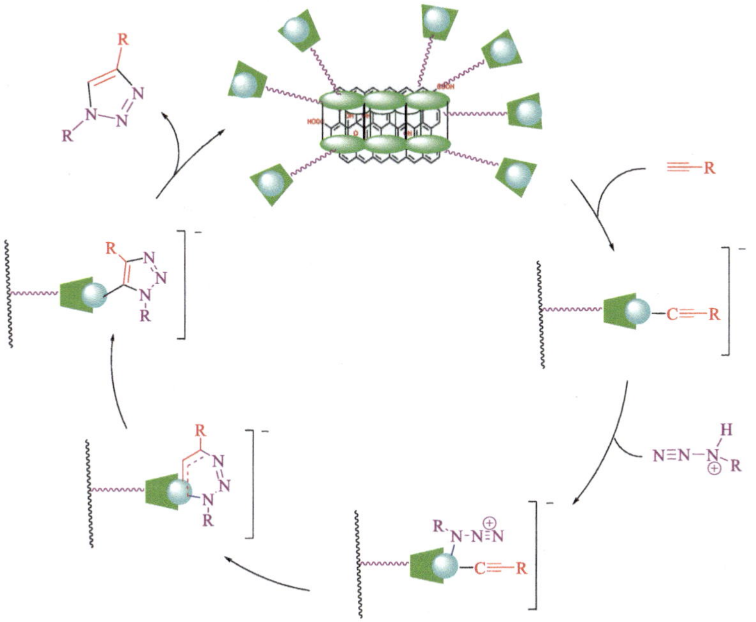

Scheme 21 The mechanism of triazole derivative preparation using this new current catalyst developed by Bahadorikhalili et al. [111]

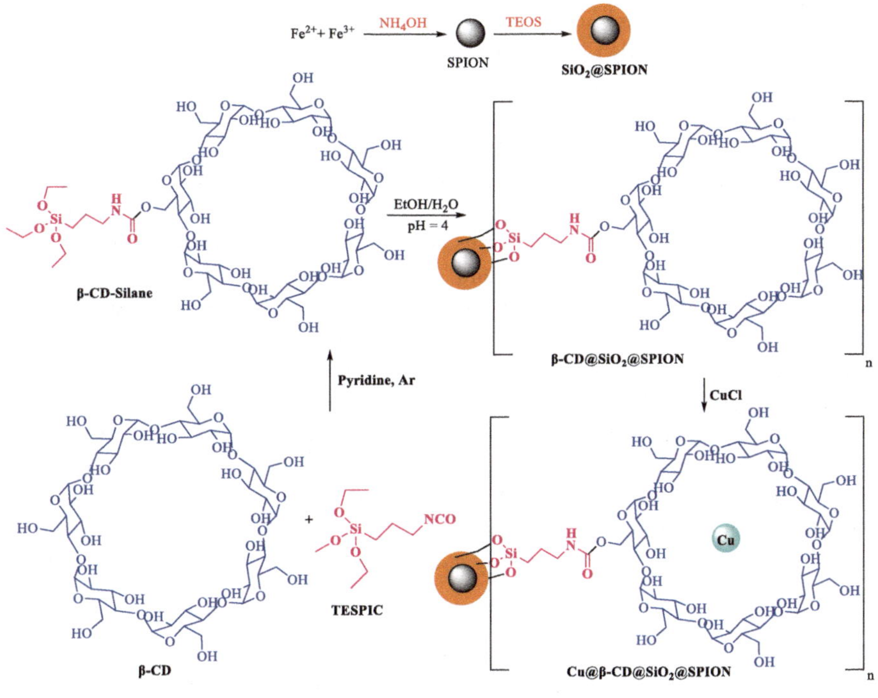

Scheme 22 The stepwise preparation pathway of Cu@β-CD@SiO_2@SPION developed by Bahadorikhalili et al. [112]

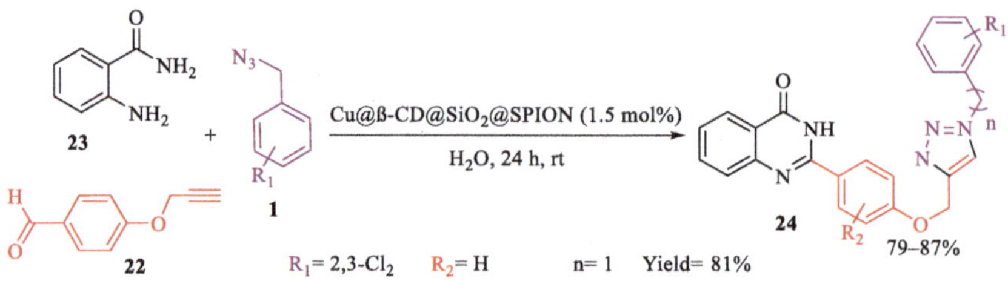

R_1= 2,3-Cl_2	R_2= H	n= 1	Yield= 81%
R_1= H	R_2= 2-OCH_3	n= 1	Yield= 88%
R_1= 4-CH_3	R_2= 2-OCH_3	n= 1	Yield= 84%
R_1= 4-F	R_2= 2-OCH_3	n= 1	Yield= 82%
R_1= 4-Me	R_2= 2-OCH_3	n= 1	Yield= 85%
R_1= 4-Cl	R_2= H	n= 0	Yield= 80%

Scheme 23 The employment of Cu@β-CD@SiO_2@SPION as a catalyst in synthesizing **23** developed by Bahadorikhalili et al. [112]

3 Conclusion

The β-CD-catalyzed click reaction is a valuable and essential method for constructing the 1,2,3-triazole framework, as evident from the reports examined in this chapter. This kind of reaction has been found to have broad uses in the synthesis of products of drug significance. This chapter focuses primarily on functionalized β-CD-Cu catalysts, both with and without ligands, which exhibit superb catalytic performance and reusability in click reactions. Thus, there is an interest in using increasingly various modified CD systems that could catalyze click reactions successfully with structurally diverse substrates. The study enables chemists to discover the potential of β-CD and related compounds as promising catalysts for organic conversion, mainly click reactions.

Immobilizing β-CD-Cu complexes onto magnetic supports, silica, or biopolymers may lead to highly recyclable, solid-supported catalysts with enhanced recovery and reusability in industrial-scale processes. Functional modification of β-CD (such as introducing ligating groups, ionic tags, or chiral moieties) may further improve regioselectivity, enantioselectivity, and substrate compatibility in click reactions, allowing for broader application in asymmetric synthesis and drug development. β-CD-based click catalysts could be incorporated into stimuli-responsive materials, hydrogels, or polymeric networks for creating self-healing, adaptive, or recyclable materials via triazole linkages. Continued innovation may yield β-CD catalysts that operate efficiently in pure water or solvent-free conditions, aligning with the principles of green chemistry and reducing environmental impact. β-CD's biocompatibility makes it an ideal scaffold for in vivo click chemistry and drug delivery systems, where mild, nontoxic conditions are essential. Future research may focus on CD-Cu nanocarriers for targeted therapeutic delivery and real-time bioconjugation.

Table 2 The catalytic activity of diverse β-CD catalysts for click reactions

Entry	Catalyst	Cu (mol%)	Solvent	Temp. (°C)	Time (h)	Yield (%)	Reference
1	$CuSO_4 \cdot 5H_2O$/β-CD	5	H_2O	Rt	5–60 min	90–99	[103]
2	Cu(II)-β-CD complex	5	H_2O	Rt	2–6	90–97	[104]
3	Cu(II)-β-CD	5	H_2O	Rt	15–70 min	90–89	[105]
4	ZrO_2Cu_2-β-CD	40	DMF	100	14–48	62–82	[106]
5	β-CD@TSC-@Cu	5	H_2O	Rt	30 min–4 h	80–95	[107]
6	h-BN@γ-CD@$Cu(OAc)_2$	0.03 mmol	H_2O	Rt	13–30	51–94	[108]
7	Fe_3O_4-β-CD-Cu(II)	15	H_2O	Rt	2–6	60–96	[109]
8	Cu@β-CD@SPIONs	1	H_2O-EtOH	Rt	24	71–87	[110]
9	Cu@β-CD-PEG-mesoGO	5	H_2O	Rt	1–2	82–90	[111]
10	Cu@β-CD@SiO_2@SPION	1.5	H_2O	Rt	24	79–88	[112]

Acknowledgements The authors would like to acknowledge the support from the Research Council of Urmia University.

References

1. Payamifar S, Abdouss M, Poursattar Marjani A (2025) An overview of porphyrin-based catalysts for sulfide oxidation reactions. Polyhedron 269:117389. https://doi.org/10.1016/j.poly.2025.117389
2. Payamifar S, Poursattar Marjani A (2023) Recent advances in β-cyclodextrin-based catalysts for reducing toxic nitroaromatic: an overview. Appl Organomet Chem 37:e7287. https://doi.org/10.1002/aoc.7287
3. Payamifar S, Abdouss M, Poursattar Marjani A (2024) The recent development of β-cyclodextrin-based catalysts system in Suzuki coupling reactions. Appl Organomet Chem 38:e7458. https://doi.org/10.1002/aoc.7458
4. Payamifar S, Abdouss M, Poursattar Marjani A (2024) Recent advances in β-cyclodextrin-based catalyst systems for the synthesis of heterocyclic compounds via multicomponent reactions (MCRs). Arab J Chem 17:105967. https://doi.org/10.1016/j.arabjc.2024.105967
5. Crini G, Fourmentin S, Fenyvesi É, Torri G, Fourmentin M, Morin-Crini N (2018) Cyclodextrins, from molecules to applications. Environ Chem Lett 16:1361–1375. https://doi.org/10.1007/s10311-018-0763-2
6. Payamifar S, Kazemi F, Kaboudin B (2021) Nickel/β-CD-catalyzed Suzuki–Miyaura cross-coupling of aryl boronic acids with aryl halides in water. Appl Organomet Chem 35:e6378. https://doi.org/10.1002/aoc.6378
7. Jicsinszky L, Cravotto G (2021) Toward a greener world-cyclodextrin derivatization by mechanochemistry. Molecules 26:5193. https://doi.org/10.3390/molecules26175193
8. Khandelwal R, Vasava M, Abhirami R, Karsharma M (2024) Recent advances in triazole synthesis via click chemistry and their pharmacological applications: a review. Bioorg Med Chem Lett 112:129927. https://doi.org/10.1016/j.bmcl.2024.129927
9. Payamifar S, Poursattar Marjani A (2024) The recent development of β-cyclodextrin-based catalysts system in click reactions: a review. Appl Organomet Chem 38:e7365. https://doi.org/10.1002/aoc.7365
10. Kolb HC, Finn M, Sharpless KB (2001) Click chemistry: diverse chemical function from a few good reactions. Angew Chem Int Ed 40:2004–2021. https://doi.org/10.1002/1521-3773(20010601)40:11<2004:AID-ANIE2004>3.0.CO;2-5
11. Amna B, Ozturk T (2021) Click chemistry: a fascinating method of connecting organic groups. Org Commun 14:120. https://doi.org/10.25135/acg.oc.100.21.03.2006
12. (a) Swords B (1999) American Chemical Society 217th National Meeting division of Medicinal Chemistry general session. 21–25 March 1999, Anaheim, CA, USA IDrugs 2:507–510; (b) Moses JE, Moorhouse AD (2007) The growing applications of click chemistry. Chem Soc Rev 36:1249–1262. https://doi.org/10.1039/B613014N
13. Rostovtsev VV, Green LG, Fokin VV, Sharpless KB (2002) A stepwise Huisgen cycloaddition process: copper(I)-catalyzed regioselective "ligation" of azides and terminal alkynes. Angew Chem Int Ed 41:2596–2599. https://doi.org/10.1002/1521-3773(20020715)41:14<2596::AID-ANIE2596>3.0.CO;2-4
14. Yadav MS, Rajkhowa S, Singh SK, Jaiswal MK, Tiwari VK (2024) Multicomponent click reaction: an indispensable tool for easy access of functionalized 1,2,3-triazoles. ChemistrySelect 9:e202400776. https://doi.org/10.1002/slct.202400776
15. Agrahari AK, Bose P, Jaiswal MK, Rajkhowa S, Singh AS, Hotha S, Mishra N, Tiwari VK (2021) Cu(I)-catalyzed click chemistry in glycoscience and their diverse applications. Chem Rev 121:7638–7956. https://doi.org/10.1021/acs.chemrev.0c00920

16. Tiwari VK, Jaiswal MK, Rajkhowa S, Singh SK (2024) Click chemistry. Springer Singapore Singapore doi:https://doi.org/10.1007/978-981-97-4596-8
17. Kolb HC, Sharpless KB (2003) The growing impact of click chemistry on drug discovery. Drug Discov Today 8:1128–1137. https://doi.org/10.1016/S1359-6446(03)02933-7
18. Sarreshtehdar Aslaheh H, Payamifar S, Poursattar Marjani A (2024) Review of the use of the nickel catalyst in azide-alkyne cycloaddition reactions. Appl Organomet Chem 38:e7692. https://doi.org/10.1002/aoc.7692
19. Zaia J (2023) The 2022 Nobel Prize in Chemistry for the development of click chemistry and bioorthogonal chemistry. Anal Bioanal Chem 415:527–532. https://doi.org/10.1007/s00216-022-04483-9
20. Finn M, Kolb HC, Sharpless KB (2022) Click chemistry connections for functional discovery. Nat Synth 1:8–10. https://doi.org/10.1038/s44160-021-00017-w
21. Leyva E, Rubén Rodríguez-Gutiérrez I, Moctezuma E, Noriega S (2022) Mechanisms, copper catalysts, and ligands involved in the synthesis of 1,2,3-triazoles using click chemistry. Curr Org Chem 26:2098–2121. https://doi.org/10.2174/1385272827666230201103825
22. Liang L, Astruc D (2011) The copper(I)-catalyzed alkyne-azide cycloaddition (CuAAC) "click" reaction and its applications. An overview. Coord Chem Rev 255:2933–2945. https://doi.org/10.1016/j.ccr.2011.06.028
23. Haldón E, Nicasio MC, Pérez PJ (2015) Copper-catalysed azide–alkyne cycloadditions (CuAAC): an update. Org Biomol Chem 13:9528–9550. https://doi.org/10.1039/C5OB01457C
24. Griffiths PR (1983) Fourier transform infrared spectrometry. Science 222:297–302. https://doi.org/10.1126/science.6623077
25. Pasini D (2013) The click reaction as an efficient tool for the construction of macrocyclic structures. Molecules 18:9512–9530. https://doi.org/10.3390/molecules18089512
26. Cabrele C, Reiser O (2016) The modern face of synthetic heterocyclic chemistry. J Org Chem 81:10109–10125. https://doi.org/10.1021/acs.joc.6b02034
27. Dheer D, Singh V, Shankar R (2017) Medicinal attributes of 1,2,3-triazoles: current developments. Bioorg Chem 71:30–54. https://doi.org/10.1016/j.bioorg.2017.01.010
28. Hsieh HY, Lee WC, Senadi GC, Hu WP, Liang JJ, Tsai TR, Chou YW, Kuo KK, Chen CY, Wang JJ (2013) Discovery, synthetic methodology, and biological evaluation for antiphotoaging activity of bicyclic[1,2,3]triazoles: in vitro and in vivo studies. J Med Chem 56:5422–5435. https://doi.org/10.1021/jm400394s
29. Aragão-Leoneti V, Campo VL, Gomes AS, Field RA, Carvalho I (2010) Application of copper (I)-catalysed azide/alkyne cycloaddition (CuAAC) 'click chemistry' in carbohydrate drug and neoglycopolymer synthesis. Tetrahedron 66:9475–9492. https://doi.org/10.1016/j.tet.2010.10.001
30. Ma Q, Qi S, He X, Tang Y, Lu G (2017) 1,2,3-Triazole derivatives as corrosion inhibitors for mild steel in acidic medium: experimental and computational chemistry studies. Corros Sci 129:91–101. https://doi.org/10.1016/j.corsci.2017.09.025
31. Duan T, Fan K, Fu Y, Zhong C, Chen X, Peng T, Qin J (2012) Triphenylamine-based organic dyes containing a 1,2,3-triazole bridge for dye-sensitized solar cells via a 'Click' reaction. Dyes Pigments 94:28–33. https://doi.org/10.1016/j.dyepig.2011.11.008
32. Borgati TF, Alves RB, Teixeira RR, Freitas RP, Perdigão TG, Silva SF, Santos AA, Bastidas AJO (2013) Synthesis and phytotoxic activity of 1,2,3-triazole derivatives. J Braz Chem Soc 24:953–961. https://doi.org/10.5935/0103-5053.20130121
33. Huisgen R (1963) 1.3-Dipolare cycloadditionen Rückschau und Ausblick. Angew Chem 75: 604–637. https://doi.org/10.1002/ange.19630751304
34. Himo F, Lovell T, Hilgraf R, Rostovtsev VV, Noodleman L, Sharpless KB, Fokin VV (2005) Copper(I)-catalyzed synthesis of azoles. DFT study predicts unprecedented reactivity and intermediates. J Am Chem Soc 127:210–216. https://doi.org/10.1021/ja0471525

35. Sarreshtehdar Aslaheh H, Payamifar S, Poursattar Marjani A (2025) A review of the latest advances in click reaction using magnetite nanocatalysts. J Organomet Chem 1036:123704. https://doi.org/10.1016/j.jorganchem.2025.123704
36. Meldal M, Tornøe CW (2008) Cu-catalyzed azide-alkyne cycloaddition. Chem Rev 108: 2952–3015. https://doi.org/10.1021/cr0783479
37. BrevØ TG, Filius M, Araman C, van der Helm MP, Hagedoorn PL, Joo C, van Kasteren SI, Eelkema R (2020) Conditional copper-catalyzed azide–alkyne cycloaddition by catalyst encapsulation. Angew Chem Int Ed 59:9340–9344. https://doi.org/10.1002/ange.202001369
38. Cram DJ, Cram JM (1974) Host-guest chemistry: complexes between organic compounds simulate the substrate selectivity of enzymes. Science 183:803–809. https://doi.org/10.1126/science.183.4127.803
39. Dietrich B, Lehn J, Sauvage J (1969) Les cryptates. Tetrahedron Lett 10:2889–2892. https://doi.org/10.1016/S0040-4039(01)88300-3
40. Dietrich B, Lehn JM, Sauvage JP, Blanzat J (1973) Cryptates-X: syntheses et proprietes physiques de systemes diaza-polyoxa-macrobicycliques. Tetrahedron 29:1629–1645. https://doi.org/10.1016/S0040-4020(01)83409-8
41. Payamifar S, Poursattar Marjani A (2023) A new β-cyclodextrin-based nickel as green and water-soluble supramolecular catalysts for aqueous Suzuki reaction. Sci Rep 13:21279. https://doi.org/10.1038/s41598-023-48603-6
42. Payamifar S, Abdouss M, Poursattar Marjani A (2025) The recent advances of copper (II)-β-cyclodextrin complex as a practical catalyst system for various organic conversions: a review. J Organomet Chem 1027:123509. https://doi.org/10.1016/j.jorganchem.2025.123509
43. Kim C, Jeong D, Kim S, Kim Y, Jung S (2019) Cyclodextrin functionalized agarose gel with low gelling temperature for controlled drug delivery systems. Carbohydr Polym 222:115011. https://doi.org/10.1016/j.carbpol.2019.115011
44. Zhou HY, Jiang LJ, Zhang YP, Li JB (2012) β-Cyclodextrin inclusion complex: preparation, characterization, and its aspirin release in vitro. Front Mater Sci 6:259–267. https://doi.org/10.1007/s11706-012-0176-2
45. Herrera BA, Bruna TC, Sierpe RA, Lang EP, Urzúa M, Flores MI, Jara PS, Yutronic NI (2020) A surface functionalized with per-(6-amino-6-deoxy)-β-cyclodextrin for potential organic pollutant removal from water. Carbohydr Polym 233:115865. https://doi.org/10.1016/j.carbpol.2020.115865
46. Abou-Elanwar AM, Shirke YM, Ingole PG, Choi WK, Lee H, Hong SU, Lee HK, Jeon JD (2018) Nanocomposite hollow fiber membranes with recyclable β-cyclodextrin encapsulated magnetite nanoparticles for water vapor separation. J Mater Chem A 6:24569–24579. https://doi.org/10.1039/C8TA09413F
47. Zhang R, Li Y, Zhu X, Han Q, Zhang T, Liu Y, Zeng K, Zhao C (2020) Application of β-cyclodextrin-modified/PVDF blend magnetic membranes for direct metal ions removal from wastewater. J Inorg Organomet Polym Mater 30:2692–2707. https://doi.org/10.1007/s10904-019-01416-5
48. Payamifar S, Abdouss M, Poursattar Marjani A (2025) A review of using β-cyclodextrin-based catalysts for Sonogashira reactions. J Organomet Chem 1031:123592. https://doi.org/10.1016/j.jorganchem.2025.123592
49. Del Valle EM (2004) Cyclodextrins and their uses: a review. Process Biochem 39:1033–1046. https://doi.org/10.1016/S0032-9592(03)00258-9
50. Uekama K, Otagiri M (1987) Cyclodextrins in drug carrier systems. Crit Rev Ther Drug Carr Syst 3:1–40
51. Wenz G (1994) Cyclodextrins as building blocks for supramolecular structures and functional units. Angew Chem Int Ed Engl 33:803–822. https://doi.org/10.1002/anie.199408031
52. Nicolazzi C, Abdou S, Collomb J, Marsura A, Finance C (2001) Effect of the complexation with cyclodextrins on the in vitro antiviral activity of ganciclovir against human cytomegalovirus. Bioorg Med Chem 9:275–282. https://doi.org/10.1016/S0968-0896(00)00247-9

53. Osa T, Szejtli J, Vogtle F (1996) Comprehensive supramolecular chemistry: cyclodextrins. Pergamon
54. Nepogodiev SA, Stoddart JF (1998) Cyclodextrin-based catenanes and rotaxanes. Chem Rev 98:1959–1976. https://doi.org/10.1021/cr970049w
55. Payamifar S, Abdouss M, Javanshir S, Poursattar Marjani A (2025) A review of the latest advancements in using β-cyclodextrin-based as supramolecular catalysts for Heck reactions. Carbohydr Res 550:109381. https://doi.org/10.1016/j.carres.2025.109381
56. Breslow R, Dong SD (1998) Biomimetic reactions catalyzed by cyclodextrins and their derivatives. Chem Rev 98:1997–2012. https://doi.org/10.1021/cr970011j
57. Payamifar S, Foroozandeh A, Pourmadadi M, Abdouss M, Hasanzadeh M (2025) Cyclodextrin-based nanocarriers as promising scaffolds for overcoming challenges of doxorubicin delivery in cancer chemotherapy. Carbohydr Polym Technol Appl 9:100677. https://doi.org/10.1016/j.carpta.2025.100677
58. Szejtli J (1998) Introduction and general overview of cyclodextrin chemistry. Chem Rev 98: 1743–1754. https://doi.org/10.1021/cr970022c
59. Crini G (2014) A history of cyclodextrins. Chem Rev 114:10940–10975. https://doi.org/10.1021/cr500081p
60. Seggio M, Payamifar S, Fraix A, Kalydi E, Kasal P, Catanzano O, Conte C, Quaglia F, Sortino S (2021) Visible light-activatable cyclodextrin-conjugates for the efficient delivery of nitric oxide with fluorescent reporter and their inclusion complexes with betaxolol. New J Chem 45: 8449–8455. https://doi.org/10.1039/D1NJ00039J
61. Payamifar S, Foroozandeh A, Abdouss M, Poursattar Marjani A (2024) Magnetic nickel nanoparticle catalyst on β-cyclodextrin-modified Fe_3O_4 for nitroarene hydrogenation. Sci Rep 14:28493. https://doi.org/10.1038/s41598-024-79709-0
62. Payamifar S, Poursattar Marjani A (2025) A review of β-cyclodextrin-based catalysts system in the chemical fixation of carbon dioxide. J CO_2 Util 91:103011. https://doi.org/10.1016/j.jcou.2024.103011
63. Payamifar S, Foroozandeh A, Pourmadadi M, Abdouss M (2024) Cyclodextrin nanocarriers in coordination chemistry: enhancing encapsulation and targeted delivery of 5-fluorouracil for cancer treatment. Results Chem 12:101878. https://doi.org/10.1016/j.rechem.2024.101878
64. Szejtli J (2004) Past, present and future of cyclodextrin research. Pure Appl Chem 76:1825–1845. https://doi.org/10.1351/pac200476101825
65. Poulson BG, Alsulami QA, Sharfalddin A, El Agammy EF, Mouffouk F, Emwas AH, Jaremko L, Jaremko M (2021) Cyclodextrins: structural, chemical, and physical properties, and applications. Polysaccharides 3:1–31. https://doi.org/10.3390/polysaccharides3010001
66. Iacovino RV, Caso J, Di Donato C, Malgieri G, Palmieri M, Russo L, Isernia C (2017) Cyclodextrins as complexing agents: preparation and applications. Curr Org Chem 21:162–176. https://doi.org/10.2174/1385272820666160909111842
67. Bai L, Xu XM, He J, Pan SZ (2009) Inclusion complexation, encapsulation interaction and inclusion number in cyclodextrin chemistry. Coord Chem Rev 253:1276–1284. https://doi.org/10.1016/j.ccr.2008.08.011
68. Kurkov SV, Loftsson T (2013) Cyclodextrins. Int J Pharm 453:167–180. https://doi.org/10.1016/j.ijpharm.2012.06.055
69. Jansook P, Ogawa N, Loftsson T (2018) Cyclodextrins: structure, physicochemical properties and pharmaceutical applications. Int J Pharm 535:272–284. https://doi.org/10.1016/j.ijpharm.2017.11.018
70. Stella VJ, He Q (2008) Cyclodextrins. Toxicol Pathol 36:30–42. https://doi.org/10.1177/0192623307310945
71. Abdellatif AA, Ahmed F, Mohammed AM, Alsharidah M, Al-Subaiyel A, Samman WA, Alhaddad AA, Al-Mijalli SH, Amin MA, Barakat H (2023) Recent advances in the pharmaceutical and biomedical applications of cyclodextrin-capped gold nanoparticles. Int J Nanomedicine 2023:3247–3281

72. Noël S, Léger B, Ponchel A, Sadjadi S, Monflier E (2021) Cyclodextrins as multitask agents for metal nano-heterogeneous catalysis: a review. Environ Chem Lett 19:4327–4348. https://doi.org/10.1007/s10311-021-01298-5
73. Payamifar S, Behrouzi L, Poursattar Marjani A (2024) The electrochemical coupling reactions of organic halides compound in a valuable and practical manner for CC and C–heteroatom formation: an overview. Arab J Chem 17:105822. https://doi.org/10.1016/j.arabjc.2024.105822
74. Zimmerman JB, Anastas PT, Erythropel HC, Leitner W (2020) Designing for a green chemistry future. Science 367:397–400. https://doi.org/10.1126/science.aay3060
75. Payamifar S, Abdouss M, Poursattar Marjani A (2025) Recent advances in the application of magnetite nanocatalysts for CO_2 fixation reactions. Appl Organomet Chem 39:e70029. https://doi.org/10.1002/aoc.70029
76. Dalal DS, Patil DR, Tayade YA (2018) β-Cyclodextrin: a green and efficient supramolecular catalyst for organic transformations. Chem Rec 18:1560–1582. https://doi.org/10.1002/tcr.201800016
77. de Marco BA, Rechelo BS, Tótoli EG, Kogawa AC, Salgado HRN (2019) Evolution of green chemistry and its multidimensional impacts: a review. Saudi Pharm J 27:1–8. https://doi.org/10.1016/j.jsps.2018.07.011
78. Payamifar S, Khalili Y, Foroozandeh A, Abdouss M, Hasanzadeh M (2025) Magnetic mesoporous silica nanoparticles as advanced polymeric scaffolds for efficient cancer chemotherapy: recent progress and challenges. RSC Adv 15:16050–16074. https://doi.org/10.1039/D5RA00948K
79. Jadhav CK, Nipate AS, Chate AV, Gill CH (2022) β-Cyclodextrin: an efficient supramolecular catalyst for the synthesis of pyranoquinolines derivatives under ultrasonic irradiation in water. Polycyclic Aromat Comp 42:4224–4239. https://doi.org/10.1080/10406638.2021.1886125
80. Payamifar S, Abdouss M, Poursattar Marjani A (2024) The application of magnetic nanoparticles based β-cyclodextrin as recoverable catalyst in various organic transformations: an overview. Arab J Chem 18:106080. https://doi.org/10.1016/j.arabjc.2024.106080
81. Molnár Á (2021) Synthetic application of cyclodextrins in combination with metal ions, complexes, and metal particles. ChemCatChem 13:1424–1474. https://doi.org/10.1002/cctc.202001610
82. Xu P, Duan XH (2021) Pd/β-cyclodextrin-catalyzed C–H functionalization in water: a greener approach to regioselective arylation of (NH)-indoles with aryl bromides. New J Chem 45:19425–19431. https://doi.org/10.1039/D1NJ03400F
83. Liu Y, Hartman RL (2019) Reaction kinetics of a water-soluble palladium-β-cyclodextrin catalyst for a Suzuki-Miyaura cross-coupling in continuous flow. React Chem Eng 4:1341–1346. https://doi.org/10.1039/C9RE00159J
84. Das A, Mishra DK, Sinha DB (2017) A Pd(II) complex of a β-cyclodextrin-based polydentate ligand: an efficient catalyst for the Suzuki reaction in aqueous media. J Coord Chem 70:3035–3047. https://doi.org/10.1080/00958972.2017.1378812
85. Bricout H, Hapiot F, Ponchel A, Tilloy S, Monflier E (2010) Cyclodextrins as mass transfer additives in aqueous organometallic catalysis. Curr Org Chem 14:1296–1307. https://doi.org/10.2174/138527210791616920
86. Lima CG, Pauli FP, Madriaga VG, Amaral AA, Graciano IA, Meira VL, da SM Forezi L, Ferreira VF, de M Lima T, de Carvalho da Silva F (2022) Supramolecular catalysts for organic synthesis: preparation and applications of cyclodextrins and calixarenes in C–C cross-coupling reactions. Eur J Org Chem 2022:e202200904. https://doi.org/10.1002/ejoc.202200904
87. Bai CC, Tian BR, Zhao T, Huang Q, Wang ZZ (2017) Cyclodextrin-catalyzed organic synthesis: reactions, mechanisms, and applications. Molecules 22:1475. https://doi.org/10.3390/molecules22091475
88. Tabushi I, Yamamura K, Nabeshima T (1984) Characterization of regiospecific AC-and AD-disulfonate capping of beta-cyclodextrin. Capping as an efficient production technique. J Am Chem Soc 106:5267–5270. https://doi.org/10.1021/ja00330a039

89. Khan AR, Forgo P, Stine KJ, D'Souza VT (1998) Methods for selective modifications of cyclodextrins. Chem Rev 98:1977–1996. https://doi.org/10.1021/cr970012b
90. Benkovics G, Bálint M, Fenyvesi É, Varga E, Béni S, Yannakopoulou K, Malanga M (2019) Homo- and hetero-difunctionalized β-cyclodextrins: short direct synthesis in gram scale and analysis of regiochemistry. Beilstein J Org Chem 15:710–720. https://doi.org/10.3762/bjoc.15.66
91. Yamamura H (2017) Chemical modification of cyclodextrin and amylose by click reaction and its application to the synthesis of poly-alkylamine-modified antibacterial sugars. Chem Pharm Bull 65:312–317. https://doi.org/10.1248/cpb.c16-00739
92. Kasal P, Jindřich J (2021) Mono-6-substituted cyclodextrins – synthesis and applications. Molecules 26:5065. https://doi.org/10.3390/molecules26165065
93. Niu H, Chen W, Chen W, Yun Y, Zhong Q, Fu X, Chen H, Liu G (2019) Preparation and characterization of a modified-β-cyclodextrin/β-carotene inclusion complex and its application in Pickering emulsions. J Agric Food Chem 67:12875–12884. https://doi.org/10.1021/acs.jafc.9b05467
94. Liu JY, Zhang X, Tian B (2020) Selective modifications at the different positions of cyclodextrins: a review of strategies. Turk J Chem 44:261–278. https://doi.org/10.3906/kim-1910-43
95. Raju N, Singh A, Benjakul S (2021) Recovery, reusability and stability studies of beta cyclodextrin used for cholesterol removal from shrimp lipid. RSC Adv 11:23113–23121. https://doi.org/10.1039/D1RA03282H
96. Dodero A, Schlatter G, Hébraud A, Vicini S, Castellano M (2021) Polymer-free cyclodextrin and natural polymer-cyclodextrin electrospun nanofibers: a comprehensive review on current applications and future perspectives. Carbohydr Polym 264:118042. https://doi.org/10.1016/j.carbpol.2021.118042
97. Szejtli J, Budai Z (1976) Acid hydrolysis of β-cyclodextrin. Acta Chim (Budapest) 91:73–80
98. Bender ML, Komiyama M (2012) Cyclodextrin chemistry. Springer Science & Business Media
99. Szejtli J (1982) Cyclodextrins and their inclusion complexes. Akademiai Kiado, Budapest
100. Page MI (1984) The chemistry of enzyme action. In: Page MI (ed), vol 14. Elsevier/North-Holland Biochemical Press B.V., Amsterdam, p 505
101. Dodziuk H (2006) Cyclodextrins and their complexes: chemistry, analytical methods, applications. John Wiley & Sons. https://doi.org/10.1002/3527608982
102. Connors KA (1997) The stability of cyclodextrin complexes in solution. Chem Rev 97:1325–1358. https://doi.org/10.1021/cr960371r
103. Shin JA, Lim YG, Lee KH (2012) Copper-catalyzed azide–alkyne cycloaddition reaction in water using cyclodextrin as a phase transfer catalyst. J Org Chem 77:4117–4122. https://doi.org/10.1021/jo3000095
104. Kaboudin B, Abedi Y, Yokomatsu T (2012) One-pot synthesis of 1,2,3-triazoles from boronic acids in water using Cu(II)–β-cyclodextrin complex as a nanocatalyst. Org Biomol Chem 10: 4543–4548. https://doi.org/10.1039/C2OB25061F
105. Gupta D, Mishra A, Kundu S (2017) Cu(II)-β-CD as water-loving catalyst for one-pot synthesis of triazoles and biofuels intermediate at room temperature without any other additive. ChemistrySelect 2:2997–3008. https://doi.org/10.1002/slct.201700020
106. Girish YR, Kumar KSS, Muddegowda U, Lokanath NK, Rangappa KS, Shashikanth S (2014) ZrO_2-supported Cu(II)–β-cyclodextrin complex: construction of 2,4,5-trisubstituted-1,2,3-triazoles via azide-chalcone oxidative cycloaddition and post-triazole alkylation. RSC Adv 4:55800–55806. https://doi.org/10.1039/C4RA09970B
107. Tajbakhsh M, Naimi-Jamal MR (2022) Copper-doped functionalized β-cyclodextrin as an efficient green nanocatalyst for synthesis of 1,2,3-triazoles in water. Sci Rep 12:4948. https://doi.org/10.1038/s41598-022-08868-9
108. Nie R, Sang R, Ma X, Zheng Y, Cheng X, Li W, Guo L, Jin H, Wu Y (2016) Copper-γ-cyclodextrin complexes immobilized on hexagonal boron nitride as an efficient

catalyst in the multicomponent synthesis of 1,2,3-triazoles. J Catal 344:286–292. https://doi.org/10.1016/j.jcat.2016.09.022

109. Kaboudin B, Mostafalu R, Yokomatsu T (2013) Fe_3O_4 nanoparticle-supported Cu (II)-β-cyclodextrin complex as a magnetically recoverable and reusable catalyst for the synthesis of symmetrical biaryls and 1,2,3-triazoles from aryl boronic acids. Green Chem 15:2266–2274. https://doi.org/10.1039/C3GC40753E
110. Mahdavi M, Lijan H, Bahadorikhalili S, Ma'mani L, Ranjbar PR, Shafiee A (2016) Copper supported β-cyclodextrin grafted magnetic nanoparticles as an efficient recyclable catalyst for one-pot synthesis of 1-benzyl-1*H*-1,2,3-triazoldibenzodiazepinone derivatives via click reaction. RSC Adv 6:28838–28843. https://doi.org/10.1039/C5RA27275K
111. Bahadorikhalili S, Ma'mani L, Mahdavi H, Shafiee A (2018) Copper supported β-cyclodextrin functionalized PEGylated mesoporous silica nanoparticle-graphene oxide hybrid: an efficient and recyclable nano-catalyst for straightforward synthesis of 2-arylbenzimidazoles and 1,2,3-triazoles. Micropor Mesopor Mat 262:207–216. https://doi.org/10.1016/j.micromeso.2017.11.046
112. Bahadorikhalili S, Ashtari A, Ma'mani L, Ranjbar PR, Mahdavi M (2018) Copper-supported β-cyclodextrin-functionalized magnetic nanoparticles: efficient multifunctional catalyst for one-pot 'green' synthesis of 1,2,3-triazolylquinazolinone derivatives. Appl Organomet Chem 32:e4212. https://doi.org/10.1002/aoc.4212

Top Heterocycl Chem (2026) 63: 69–122
https://doi.org/10.1007/7081_2025_94

Published online: 15 February 2026

[2 + 2] Cycloaddition-Retroelectrocyclization (CA-RE) Click Chemistry

Xu Chen and Tsuyoshi Michinobu

Contents

1 Introduction 70
1.1 Early Exploration of CA-RE Reaction 70
1.2 Unexpected Reactions Observed During the CA-RE Transformation 73
1.3 Comparison to Other Click Reactions 75
2 Historical Development of the [2 + 2] CA-RE Reaction 76
2.1 Exploration of Donor and Acceptor Units 76
2.2 Synthesis of Molecular Push-Pull Chromophores 84
2.3 Synthesis of Polymeric Push-Pull Chromophores 91
3 Application to Organic Functional Materials 97
3.1 Nonlinear Optical Materials 97
3.2 Photovoltaic Materials 101
3.3 p-Dopants for Organic Semiconductors 105
3.4 Colorimetric Sensors 106
3.5 Other Applications 106
4 Conclusion 107
References 108

Abstract The [2 + 2] cycloaddition-retroelectrocyclization (CA-RE) reaction, which generally occurs between electron-rich alkynes and electron-deficient alkenes, has been extensively investigated because of the characteristics of click chemistry, such as high yield, no catalysts, no side reactions, and mild conditions. Therefore, it has been employed as a powerful tool for the construction of donor-acceptor (push-pull) chromophores with tunable energy levels and absorption properties, as well as for the postfunctionalization of polymers. In this chapter, the development and mechanism of the [2 + 2] CA-RE reaction are summarized, and its application to the synthesis of molecular and polymeric push-pull chromophores is reviewed. In

X. Chen and T. Michinobu (✉)
Department of Materials Science and Engineering, Institute of Science Tokyo, Meguro-ku, Tokyo, Japan
e-mail: michinobu.t.aa@m.titech.ac.jp

addition, functional applications of CA-RE products such as nonlinear optical materials, photovoltaic materials, p-dopants for organic semiconductors, and colorimetric sensors are briefly presented.

Keywords Charge-transfer · Click reaction · Energy levels · Postfunctionalization of polymers · Push-pull chromophores

1 Introduction

1.1 Early Exploration of CA-RE Reaction

The first report of the [2 + 2] cycloaddition-retroelectrocyclization (CA-RE) reaction dates back to 1981 by Bruce et al., involving the reaction of an electron-deficient olefin with a metal-substituted acetylide under mild conditions [1]. In this study, tetracyanoethylene (TCNE) was mixed with Ru-acetylide in benzene, and after several hours, orange crystals were obtained with a yield of 82%. The molecular structure was confirmed by X-ray single-crystal analysis to be 1,1,4,4-tetracyanobuta-1,3-diene (TCBD) derivative **1**, formed through the cleavage of the olefin bond in TCNE on reaction with an alkyne (Fig. 1). Subsequently, the research was extended to other transition-metal-substituted (W, Fe, Mn, Ni) acetylides, and the reaction mechanism was proposed after the cyclobutene intermediates were successfully isolated and characterized (Scheme 1) [2].

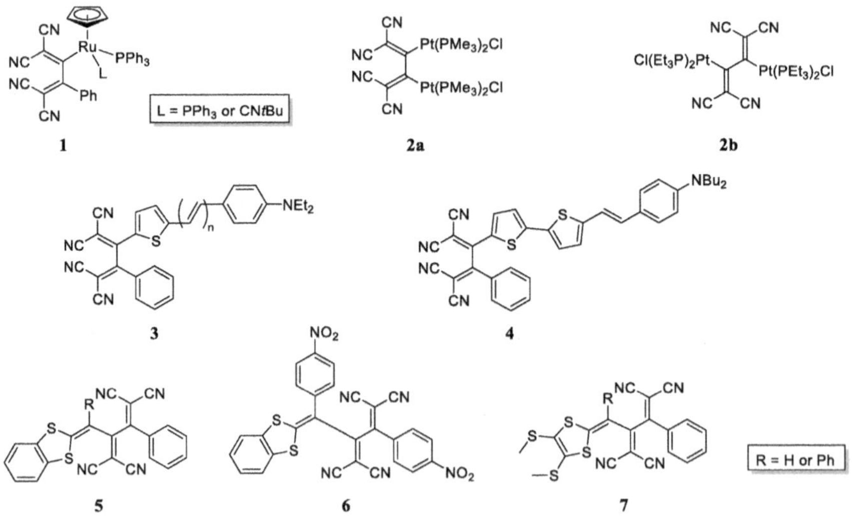

Fig. 1 Chemical structures of early reported TCBD derivatives

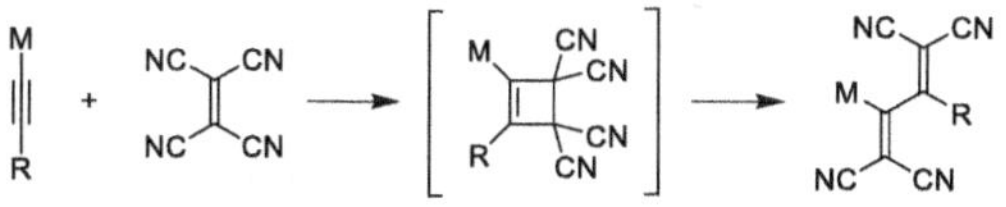

Scheme 1 Mechanism of the [2 + 2] CA-RE reaction between metal acetylide and TCNE

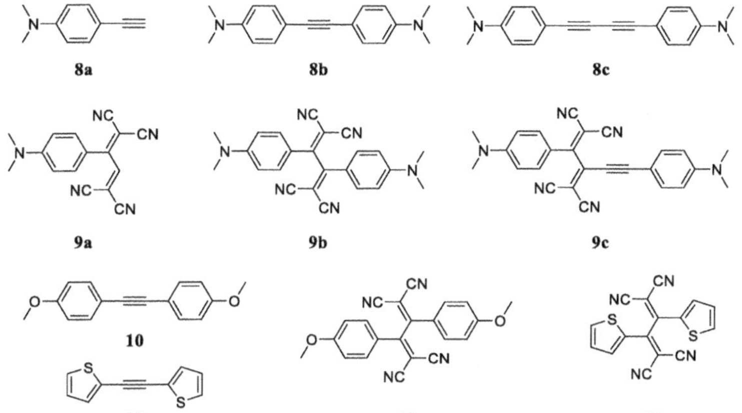

Fig. 2 Examples of initially reported EDGs for activating alkyne units in the [2 + 2] CA-RE transformation

Later, Takahashi et al. investigated the [2 + 2] CA-RE reaction between Pt (II) acetylides and TCNE, yielding TCBD derivatives **2a** and **2b** with s-cis and s-trans conformations, respectively, depending on the bulkiness of the substituents at the metal center [3]. In 1999, Jen et al. and Suter et al. independently reported the synthesis of purely organic TCBD derivatives **3** and **4**, marking the first examples of such compounds [4]. Yamashita et al. synthesized a series of donor-acceptor compounds **5–7** by reacting TCNE with 1,3-dithiol-2-ylidene-substituted alkynes in 2004 [5]. Thus far, reports on the [2 + 2] CA-RE reaction have been sporadic and fragmented, primarily focusing on the synthesis of new molecules, with limited attention given to tuning the electron deficiency of the alkyne moiety and improving reaction efficiency.

Starting in 2005, Diederich and Michinobu et al. published research on [2 + 2] CA-RE reactions, primarily focusing on adjusting alkyne reactivity by replacing the electron-donating group (EDG) (Fig. 2). Initially, they used the *N*,*N*-dimethylaniline group as a donor moiety to activate the alkyne unit. The precursor alkynes **8a-8c** were then mixed with TCNE in benzene or THF at 20 °C to undergo the [2 + 2] CA-RE transformation, generating the corresponding TCBD derivatives **9a-9c**, respectively, in nearly quantitative yields (96–100%) [6]. This was followed by *p*-methoxyphenyl and thienyl groups as EDG. Although these two groups were considered strong electron-donating units, they did not exhibit the anticipated ability to activate the alkyne unit. The *p*-methoxyphenyl-substituted alkyne **10** underwent

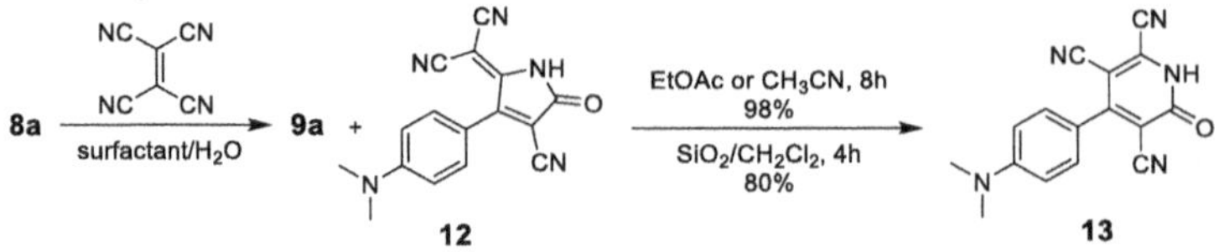

Scheme 2 [2 + 2] CA-RE reaction in water and further conversion from **12** to **13**

the [2 + 2] CA-RE transformation neat at 110 °C, yielding TCBD derivative **10a** in 42% yield. Meanwhile, the thienyl-substituted alkyne **11** required 150 °C to produce **11a** with an 80% yield [7]. Despite the variation in yield, it is noteworthy that no side reactions or by-products occur, even in the presence of an excess of TCNE. Furthermore, the reaction smoothly proceeds without any catalysts, allowing it to be classified as a "click" reaction [8]. Their initial research was summarized in three reviews [9].

The [2 + 2] CA-RE reaction proceeds through an initial cycloaddition step, forming a cyclobutene intermediate, followed by retro-electrocyclization to yield the TCBD product. Experimental and computational studies confirmed that the CA step is the rate-determining step, proceeding with bimolecular, second-order kinetics, as observed in the reaction between 4-(*N*,*N*-dimethylamino)phenylacetylene and *para*-substituted benzylidenemalononitriles in polar solvents [10]. Further research revealed that the first step is not a standard second-order reaction. The final products promote the formation of charge-transfer complexes, which generate the cyclobutene intermediates and exert autocatalytic effects on the first step. The RE step follows a typical first-order reaction mechanism, with no polar intermediates involved. It was found that this reaction is inhibited in acetic acid, indicating the effect of the aniline-cyano push-pull interaction.

Water as a solvent in organic synthesis is an important strategy that meets the requirements of green chemistry [11] and offers different reactivity and selectivity. Recently, Jayamurugan et al. reported the [2 + 2] CA-RE reaction in a water/surfactant medium and found that TCBD derivatives could be obtained in excellent yields (85–98%) at higher concentrations of various surfactants, including ionic (Tween 20 and Brij 30), anionic (sodium dioctylsulfosuccinate (SDOSS) and sodium dodecyl sulfate (SDS)), and cationic (cetyltrimethylammonium bromide (CTAB)) types (Scheme 2) [12]. TCBD formation exhibits unique reactivity in nonionic surfactants at lower concentrations, due to the presence of vesicle-based transient nanoreactors that facilitate H_2O addition in the hydrophilic core. This process can be classified as substrate-induced generation of transient self-assembled catalytic systems. [13] Besides the formation of TCBD products, amidopentafulvene **12** was formed (23–47%), and it was transformed *via* keto-enol tautomerism into **13** in polar aprotic solvents with high yield. Subsequently, disubstituted alkynes were also investigated, and only TCBD products were obtained.

1.2 Unexpected Reactions Observed During the CA-RE Transformation

The simultaneous formation of **9a** and **12** in a one-pot reaction suggests the presence of competing reactions in the [2 + 2] CA-RE transformation. Most examples of these competing reactions involve dicyanovinyl (DCV) acceptors. In the case of the reaction between alkyne **14** and 2-(dicyanomethylene)indan-1,3-dione **15**, [4 + 2] hetero-Diels-Alder (HDA) reaction occurred as a side reaction (Scheme 3) [14]. The involvement of Lewis acids, such as $LiClO_4$, significantly facilitates the HDA reaction. Depending on the donor-substituted alkynes, the yield ratio between the CA-RE product **16** and the HDA product **17** was from 98:2 to 80:20 when $LiClO_4$ was not added. This ratio dramatically changed upon adding $LiClO_4$, ranging from 4: 96 to 13:87. Similar to the HAD reaction, the [4 + 2] Diels-Alder reaction can also compete with the CA-RE reaction [15].

When 2-(dicyanomethylene)malonate **18** reacted with electronically neutral alkynes, such as phenylacetylene **19**, a formal [3 + 2] cycloaddition-rearrangement cascade reaction occurred, forming **21** through the intermediates **20a** and **20b** with a yield of 65% (Scheme 4) [16]. This reaction also occurred with many other unactivated or moderately activated alkynes, yielding a variety of substituted penta-2,4-dien-1-ones in very good yields [17]. Additionally, the [3 + 2] cycloaddition-rearrangement was observed when mixing triester-substituted cyanoalkene **22** with donor-activated alkynes, such as **8a**, although the product **24** was obtained through the intermediates **23a** and **23b** in a low yield of 4% [16]. However, the yield increased to 29% when tetra-ester alkenes were used. It is worth noticing that the yield of the [3 + 2] cycloaddition-rearrangement reaction can be further improved by replacing some ethyl groups in the ester substituents with other alkyl groups.

Some CA-RE products undergo further transformation [18]. The 1,3-di-*tert*-butyl-1,3-diethynylallene (DEA) was developed to investigate multiple cascade pericyclic reactions, which rapidly proceeded in a one-pot fashion. The reaction between racemic DEA **(±)-25** and TCNE produced butadienes **(±)-26** via a [2 + 2] cycloaddition, followed by cycloreversion (CA/CR). Then, a 4π electrocyclization

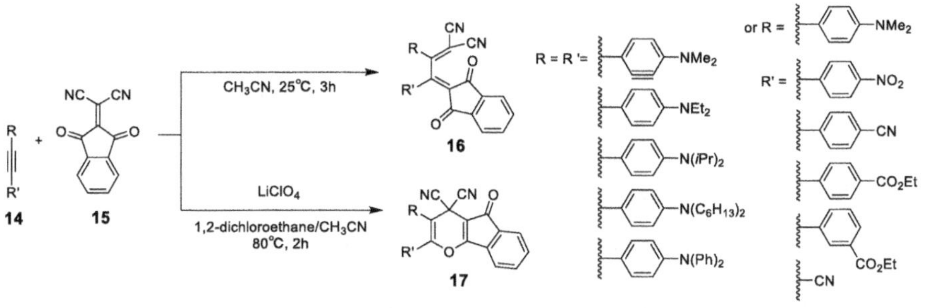

Scheme 3 [2 + 2] CA-RE and [4 + 2] HAD reactions between donor-substituted alkynes **14** and **15**

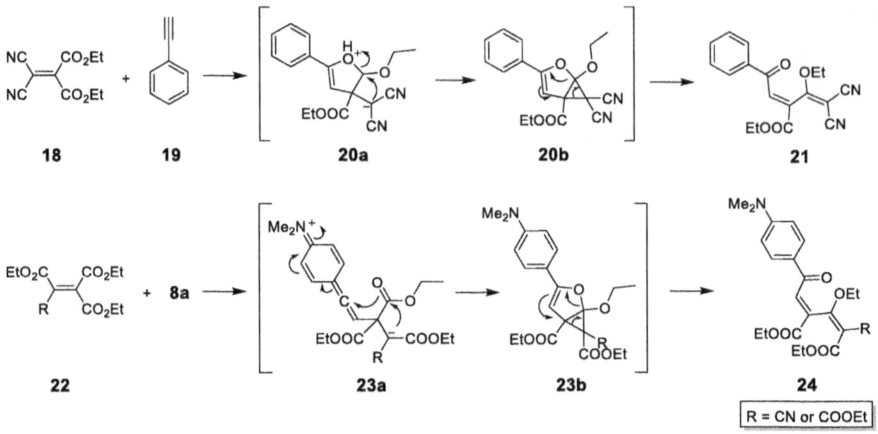

Scheme 4 Proposed mechanism for the formal [3 + 2] cycloaddition-rearrangement cascade reaction

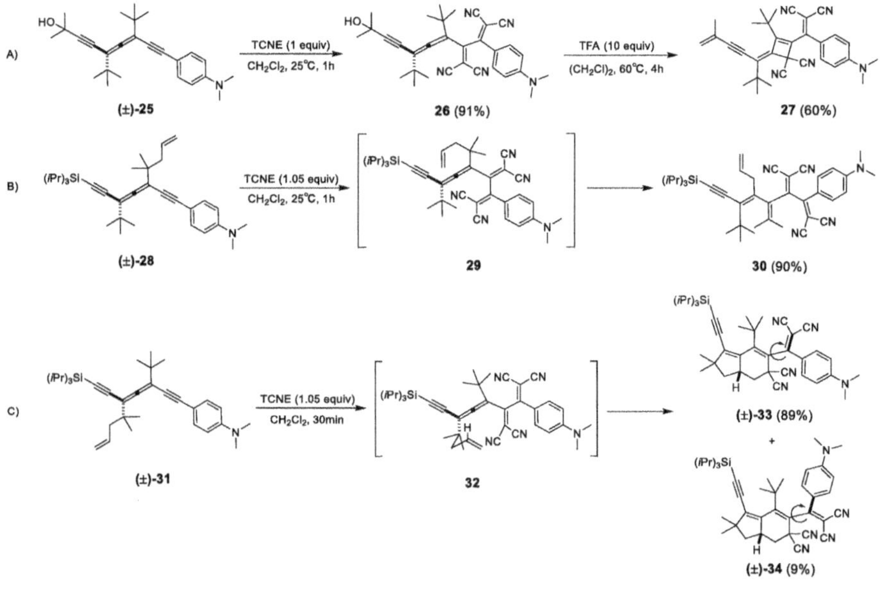

Scheme 5 CA-RE initiated cascade reactions: (**a**) CA/CR/4π EC reaction, (**b**) CA/CR/allenyl-Cope EC reaction, and (**c**) CA/CR/Diels-Alder reaction

(EC) occurred in the [2 + 2] product between the allene core and proximal dicyanovinyl moiety, resulting in the formation of cyclobutene **27** (Scheme 5a). Subsequently, exposure of **(±)-28** to TCNE formed **(±)-29** via a CA/CR reaction, and the subsequent allenyl-Cope [3] sigmatropic rearrangement yielded the [4] dendralene **30** (Scheme 5b). Furthermore, mixing DEA **(±)-31** with TCNE at

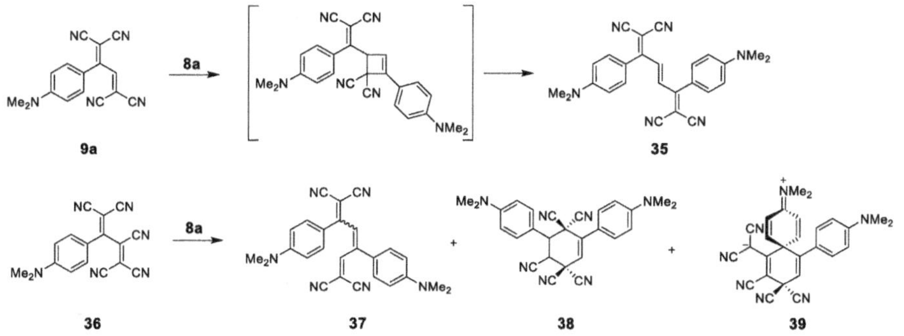

Scheme 6 Post-reactions of TCBD derivatives

room temperature resulted in the formation of the new intermediate **(±)-32**, and the cascade transformation proceeded via a [4 + 2] reaction mechanism. It is noteworthy that the final product was a mixture of **(±)-33** and **(±)-34** due to the hindered rotation around the butadiene axis (Scheme 5c). Additionally, many other reports have described similar cascade reactions. For instance, a C_{60} sphere directly connected to a TCBD moiety underwent thermal rearrangement to form chiral 1,2,9,12-tetrakis-adducts [19]. The CA-RE product of 6,6-bis[(4-*N*,*N*-dimethylanilino)ethynyl]-substituted pentafulvene with TCNE transformed into dihydropentalene as the sole product in very high yield [20]. Moreover, the synthesis of 2-aminofurans has been achieved through an initial [2 + 2] CA-RE reaction, followed by an intramolecular nucleophilic addition of the internal hydroxyl groups to the TCBD moiety [21]. Interestingly, the 2-aminofuran derivatives could be further transformed to 6-aminopentafulvenes in amine solvents under mild conditions.

The DCV moiety of TCBD derivatives can undergo CA-RE transformation with *N*,*N*-dimethylanilino (DMA)-substituted alkynes (Scheme 6). TCBD **9a** reacted with alkyne **8a** at room temperature to afford the 1,1,6,6-tetracyano-1,3,5-hexatriene derivative **35** in 88% yield [22]. DMA-substituted 1,1,2,4,4-pentacyanobuta-1,3-diene (PCBD) **36** also underwent the same transformation to yield **37**, along with the unexpected [4 + 2] cycloadducts **38** and **39** [23].

1.3 Comparison to Other Click Reactions

The [2 + 2] CA-RE reaction is a more straightforward method that is closer to the principles of green chemistry than other typical click reactions. It smoothly proceeds at room temperature without the use of a catalyst when appropriate EDGs are selected to activate alkynes [24]. On the other hand, copper(I)-catalyzed azide-alkyne cycloaddition (CuAAC), a typical example of click reactions, has the limitation of requiring a toxic copper catalyst. [25] Subsequently, strain-promoted azide-alkyne cycloaddition (SpAAC) reactions, which do not require toxic catalysts, have

Table 1 Comparison of [2 + 2] CA-RE reaction with other click reactions. EDG = Electron-donating groups

	Substrates	Catalysts	Conditions	Products
CA-RE	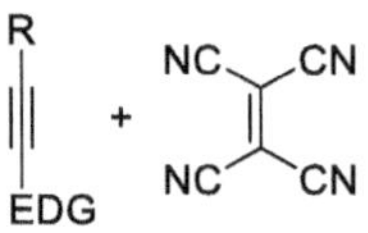	–	r. t. or mild heating	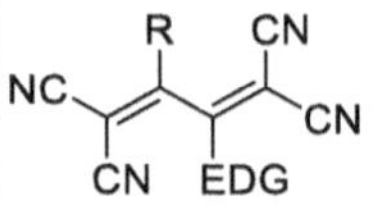
CuAAC		Cu(I)	r. t.	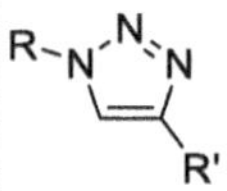
SpAAC	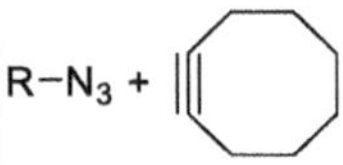	–	r. t.	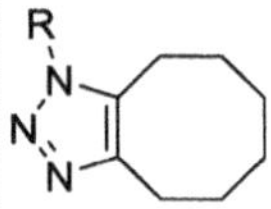
Thiol-ene	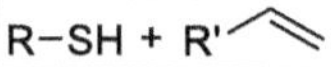	hν or mild base	r. t.	
Diels-Alder	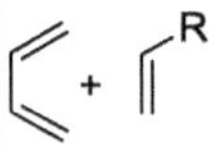	–	Heating	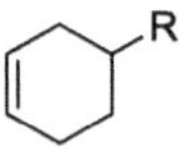

been developed [26], but they produce a mixture of triazole regioisomers [27a], and the cyclooctynes used in the reaction are relatively unstable [27b]. Similar to this example, thiol-ene click reactions generally require either a mild base as a catalyst or light to initiate the reaction [28]. In addition, the products are limited to sulfur-containing compounds due to the use of thiol groups. Diels-Alder cycloaddition reactions are also regarded as click reactions, and they are powerful tools for constructing hexacyclic compounds, but temperature control is required for efficient conversion [29]. Furthermore, none of the conventional click reactions do not directly produce push-pull chromophores. Collectively, the [2 + 2] CA-RE reaction is not only excellent as a click reaction, but also extremely promising in terms of creating functional molecules (Table 1).

2 Historical Development of the [2 + 2] CA-RE Reaction

2.1 Exploration of Donor and Acceptor Units

Inspired by the research of Diederich and Michinobu et al. on adjusting electron-donating groups (EDGs) to activate the alkyne unit, an increasing number of EDGs have been explored to broaden the application scope of the [2 + 2] CA-RE reaction. Besides aniline groups with various alkyl substituents [30], many aromatic amines have demonstrated excellent effectiveness in activating the alkyne unit. The alkyne reactivity depends on the solvent species and temperature. The enantiopure *N*-arylated-3,5-dihydro-4*H*-dinaphtho[2,1,-*c*:1′,2′-*e*]azepine effectively activated the alkyne unit for the [2 + 2] CA-RE reaction with TCNE at 20 °C, yielding the corresponding TCBD derivative in 95% yield (Table 2, entry 1) [31]. Recent studies have shown that urea activates the alkyne unit, leading to the formation of TCBD derivatives with good yields (90%, Table 2, entry 2) [32]. Another effective candidate for activating the alkyne was the triarylamine moiety, which allowed the TCBD yield to reach 100% (Table 2, entry 3) [33]. Quinoxalines, typically known as electron-withdrawing units, can act as electron donors when substituted with two triarylamine moieties, and the corresponding alkyne produced TCBD derivatives with a 75% yield (Table 2, entry 4) [34], further highlighting the strong electron-donating properties of triarylamine. Carbazole and indole are common *N*-heterocyclic compounds, and it is noteworthy that their derivatives exhibit varying reactivities depending on the substitution position, including both the C- and N-positions (Table 2, entries 5–10) [34–37]. Azuleno[10,20:4,5]pyrrolo[2,1-*b*]quinazoline-6,14-dione has been reported as a donor unit for activating alkynes, achieving the CA-RE transformation with a 97% yield (Table 2, entry 11) [38]. The alkyne substituted at the 10-position of phenothiazine reacted with TCNE at 20 °C, yielding the corresponding TCBD with a 62% yield (Table 2, entry 12) [39]. The yield increased to 83% when the substitution was shifted to the 3-position (Table 2, entry 13) [40].

Oligothiophene-substituted alkynes reacted with TCNE at 20 °C to afford the corresponding TCBD derivatives in moderate yields of 45–59% (Table 2, entry 14–15) [41, 42]. Tetrathiafulvalene (TTF) and its extended derivatives did not exhibit significant electron-donating ability, necessitating heating to reflux in 1,2-dichloroethane to obtain the TCBD products in moderate yields of 41–70% (Table 2, entries 16–17) [43, 44]. Azulene-substituted alkynes, which are known for their strong electron-donating properties, could react with TCNE at 20 °C to yield over 89% (Table 2, entries 18–19) [31b, 45, 46]. 2*H*-Cyclohepta[*b*]furan-2-one derivatives could also activate alkynes, but heating to reflux in EtOAc was necessary to achieve a high yield in the CA-RE transformation (Table 2, entry 20) [47].

B,N-heterocyclic donor-substituted alkynes, including boron dipyrromethene (BODIPY) [48] and subphthalocyanines (SubPcs) [49], reacted with TCNE at 20 ° C (Table 2, entries 21–22). Systematic investigations of γ-pyranylidene derivative-substituted alkynes have been reported, showing that their CA-RE reaction with TCNE proceeds at 20 °C with a moderate yield of 69% (Table 2, entry 23) [50]. Both free-base and metalloporphyrins (Zn or Cu), as well as chlorophyll derivatives, demonstrated sufficient electron-donating power to activate the alkyne unit, leading to CA-RE reactions with TCNE that achieved yields of up to 80% (Table 2, entries 24–25) [48c, 51, 52].

Several reports have described alkynes directly attached to the nitrogen atom of donor moieties undergoing [2 + 2] CA-RE transformations at 20 °C. Triazene derivatives could activate the alkyne, resulting in good yields in the [2 + 2] CA-RE reaction (Table 2, entry 26) [53]. Alkynes activated by ynamides could also react with TCNE to afford TCBD derivatives in very high yields (Table 2, entry 27) [54], even when one of the substituents on amine was electron-withdrawing groups (Table 2, entry 28) [55]. The reaction between cyclic amine end-capped

Table 2 Electron donors on alkynes used in the [2 + 2] CA-RE reaction with TCNE. EWG = electron-withdrawing group

Entry	Donor structure	Typical conditions	Yield (%)	Ref.
1	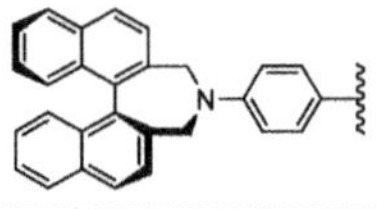	CH_2Cl_2, 20 °C	95	[31]
2	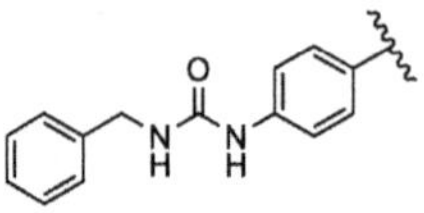	CH_2Cl_2, 25 °C	90	[32]
3	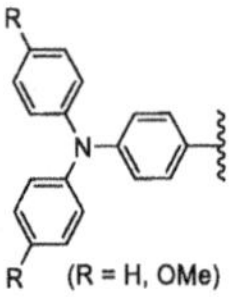	CH_2Cl_2, 20 °C	100	[33]
4	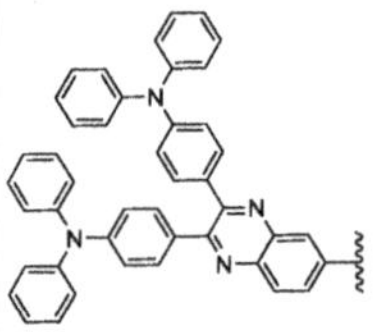	1,2-Dichloroethane, 20 °C	75	[34]
5	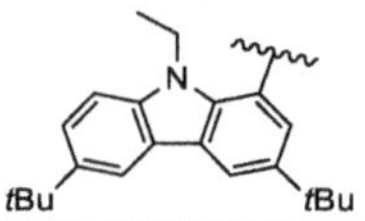	1,2-Dichloroethane, 20 °C	100	[35]
6	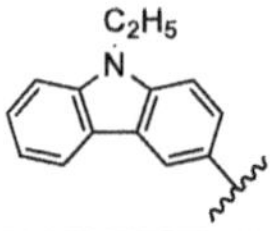	1,2-Dichloroethane, 20 °C	90	[35]
7	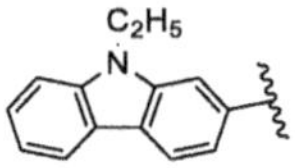	1,2-Dichloroethane, reflux	85	[35]
8	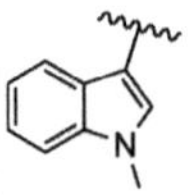	1,2-Dichloroethane, 25 °C	40–96	[34, 36]
9	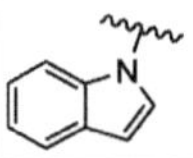	CH_2Cl_2, 20 °C	59	[37]
10	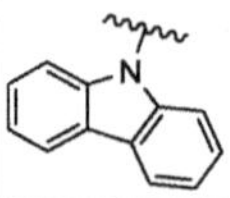	CH_2Cl_2, 20 °C	83	[35]
11	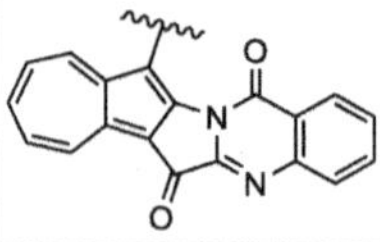	1,2-Dichloroethane, reflux	97	[38]

(continued)

Table 2 (continued)

Entry	Donor structure	Typical conditions	Yield (%)	Ref.
12	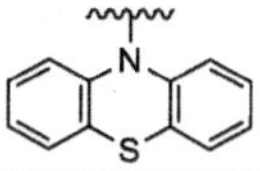	CH_2Cl_2, 20 °C	62	[39]
13	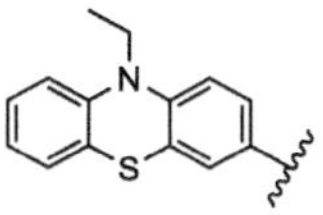	CH_2Cl_2, 20 °C	83	[40]
14	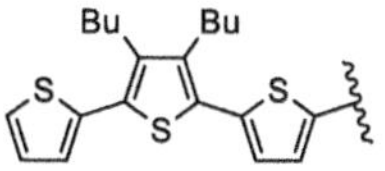	Benzene, 20 °C	59	[41]
15	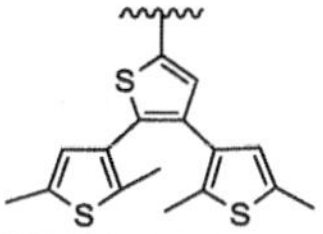	CH_2Cl_2, 20 °C	45	[42]
16	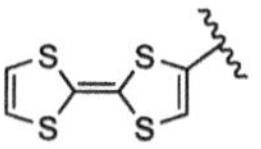	1,2-Dichloroethane or benzene, 80 °C	41	[43]
17	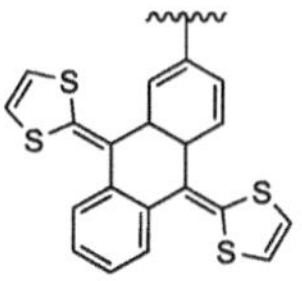	1,2-Dichloroethane, reflux	70	[44]
18	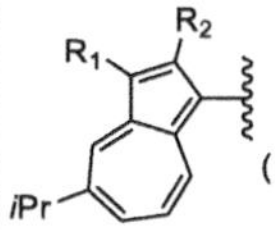 (R_1 = COOMe, Ph, R_2 = H, Me)	AcOEt or CH_2Cl_2, 20 °C	89–98	[38b, 45]
19	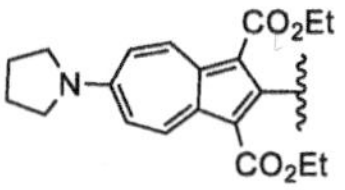	CH_2Cl_2, 20 °C	94	[46]
20	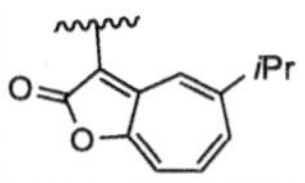	AcOEt, reflux	89–99	[47]
21	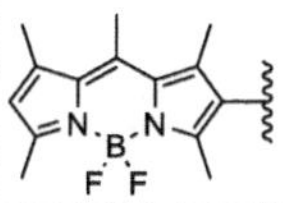	CH_2Cl_2, 20 °C	51–77	[48]
22	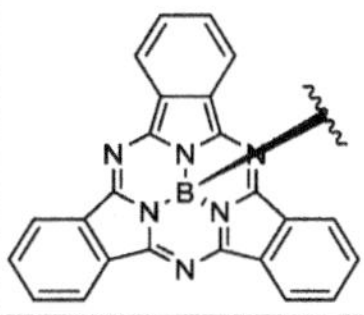	THF, 20 °C	81	[49]

(continued)

Table 2 (continued)

Entry	Donor structure	Typical conditions	Yield (%)	Ref.
23	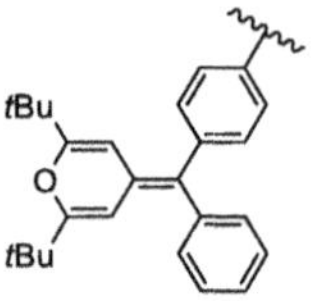	CH_2Cl_2, 20 °C	69	[50]
24	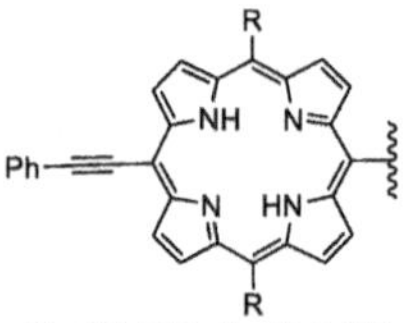 (R = 3,4,5-$(C_{10}H_{21}O)_3C_6H_2$)	CH_2Cl_2, 20 °C	83	[48c, 51]
25	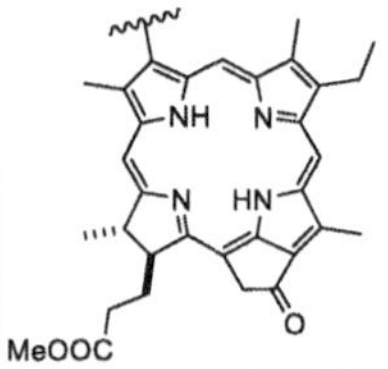	$CHCl_3$, 20 °C	80	[52]
26	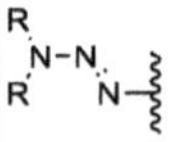 (R = Et, *i*Pr)	CH_2Cl_2, 20 °C	40–64	[53]
27	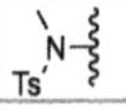	CH_2Cl_2, 20 °C	67–95	[54]
28	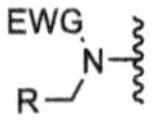 (R= Ph, Bn, allyl, propargyl; EWG = -COOMe, *p*-tosyl, 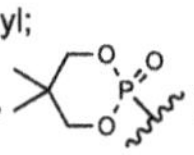)	CH_2Cl_2, 20 °C	64–99	[55]
29	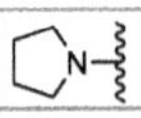	CH_2Cl_2, 20 °C	60–62	[56]
30	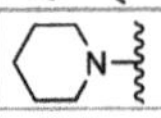	CH_2Cl_2, 20 °C	73	[56]
31	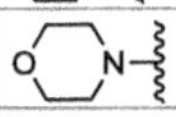	CH_2Cl_2, 20 °C	33–82	[56]
32	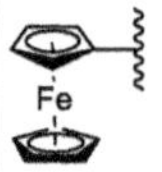	$CHCl_3$, 20 °C	60–99	[33b, 33d, 40a, 48d, 57]
33	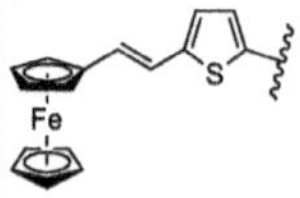	$CHCl_3$, 20 °C	53	[58]
34	$(PEt_3)_5Se_8Re_6$–	CD_2Cl_2, 20 °C	58	[59]

oligoynes and TCNE proceeded at 20 °C, even in the presence of an electron-withdrawing unit on the opposite side (Table 2, entries 29–31) [56].

In addition to pure organic substituents, ferrocene was a widely used unit for activating the alkynes, with the CA-RE yield reaching up to 99% after optimizing the conditions and structure (Table 2, entry 32) [33b, 33d, 40a, 48d, 57]. Monothiophene-substituted alkynes also reacted with TCNE at 20 °C when linked to a ferrocene moiety through alkene units (Table 2, entry 33) [58]. Most surprisingly, $[Re_6Se_8(PEt_3)_5]^+$, a type of non-carbonyl cluster, could also activate the attached alkyne unit for [2 + 2] CA-RE transformation (Table 2, entry 34) [59].

Compared to the rapid diversification of donor moieties, the development of acceptors has lagged somewhat. Among these, cyano groups were widely used due to their strong electron-withdrawing properties. As a result, nearly all alkenes suitable for CA-RE transformations contain cyano groups. As highlighted in the previous section, the most widely used alkene is TCNE. TCNQ, an extended form of TCNE, is another effective acceptor that undergoes the [2 + 2] CA-RE transformation with DMA-substituted alkyne **8a** in high yield at 20 °C (Table 3, entry 1) [60]. TCNQ derivatives can also be used for [2 + 2] CA-RE reactions. 2,3,5,6-Tetrafluoro-TCNQ (F_4TCNQ), a well-known strong acceptor, yielded the corresponding push-pull chromophore in 65% yield at 20 °C (Table 3, entry 2) [61]. However, F_4TCNQ adducts were unstable and gradually decomposed when exposed to glass surfaces. Additionally, 2,5-dialkoxy-substituted TCNQ could undergo [2 + 2] CA-RE transformation, but heating to 60 °C was required for a clean transformation (Table 3, entry 3) [62].

Dicyanovinyl (DCV) derivatives were synthesized to study their reactivity in [2 + 2] CA-RE reactions in DMF at 100 °C. With the exception of *p*-azidephenyl-substituted DCV, which experienced decomposition issues due to the instability of the N_3 group, most DCV derivatives produced the CA-RE products in high yields, ranging from 85% to 98% (Table 3, entry 4) [63]. In addition to aromatic substituents, other groups such as trimethylsilylethynyl, cyclohexyl, and adamantyl units were introduced into the DCV derivatives, which underwent [2 + 2] CA-RE transformations with moderate to high yields (Table 3, entries 5–7). Tricyanovinyl (TCV) derivatives also reacted with DMA-substituted alkynes in high yield (Table 3, entry 8), but the steric hindrance slightly reduced the reaction efficiency. 6,6-Dicyanopentafulvene derivatives have received significant attention due to their strong electron-accepting properties [64]. The [2 + 2] CA-RE reaction of these derivatives produced the corresponding adducts in 64% yield at 80 °C (Table 3, entry 9). It is noteworthy that the adducts displayed regioselectivity based on the substituent pattern [65]. There have also been reports of extended DCV derivatives participating in [2 + 2] CA-RE reactions; however, their product yields remained low, even under reflux, due to insufficient reactivity (Table 3, entries 10–11) [66]. Recent studies have revealed that [2 + 2] CA-RE transformations are not limited to C=C bonds. *N,N′*-Dicyanoquinone diimides (DCNQIs), which contain C=N bonds, have been identified as strong acceptors capable of undergoing [2 + 2] CA-RE reactions to produce push-pull chromophores in high yields (Table 3, entry 12) [67].

Table 3 Olefinic acceptors used in [2 + 2] CA-RE reactions with DMA-substituted alkyne

Entry	Acceptor structure	Typical conditions	Yield (%)	Ref.
1	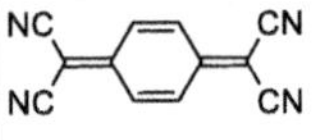 TCNQ	CH_2Cl_2, 20 °C	81	[60]
2	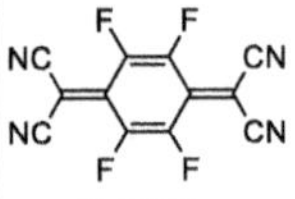 F_4TCNQ	CH_2Cl_2, 20 °C	65	[61]
3	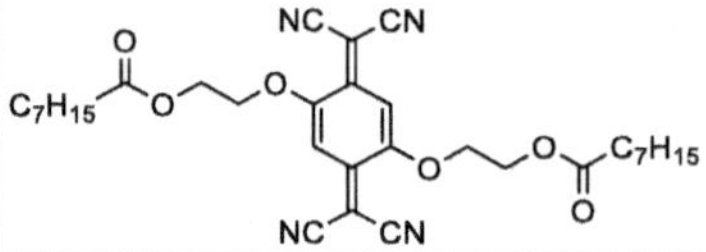	1,2-Dichloroethane, 60 °C	100	[62]
4	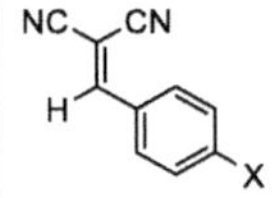 X = NO_2, CN, COOMe, CHO, Br, CCH, H, Ph, Me, OMe, NMe_2, N_3	DMF, 100 °C	85–98 35 (X = N_3)	[63]
5	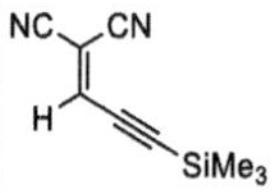	DMF, 100 °C	63	[63]
6	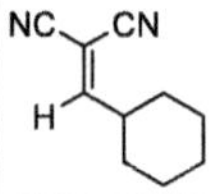	DMF, 100 °C	90	[63]
7	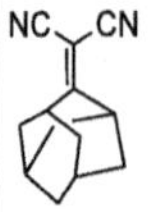	DMF, 100 °C	80	[63]
8	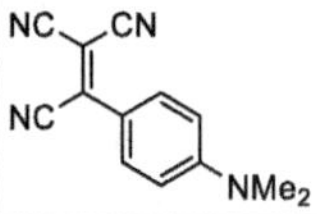	DMF, 100 °C	89	[63]
9	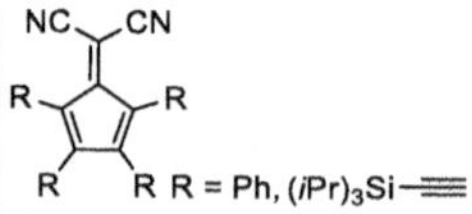	CH_3CN, 80 °C	64	[65]
10	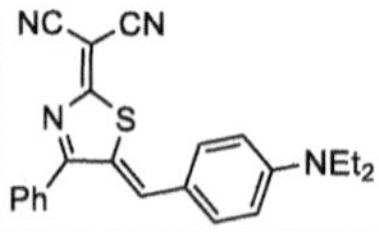	1,2-Dichloroethane, reflux	29	[66]

(continued)

Table 3 (continued)

Entry	Acceptor structure	Typical conditions	Yield (%)	Ref.
11	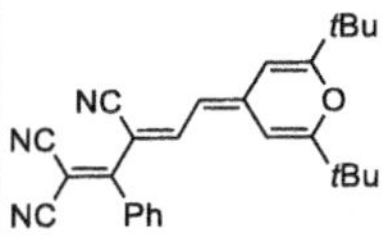	CH_3CN, reflux	5	[66]
12	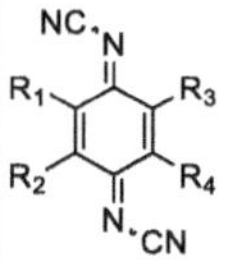 R_{1-4} = H R_1 = R_2 = -(C_4H_4)-, R_3 = R_4 = H R_1 = R_2 = Me, R_3 = R_4 = H DCNQI	1,2-Dichloroethane, 20–80 °C	85–95	[67]

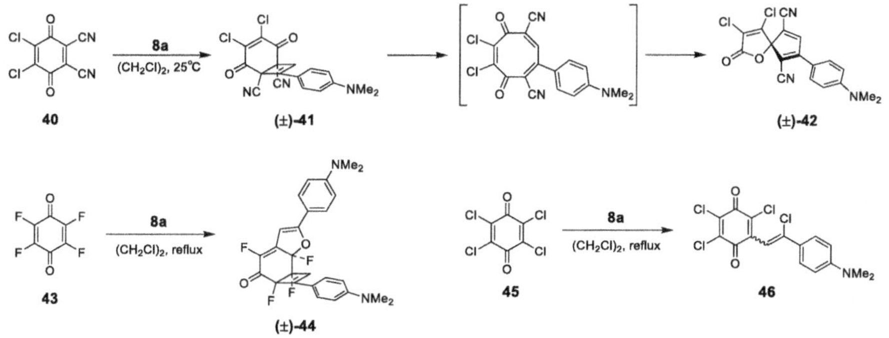

Scheme 7 Reactions of quinone acceptors with alkyne **8a**

The CA-RE reactivity of halo-*p*-benzoquinones has been extensively investigated (Scheme 7). For example, 2,3-dichloro-5,6-dicyano-1,4-benzoquinone (DDQ, **40**) reacted with alkyne **8a** to afford the CA-RE product **(±)-41** in a high yield of 85% at 25 °C. The product underwent rearrangement upon heating, resulting in the formation of **(±)-42** in a yield of 16% [68]. This reaction can also proceed with alkynes bearing other substituents [69]. Tetrafluoroquinone **43** reacted with excess **8a** in 1,2-dichloroethane under reflux conditions, yielding **(±)-44** in 20% yield [70]. In contrast, tetrachloroquinone **45** did not smoothly undergo the [2 + 2] CA reaction under similar conditions with an equivalent amount of **8a**. Instead, it produced **46** in 23% yield as a 10:1 (*Z*)/(*E*) mixture.

The reaction of cyanoethenes bearing ester groups often involves competing reactions, leading to a mixture of several isomers as the final products. Reekie et al. conducted a systematic study on the reaction between **8a** and cyanoethenes containing one, two, or three ester groups (Scheme 8) [71]. The monoester-substituted cyanoethene yielded three CA-RE products **48–50**. Since the cyano group is a stronger electron acceptor than the ester group, **49** and **50** are the expected products. However, the unexpected **48** was also obtained in 13% yield. There are

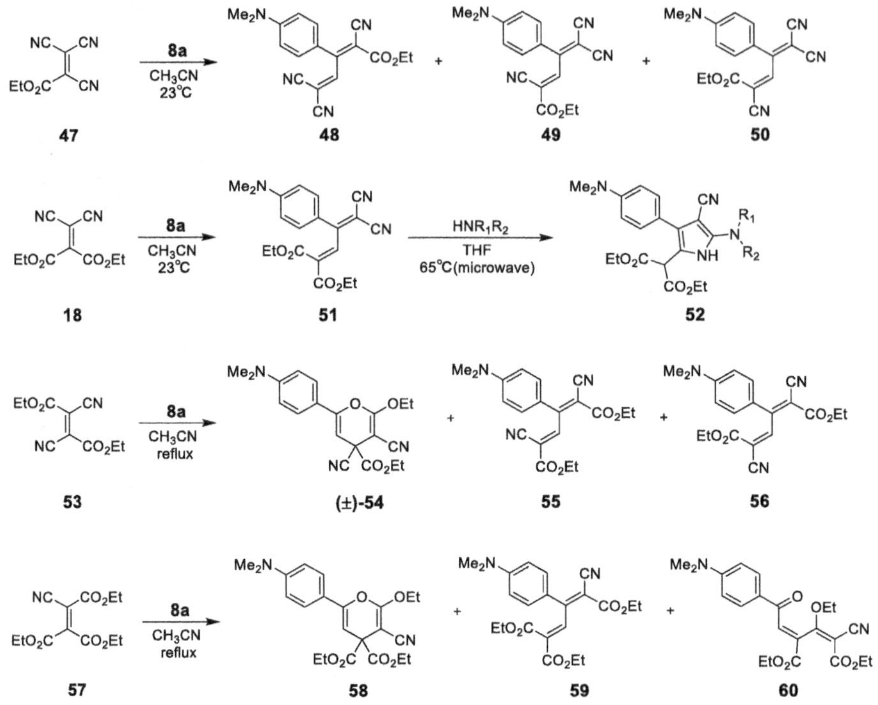

Scheme 8 Reactions of ester-substituted cyanoethenes

two isomers, **18** and **53**, for diester-substituted cyanoethenes, each exhibiting different reactivities. The DCV derivative **18** afforded the CA-RE product **51** in 91% yield, which could be transformed into **52** upon microwave heating with a secondary amine. In contrast, isomer **53** produced three products: a pair of *E/Z* isomers, **55** and **56**, through the [2 + 2] CA-RE reaction, and **(±)-54** via an HDA reaction. The triester-substituted cyanoethene **57** produced **59** via the [2 + 2] CA-RE reaction and **58** via the HDA reaction. An unexpected pentadienone product **60** was also obtained in 4% yield.

2.2 *Synthesis of Molecular Push-Pull Chromophores*

The reactivity of alkynes in [2 + 2] CA-RE transformations can be significantly increased by ring strain. Methoxy-substituted cyclooctadiyne **61** underwent a stepwise [2 + 2] CA-RE reaction with TCNE as the temperature increased (Scheme 9) [72]. The CA reaction occurred at 20 °C to yield **62**, but the further RE transformation did not proceed at this temperature. Upon raising the temperature to 40 °C, a second TCNE could undergo a [2 + 2] CA-RE transformation with **62** to afford **63**.

After increasing the temperature to 80 °C, the cyclobutene moiety in **63** underwent RE transformation, resulting in the formation of tetraene **(±)-64**. The strain-induced acceleration of the first CA reaction can reach up to 5,700-fold, according to experimental rate constants and theoretical calculations. As a chiral compound, the enantiomerically enriched products were separated using a chiral stationary phase, and their ECD spectra were measured.

Several studies have reported the synthesis of chiral TCBD derivatives and the investigation of their racemization behavior. Enantiomerically pure molecules were successfully synthesized through [2 + 2] CA-RE reactions (Fig. 3). For example, sterically congested C_{60} derivatives reacted with TCNE to produce TCBD derivatives in 80% yield. Their enantiomers with axially chiral ***(M)*-65** and ***(P)*-65** were isolated by a chiral HPLC [73]. ***(S)*-66** and ***(M)*-67** were synthesized from optically active precursors, specifically 1,3-diethynylallene and 2,2′-diethynyl-1,1′-binaphthalene derivatives, with high yield [74]. A pair of enantiomers, **(±)-68**, was obtained from spirobifluorene derivative precursors and TCNE [75]. It was also reported that SubPcs derivatives (Table 1, entry 22) could yield chiral products after the [2 + 2] CA-RE reaction with TCNE [76]. All of the described enantiomers exhibited a Cotton effect in their ECD spectra, which was associated with the intramolecular CT absorptions.

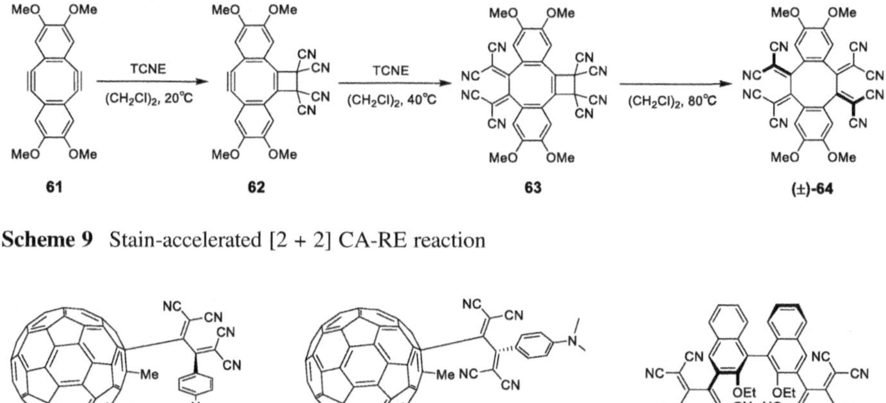

Scheme 9 Stain-accelerated [2 + 2] CA-RE reaction

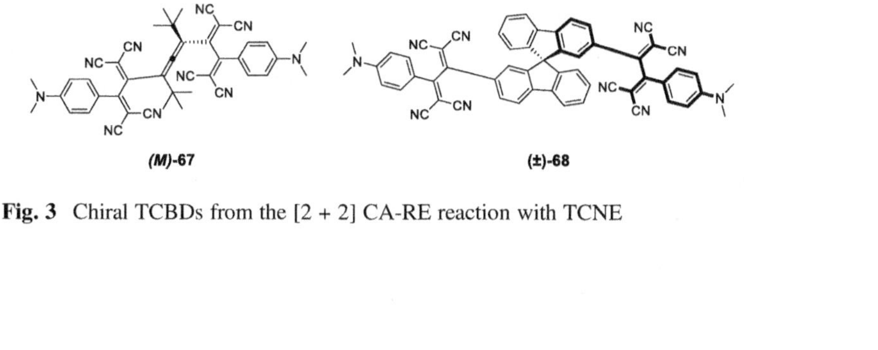

Fig. 3 Chiral TCBDs from the [2 + 2] CA-RE reaction with TCNE

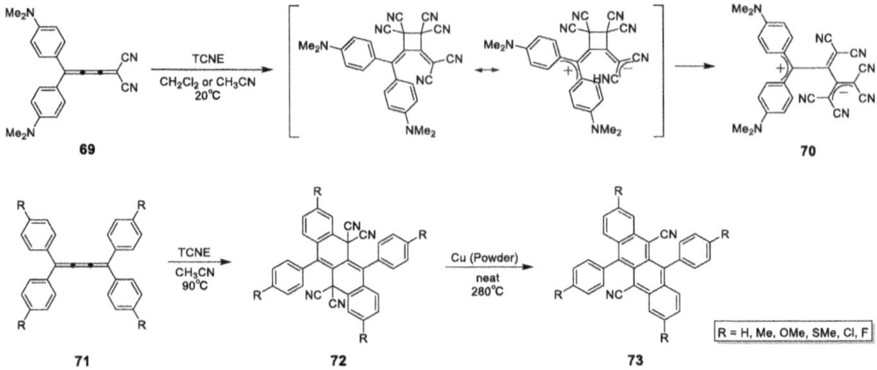

Scheme 10 Reaction of tetraaryl[3]cumulenes with TCNE

It has been reported that the length of the central alkene in donor-acceptor-substituted [3]cumulenes is similar to that of the triple bond in alkyne (1.21–1.22 Å), so their CA-RE reactivity with TCNE was investigated (Scheme 10) [77]. The asymmetric [3]cumulenes **69** reacted with TCNE in CH_2Cl_2 or CH_3CN at 20 °C to yield **70** almost quantitatively (99%). Symmetric [3]cumulenes **71** also displayed the CA-RE reactivity, but heating was required to initiate the transformation due to the reduced proacetylenic reactivity of the central alkene (1.24–1.25 Å). Unexpected tetracene derivatives **72** were obtained as a result of cascade reactions [78]. The initial products **72** were further transformed into tetracenes **73**, which have potential applications in organic electronics due to their favorable exoergic singlet exciton fission and high triplet exciton yield [79]. Furthermore, tetraaryl[5] cumulenes were also shown to react with TCNE [80].

The [2 + 2] CA-RE reaction is a highly efficient method for constructing dendritic systems (Fig. 4). Starting with aromatic cores, several alkyne-containing side chains were attached, followed by a [2 + 2] CA-RE reaction with TCNE to yield dendrimers **74**–**81** [81]. This synthetic pathway takes advantage of the click chemistry feature of the [2 + 2] CA-RE reaction. The resulting TCBD-based dendrimers exhibited unique absorption and redox properties, highlighting their potential applications in organic electronic devices.

The nonplanar configurations of the CA-RE products make them suitable as stoppering motifs for rotaxane synthesis [82]. Li et al. initially synthesized rotaxane **84** in 30% yield by threading the rod precursor **82** through the macrocycle **83**, followed by the [2 + 2] CA-RE reaction with TCNQ as the stoppering agent (Scheme 11). Since Cu^{I} can help immobilize the macrocycle moiety on the rod molecules, Diederich et al. reported the synthesis of a Cu^{I} bis-phenanthroline-based rotaxane **88** in 84% yield through the [2 + 2] CA-RE reaction of **87** formed from **85** and **86** (Scheme 11).

Recently, the applications of the [2 + 2] CA-RE reaction in organic frameworks have been reported. Xu et al. introduced TCBD moieties into metal-organic frameworks (MOFs) by reacting TCNE with alkyne side arms on a microporous Zr^{-}

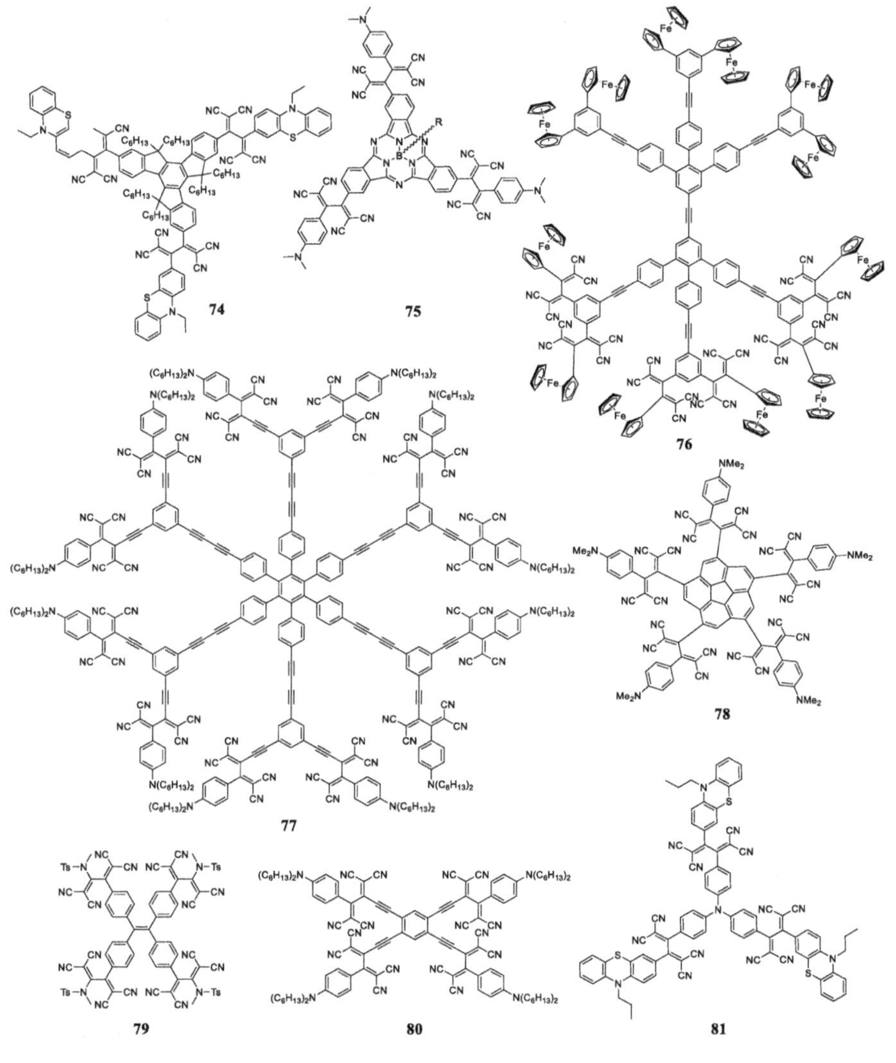

Fig. 4 Dendrimers containing TCBD moieties

organic framework [83]. The resulting black microporous crystallites maintained the crystallinity and framework rigidity, while their redox-active and photochemical properties were significantly enhanced. Shustova et al. inserted DCNQ moieties into porous covalent-organic frameworks, resulting in an eight-orders-of-magnitude increase in conductivity and significant changes in redox behavior [84].

Interestingly, the [2 + 2] CA-RE reaction is not limited to the combination of electron-rich alkynes and electron-deficient alkenes. The opposite combination, namely, electron-rich alkenes and electron-deficient alkynes, can also undergo the [2 + 2] CA-RE transformation (Scheme 12) [85]. Tetrathiafulvalene (TTF), an

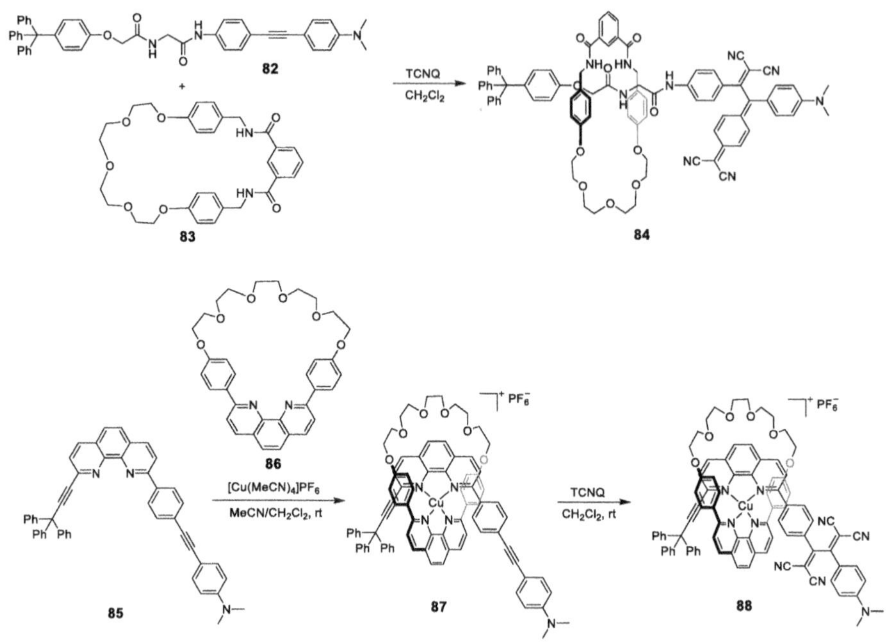

Scheme 11 Synthesis of rotaxanes via [2 + 2] CA-RE reaction

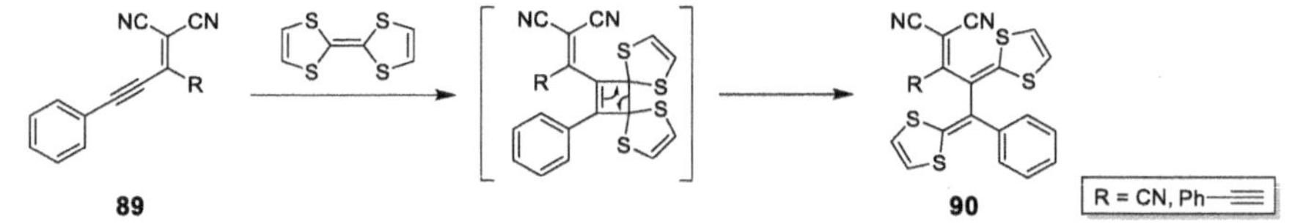

Scheme 12 [2 + 2] CA-RE reaction with the opposite electronic combination

electron-rich alkene, reacted with electron-deficient alkyne **89** in refluxing toluene to yield the CA-RE product **90** in 77% yield. This reaction was applied to construct dendrimer **91** with [AB]-type oligomers (Fig. 5) [81b]. Additionally, the sequential addition of TCNE and TTF to a polyyne backbone was demonstrated, and the final CA-RE adduct **92** was obtained in 47% yield (Fig. 5) [74a, 86].

TCBD moieties have been used to construct photoactive molecular dyad systems, taking advantage of their electron-accepting characteristics. Several multicomponent systems **93**–**97** have been reported, consisting of a zinc-porphyrin donor and donor-substituted pentacyanobuta-1,3-diene (PCBD), TCBD, or DCNQ, with or without a bridging spacer (Fig. 6) [87]. The PCBD-substituted molecule **93** exhibited ultrafast intrinsic deactivation within 10 ps. In contrast, the charge separation (CS) state of the TCBD-substituted analog **94** lasted 2.3 μs. For another molecular system **96**, which was composed of a different spacer and acceptor, the CS state lasted 931 ps, and the recombination lifetime reached 23,000 ps. However, for molecule **95**, which lacked

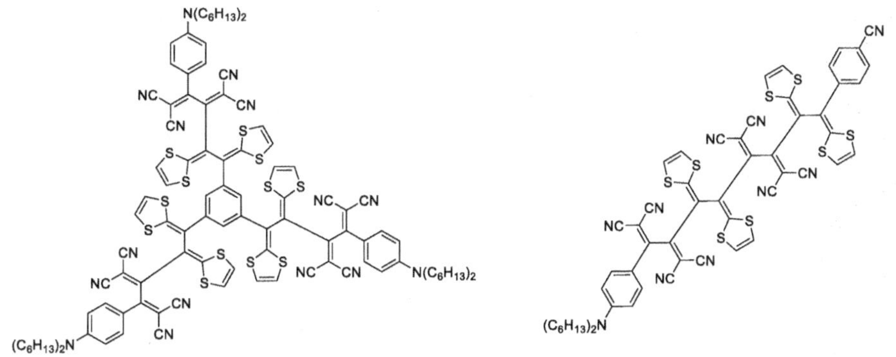

91 92

Fig. 5 Molecules synthesized by the electronically opposite [2 + 2] CA-RE reaction

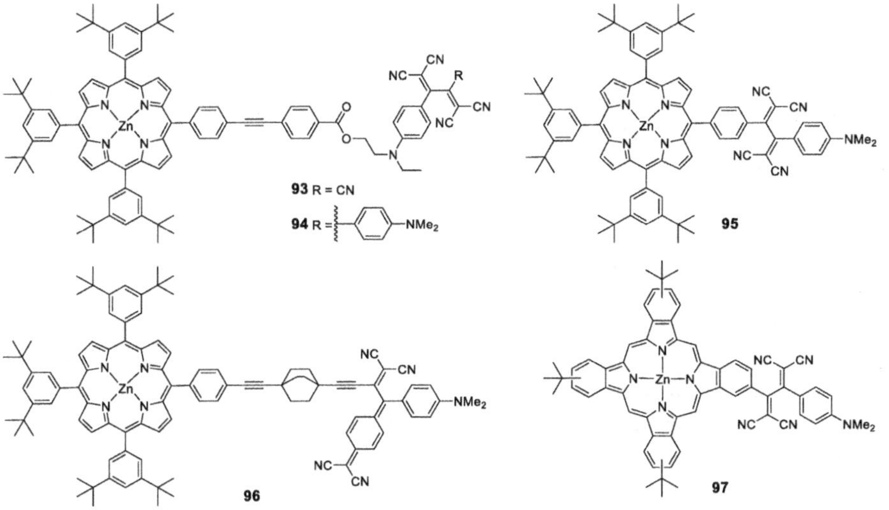

Fig. 6 Structure of molecular dyad systems

spacers, the CS state and subsequent recombination occurred within 20 ps. These results indicate that the presence of a spacer is essential for maintaining a long-lasting CS state. Molecular dyad **97** also exhibited a CS state due to extremely strong ground-state charge transfer (CT) interactions [87d]. A series of CA-RE-chromophore-C_{60} conjugates and triphenylamine-CA-RE-chromophore-BODIPY triads were synthesized, and their CS states were observed. [88] Additionally, the charge recombination kinetics in the inverted Marcus region were investigated by Diederich et al. [89] Furthermore, TCBD-containing dendrimers also displayed photoactive CS states, with electron stabilization achieved through electron exchange among different acceptor entities in multimodular donor-acceptor conjugates [90].

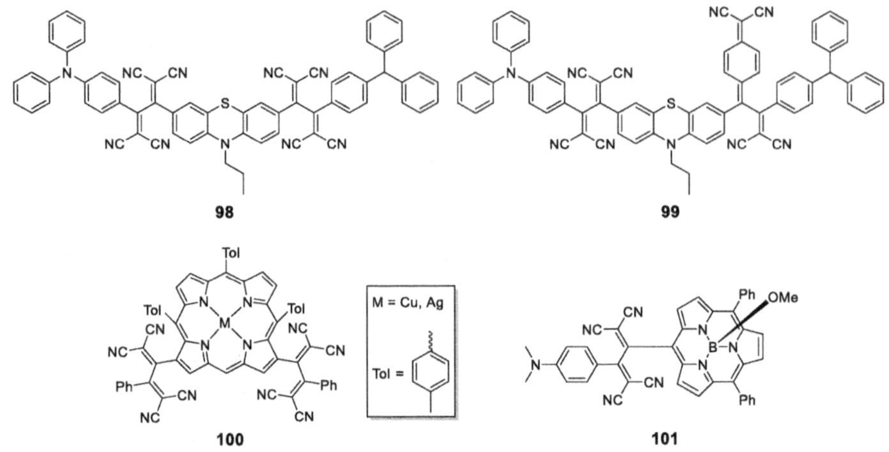

Fig. 7 Examples of push-pull chromophores for modulating CT states

The [2 + 2] CA-RE reaction is an effective method for modulating the CT state of π-conjugated systems (Fig. 7). Due to the direct conjugation of donor and acceptor units, efficient intramolecular CT was formed in the ground state of both symmetric and asymmetric chromophores **98**–**99**, resulting in their absorption bands extending into the near-infrared region [91]. Copper and silver corrole-based bis-TCBD push-pull chromophores **100** were synthesized in excellent yields, and their triplet states were effectively populated through CT [92]. Interestingly, subporphyrin-based TCBD derivative **101** displayed an emissive CT state in almost all environments [93]. Very recently, donor-acceptor complexes were constructed using a metal-ligand axial coordination self-assembly approach, with TCNE- or TCNQ-adduct derivatives as acceptors and porphyrin derivatives as donors [94]. These complexes exhibited both ground- and excited-state CTs, demonstrating potential for optoelectronic device applications.

Many push-pull chromophores obtained through the [2 + 2] CA-RE reaction show optical absorption in the near-infrared region. A series of near-infrared absorbing porphyrin [95a] or fullerene derivatives [95b] **102**–**104** were synthesized, and their photoacoustic effects were characterized (Fig. 8). The use of F_4TCNQ resulted in a significant red shift in the absorption band. Additionally, all these molecules exhibited low toxicity in cells, indicating their potential applications in *in vivo* imaging and detection. Leveraging the extended absorption, near-infrared luminescent molecules were synthesized by linking TCBD units with luminescent motifs such as pyrene and perylene. These two chromophores emit light in the solid state in the near-infrared region, up to approximately 1,350 nm, thus reaching the near-infrared-II window [96].

The structure of Aviram-Ratner-type molecular rectifiers **105** was proposed in 1974, but their synthesis has not been reported due to synthetic difficulties [97a]. Based on the proposed structure, a new rectifier **106** was designed and synthesized

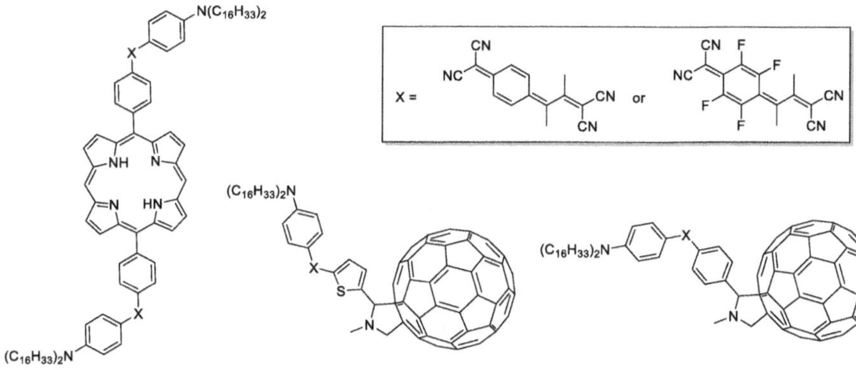

102 103 104

Fig. 8 Near-infrared absorbing porphyrin or fullerene derivatives with photoacoustic effects

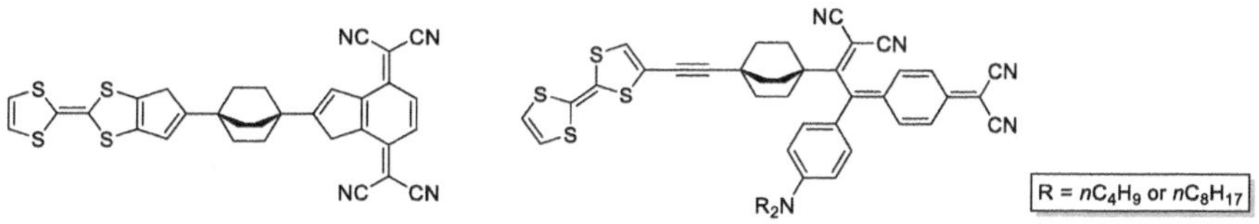

105 106

Fig. 9 Aviram-Ratner-type molecular rectifiers

using the [2 + 2] CA-RE reaction at 120 °C in 75% yield [97b]. Langmuir–Blodgett films of **106** were prepared and used in the fabrication and electrical testing of junction devices, which exhibited an asymmetric current–voltage (I–V) curve characteristic of rectification (Fig. 9).

2.3 *Synthesis of Polymeric Push-Pull Chromophores*

Postfunctionalization of polymers is a crucial method in polymer synthesis and functionalization. Click postfunctionalization is an ideal approach due to its advantages, including high yield and the absence of side reactions, which are essential for obtaining polymers with well-defined chemical structures. As already described, the [2 + 2] CA-RE reaction is an efficient, high-yielding addition reaction when appropriate donors, such as DMA, and symmetrical acceptors, such as TCNE and TCNQ, are employed. Compared with other click reactions, the [2 + 2] CA-RE reaction can modulate the intramolecular CT absorption and redox activities, making it especially suitable for optoelectronic applications.

Side-chain postfunctionalization of polymers using the [2 + 2] CA-RE has been extensively studied by Michinobu et al. [98] A typical example is a polystyrene

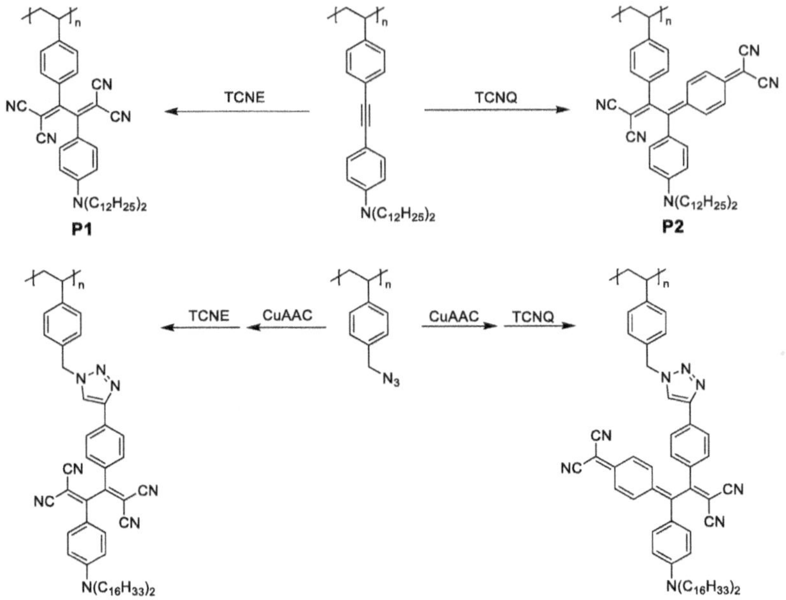

Scheme 13 Postfunctionalization of polystyrene derivatives bearing alkyne side chains

derivative bearing dialkylanilino-substituted alkynes as side chains, which efficiently undergo the [2 + 2] CA-RE with TCNE and TCNQ to yield **P1** and **P2**, respectively (Scheme 13) [99]. Additionally, alkyne-containing side chains were also introduced by Cu^{I}-catalyzed azide-alkyne cycloaddition (CuAAC) click reaction. This was followed by the [2 + 2] CA-RE click postfunctionalization to afford **P3** and **P4** (Scheme 13) [100]. This result highlights the advantages of using click chemistry in polymer synthesis. Since click chemistry reactions have orthogonal reactivities, multiple different click chemistry reactions can be combined to synthesize complex polymer structures in a well-defined manner. The colorimetric ion-sensing phenomenon of **P1–P4** was observed based on the principle of hard and soft acids and bases (HSAB) [101]. The hard basic aniline nitrogen atoms interacted with hard acidic metal ions, such as Fe^{3+}, resulting in the decolorization of polymer solutions due to the decreased CT band intensity. Conversely, a bathochromic shift in the CT band occurred while the soft basic cyano nitrogen atoms interacted with soft acidic metal ions, such as Ag^{+}. Therefore, these polymers could be used as sensors to distinguish soft and hard metal ions based on the different color changes. Detailed studies revealed that the color of **P1** and **P3** changes from orange to pink, while **P2** and **P4** shift from green to yellowish-green when mixed with Ag^{+}. It is also important to note that the interaction between the cyano group and Ag^{+} requires multivalent coordination, indicating that monomeric chromophores cannot capture Ag^{+}. This bathochromic shift was also observed in polyurethane-based push-pull chromophores **P5–P6** (Scheme 14) [102].

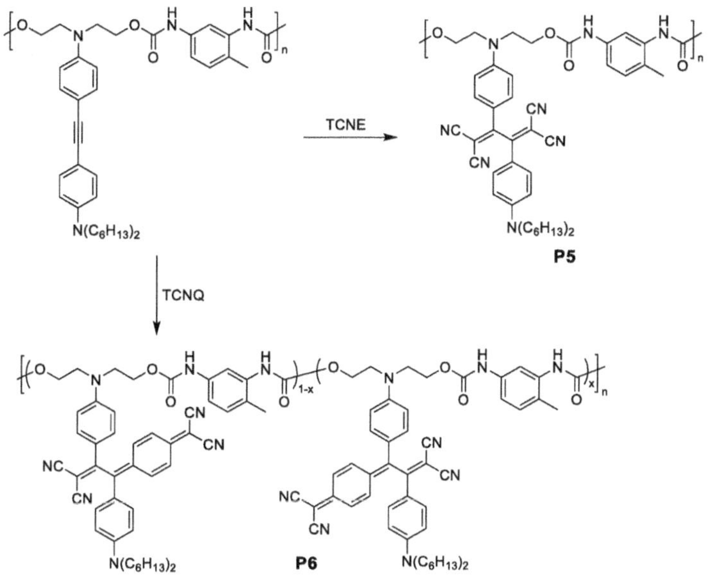

Scheme 14 Polyurethane-based push-pull chromophores **P5**–**P6** synthesized by the [2 + 2] CA-RE postfunctionalization

Side-chain postfunctionalization has also been applied in the synthesis of semi-conducting polymers. For example, aromatic polyamine **P8** was synthesized by the postfunctionalization of polyamine **P7** via the [2 + 2] CA-RE reaction with TCNE at 20 °C (Scheme 15) [103]. It was found that the first reductive potential of the polyamine decreased with the increasing added TCNE amount, and the estimated LUMO energy level reached −4.2 eV when the postfunctionalization yield was 100% ($x = 1$). Furthermore, the performance of their bottom-gate top-contact thin-film transistors (TFTs) was also investigated. [104] The hole mobility of **P7** gradually decreased with the increasing amount of TCNE, partly due to hindered hole injection. However, n-channel TFT performance was not observed for **P8**, likely due to the nonplanar TCBD moieties. It was found that **P7** underwent the [2 + 2] CA-RE functionalization with TCNQ under mild heating conditions to yield **P9** (Scheme 15) [105]. The estimated LUMO energy level of **P9** reached −4.0 eV when the alkyne unit was completely converted into the acceptor unit ($x + y = 1$). Additionally, the side-chain alkynes of other conjugated polymers were postfunctionalized by the [2 + 2] CA-RE reaction with TCNE to yield **P10**–**P13** (Fig. 10) [106]. It was shown that the regioregular poly(3-hexylthiophene) (P3HT) derivative **P10** with the side chain TCBD units becomes purely organic high-spin poly(anionic radical)s with a half-life exceeding 1 day at 20 °C [106b]. Conjugated polymers, **P11**–**P13**, exhibited longer wavelength absorption spectra, deeper HOMO and LUMO energy levels, and narrow bandgaps, as compared to the corresponding precursor polymers.

In addition to side-chain postfunctionalization, main-chain alkynes also underwent the [2 + 2] CA-RE postfunctionalization (Fig. 11). For example,

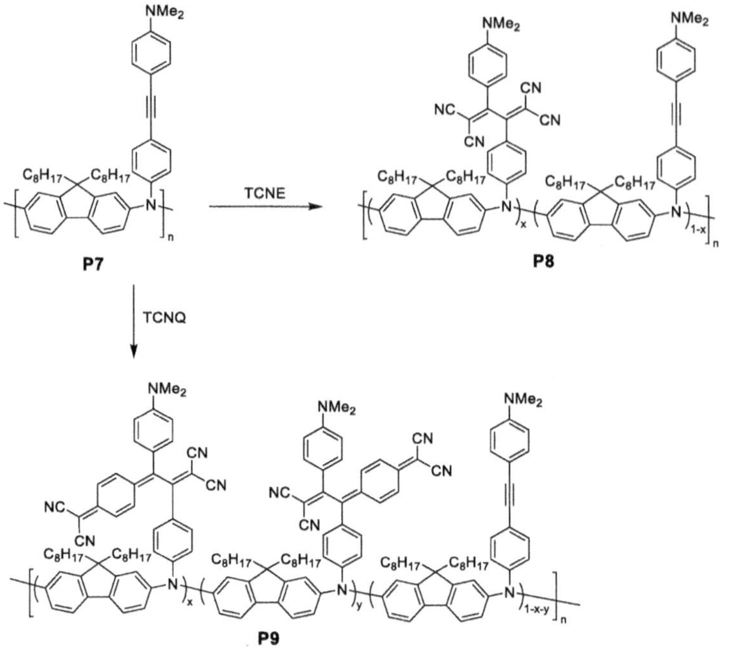

Scheme 15 Postfunctionalization to polyamine **P7** to yield TCBD-containing polyamine **P8** and DCNQ-containing polyamine **P9**

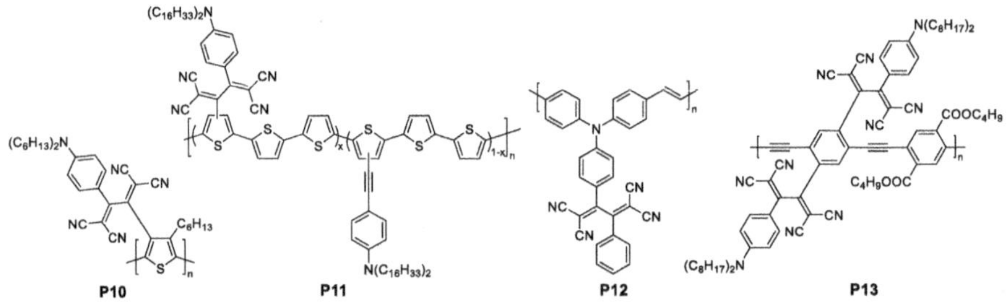

Fig. 10 Conjugated polymers postfunctionalized on the side chain with TCNE via the [2 + 2] CA-RE reaction

ferrocene-containing poly(arylene ethylene)s reacted with TCNE at 120 °C to yield the donor-acceptor alternating polymer **P14** [107]. Since no side reactions occurred, purification required only the removal of excess TCNE, which can be achieved by sublimation under vacuum. This novel synthetic method has also been applied to synthesize other donor-acceptor polymers based on the carbazole donor (**P15**), dialkylamino-benzene donor (**P16**), and alkylthiophene donor (**P17–P18**) [108]. Although the main-chain postfunctionalization was not quantitative, these polymers featured well-defined intramolecular CT absorption, narrow bandgaps, and

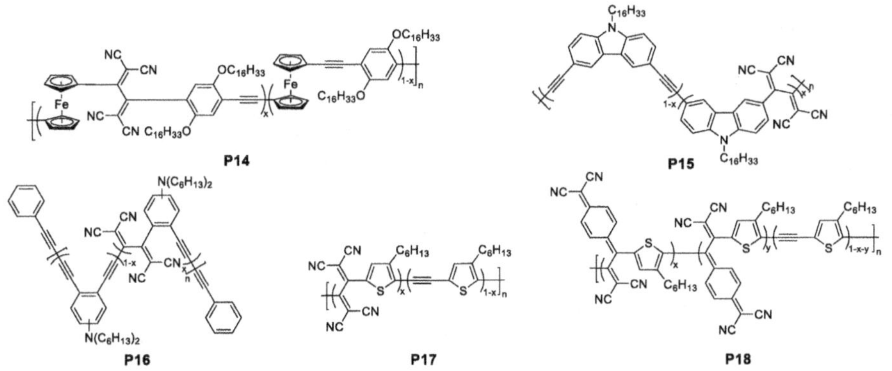

Fig. 11 Donor-acceptor alternating polymers synthesized by the [2 + 2] CA-RE postfunctionalization

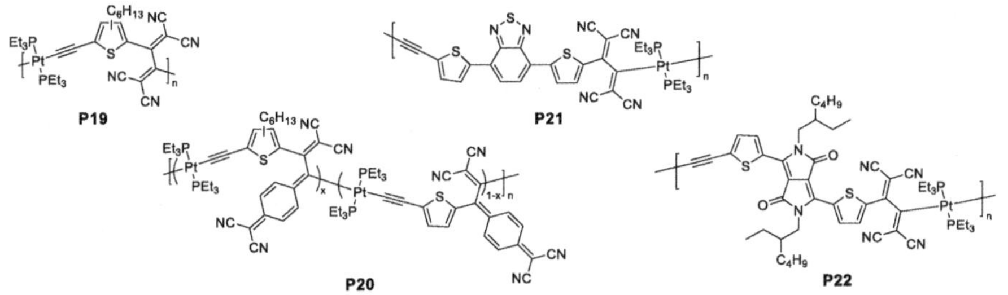

Fig. 12 Donor-acceptor polymers obtained from the [2 + 2] CA-RE postfunctionalization of platinum(II) polyynes

redox activities. Notably, the thermal stability of some functionalized polymers (**P15**, **P17**–**P18**) was improved compared to their precursor polymers, probably due to the enhanced intermolecular interactions resulting from the abundant cyano groups.

Metal-acetylide compounds were the first example of [2 + 2] CA-RE reactions, and metal-polyyne polymers have also been used in [2 + 2] CA-RE postfunctionalization (Fig. 12) [109]. Platinum(II) polyynes underwent the [2 + 2] CA-RE postfunctionalization with TCNE and TCNQ to yield the corresponding TCBD polymers **P19** and **P20**, respectively. These polymers have been investigated as n-type semiconductors for photovoltaic devices because of their deep LUMO energy levels. Additionally, electron-withdrawing aromatic units, such as benzothiadiazole and diketopyrrolopyrrole, have been incorporated into **P19** to produce **P21** and **P22** with the significantly lowered LUMO energy level of −4.5 eV [110]. The resulting polymers were tested in all-polymer solar cells in combination with P3HT, but their photoconversion efficiencies remained very low.

Furthermore, there are numerous reports about [2 + 2] CA-RE that discuss the synthesis and properties of main-chain functionalized donor-acceptor polymers

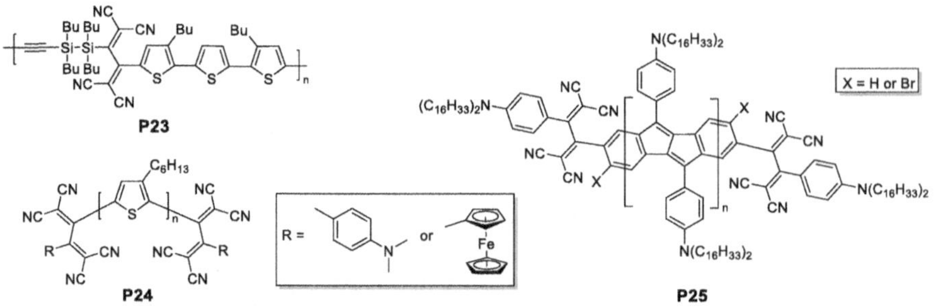

Fig. 13 Other examples of postfunctionalized donor-acceptor oligomers and polymers by the [2 + 2] CA-RE reaction

(Fig. 13). Ohshita et al. synthesized donor-acceptor-type organosilicon polymers **P23**, which exhibited red-shifted absorption bands with an absorption edge extending to approximately 800 nm [111]. This polymer exhibited solvatochromic behavior due to the push-pull chromophores. TCBD units were incorporated at the terminals of conjugated oligo(3-hexylthiophene)s **P24**. These oligomers were synthesized by Kumada catalyst-transfer polymerization and exhibited potential application in molecular electronics [112]. Michinobu et al. developed ladder-type conjugated polymers, specifically fused poly(benzopentalene) derivatives, and utilized the [2 + 2] CA-RE postfunctionalization with TCNE to conduct titration experiments for estimating the terminal-alkyne content [113]. The estimated terminal amount was employed to calculate the molecular weight of the polymers, which matched the values obtained from GPC measurements. They referred to this technique as the "exact quantification of specific alkyne amounts in polymers." As expected, the functionalized polymers **P25** exhibited decreased bandgaps.

To conduct polymerization based on [2 + 2] CA-RE, a TCNQ-based bifunctional monomer was synthesized (Fig. 14). **P26** was synthesized from TCNQ-based monomer and comonomer with two electron-rich alkynes in $CHCl_3$ at 60 °C [114]. The molecular weight (M_n) of **P26**, determined by GPC with THF as the eluent, was 19,000, indicating successful polymerization. The CA-RE polymerization technique could be combined with other click reactions, such as CuAAC and Diels-Alder reactions, leading to the formation of **P27** and **P28** [115]. Similar to the previously reported side-chain functionalized polymers, these polymers exhibited specific metal-ion-sensing behavior. For example, when Ag^+ was added, the solution color changed from green to yellowish-green.

The TCNQ-based monomer could be used to prepare cross-linked polymer films (Scheme 16) [116]. TCNQ moieties in **P29** underwent radical copolymerization with styrene derivatives, such as 4-chloromethylstyrene, resulting in the formation of **P30**. The unreacted TCNQ moieties in **P30** were further modified via a [2 + 2] CA-RE reaction with DMA-substituted alkyne **8b**, forming **P31** in 76% yield.

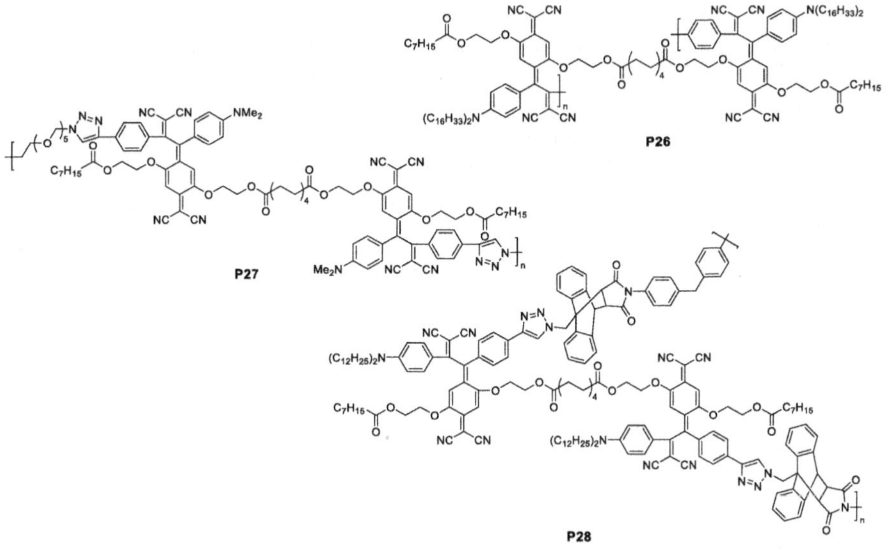

Fig. 14 TCNQ-based polymers by [2 + 2] CA-RE polymerization

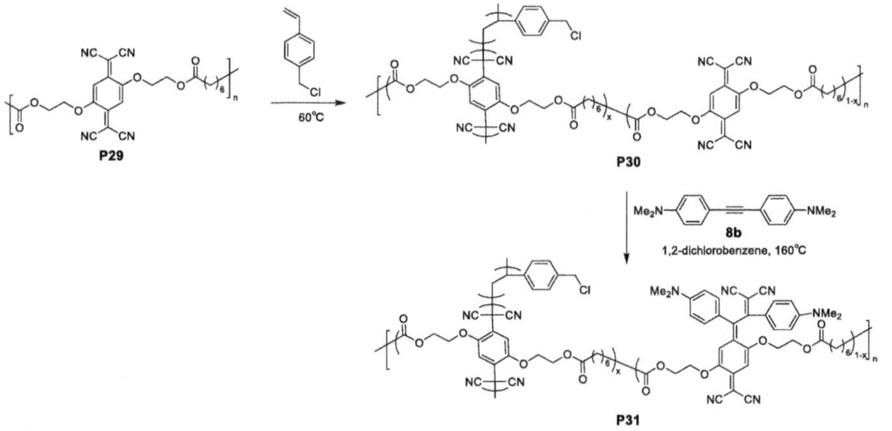

Scheme 16 Synthesis and postfunctionalization of cross-linked TCNQ-based polymers

3 Application to Organic Functional Materials

3.1 Nonlinear Optical Materials

Donor-acceptor chromophores demonstrate excellent potential for organic nonlinear optical (NLO) applications due to their highly polarizable π-electron systems [4a, 4b, 117]. Therefore, NLO properties of various push-pull chromophores synthesized via the [2 + 2] CA-RE reaction have been investigated. Small molecular NLO materials have received significant attention (Fig. 15). TCBD derivative **9c** (Fig. 2) exhibited a γ value of $(12 \pm 2) \times 10^{-48}$ m^5 V^{-2}, which is notable for its molecular size being close to the theoretical limit [118]. The nonplanar structure of **9c** ($\theta = 96.7°$) reduced intermolecular interactions, which is a critical factor for achieving high-quality amorphous optical films. In addition, **9c** has a decomposition temperature of 451 °C and a melting point of 191–193 °C, allowing for sublimation before decomposition. This property facilitates device fabrication through vapor deposition. Based on these characteristics, **9c** was used to fabricate a 4-mm-long silicon-organic hybrid

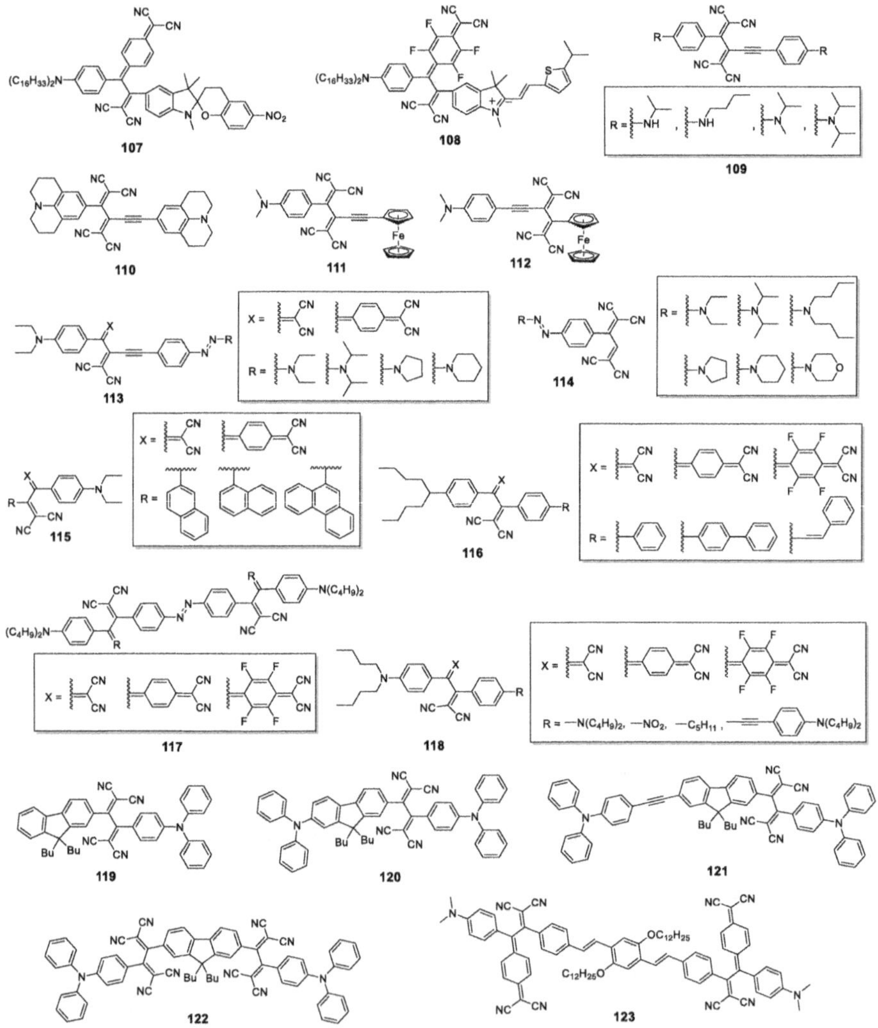

Fig. 15 Small molecular CA-RE products for NLO applications

waveguide with a record nonlinearity coefficient of $\gamma \approx 1 \times 10^5\ W^{-1}\ km^{-1}$, achieving all-optical demultiplexing from 170.8 Gb s^{-1} to 42.7 Gb s^{-1}. This marked the fastest silicon photonic optical signal processing at that time [119]. The TCNQ adduct **107** and F_4TCNQ adduct **108** exhibited third-order optical susceptibilities ($\chi^{(3)}$) of 2.03×10^{-13} and 1.51×10^{-13} esu, respectively [120]. TCBD derivatives **109–112**, inspired by the structure of **9c**, were developed for optical switching and frequency conversion on silicon photonics platforms. [121] A series of triazene-based push-pull chromophores **113–114** were synthesized, and computational calculations revealed significant first-order hyperpolarizabilities [122]. Similar to these results, polycyclic aromatic hydrocarbon-substituted chromophores **115** were computationally analyzed, showing first hyperpolarizability (β(tot)) values ranging from 198 to 538×10^{-30} esu [123]. Symmetrical and asymmetrical multi-phenylene rings **116–122**, prepared through the [2 + 2] CA-RE reaction, exhibited tunable NLO properties that varied depending on the substituents. These properties were demonstrated through Z-scan measurements [124]. The intramolecular CT compound **123** showed both third-order and fifth-order optical nonlinearities, as revealed by nonlinear optical imaging [125]. DDQ-based CA-RE products also exhibited excellent third-order polarizabilities, expanding the design scope of NLOphores [68a]. Furthermore, studies on molecules with varying acetylene spacer lengths between the anilino group and the TCBD moiety revealed that longer spacers correlate with higher off-resonant third-order polarizability values [126]. These findings underscore the versatility of [2 + 2] CA-RE chromophores in advancing the research of NLO materials.

The Z-scan measurements of the fullerene-based CA-RE adduct revealed a third-order NLO response [127]. Notably, compound **103** exhibited a unique NLO phenomenon when pumped by 21 ps pulses at 532 nm, indicating the presence of different types of reversible photoisomerization [127b]. Porphyrin derivatives **124–125** synthesized via the [2 + 2] CA-RE reaction also exhibited high third-order NLO susceptibilities ($\chi^{(3)}$) [128]. Remarkably, the third-order nonlinear absorption of these compounds displayed a reverse saturable absorption (RSA)-saturable absorption (SA) transition. The Z-scan measurements of star-shaped molecules **126–128** containing CA-RE chromophores demonstrated enhanced third-order NLO properties [129]. Compound **126** achieved an $\chi^{(3)}$ value of $8.30 \times 10^{-21}\ m^2\ V^{-2}$, representing a breakthrough as the first pyrene derivative with such high third-order susceptibility. Compound **127** exhibited third-order NLO performance with typical RSA-SA behavior. The nonlinear susceptibilities of this compound with different push-pull chromophores were measured to be 1.50×10^{-12} esu. Compound **128** exhibited even better performance, with an $\chi^{(3)}$ value reaching 1.3×10^{-12} esu. In addition, the incorporation of push-pull chromophores into polymers through CA-RE postfunctionalization—whether as side chains or end caps—has been shown to endow these polymers with considerable third-order NLO properties [130] (Fig. 16).

NLO dendrimers exhibited an efficient translation of large molecular hyperpolarizabilities ($\mu\beta$) into substantial macroscopic electro-optic (E-O) activities (r_{33}), which are essential for high-speed E-O device applications (Fig. 17)

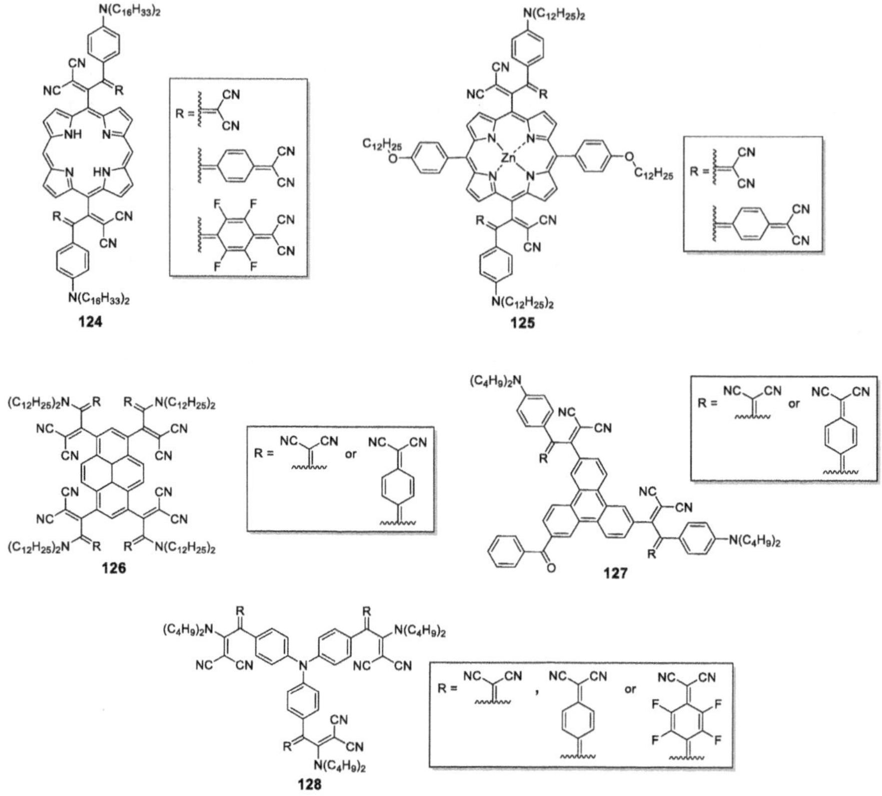

Fig. 16 Conjugated molecules and polymers with [2 + 2] CA-RE chromophores for NLO applications

[131]. Dendrimer **129** exhibited an impressive r_{33} value of 60 pm V^{-1} at 1.55 μm and maintained over 90% of its original value at 85 °C for more than 1000 h, indicating its remarkable thermal stability [131b]. Similar to this result, dendrimer **130** encapsulated by highly fluorinated dendrons achieved an r_{33} value of 30 pm V^{-1} at 1.3 μm, with the added advantage of low optical loss [131c]. In addition, TCBD-based hyperbranched polymers were prepared by [2 + 2] CA-RE postfunctionalization (Fig. 18) [132]. The NLO property of the resulting hyperbranched polymers **131–132** was investigated, revealing a macroscopic NLO activity with a second harmonic generation (SHG) coefficient (d_{33}) value of 55.6 pm V^{-1} and a depoling onset temperature of 125 °C. Furthermore, the d_{33} value could be further enhanced to 96.6 pm V^{-1} when the depoling onset temperature reached 137 °C.

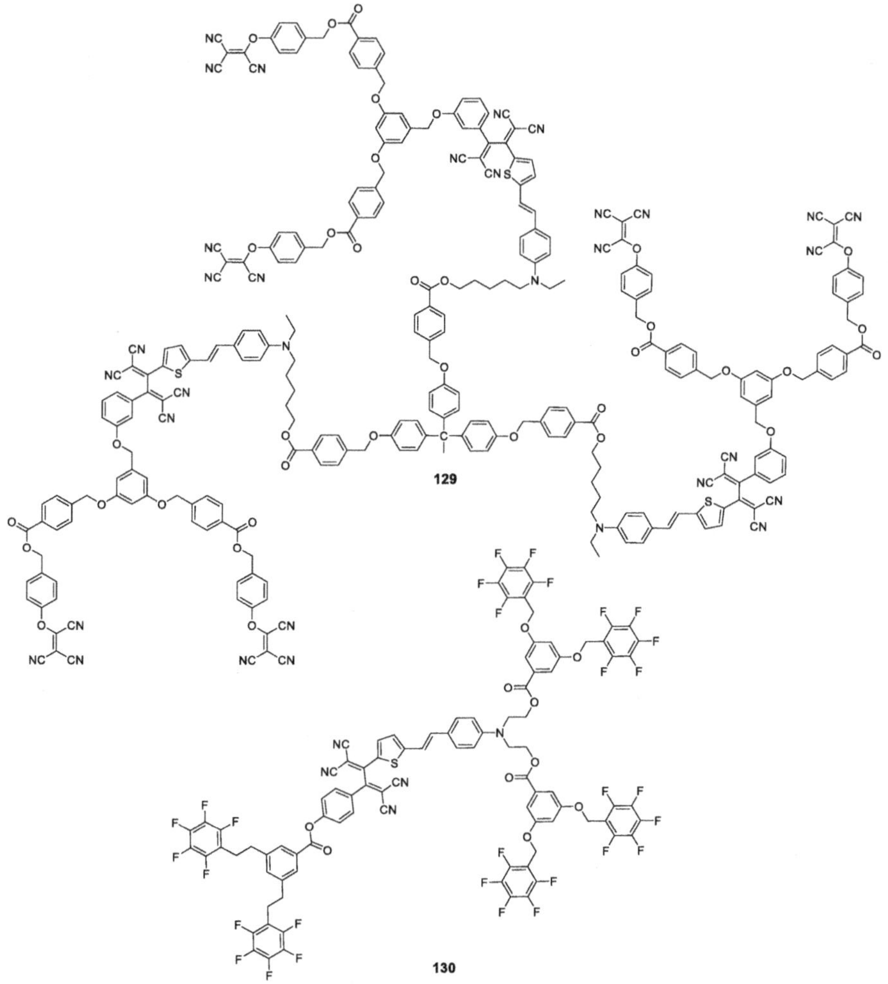

Fig. 17 Dendrimers with [2 + 2] CA-RE chromophores for NLO applications

3.2 *Photovoltaic Materials*

Dye-sensitized solar cells (DSSCs) provide a promising alternative to conventional silicon-based solar cells due to their low cost, ease of fabrication, and high overall power conversion efficiency (PCE) [133]. Dye sensitizers are a critical component of DSSCs, playing an essential role in light harvesting and electron injection into the conduction band of TiO_2. Although ruthenium- and porphyrin-based dyes have shown excellent performance, their low molar extinction coefficients, high cost, and environmental concerns have limited their wider adoption. This has shifted attention toward metal-free organic dyes as viable alternatives (Fig. 19)

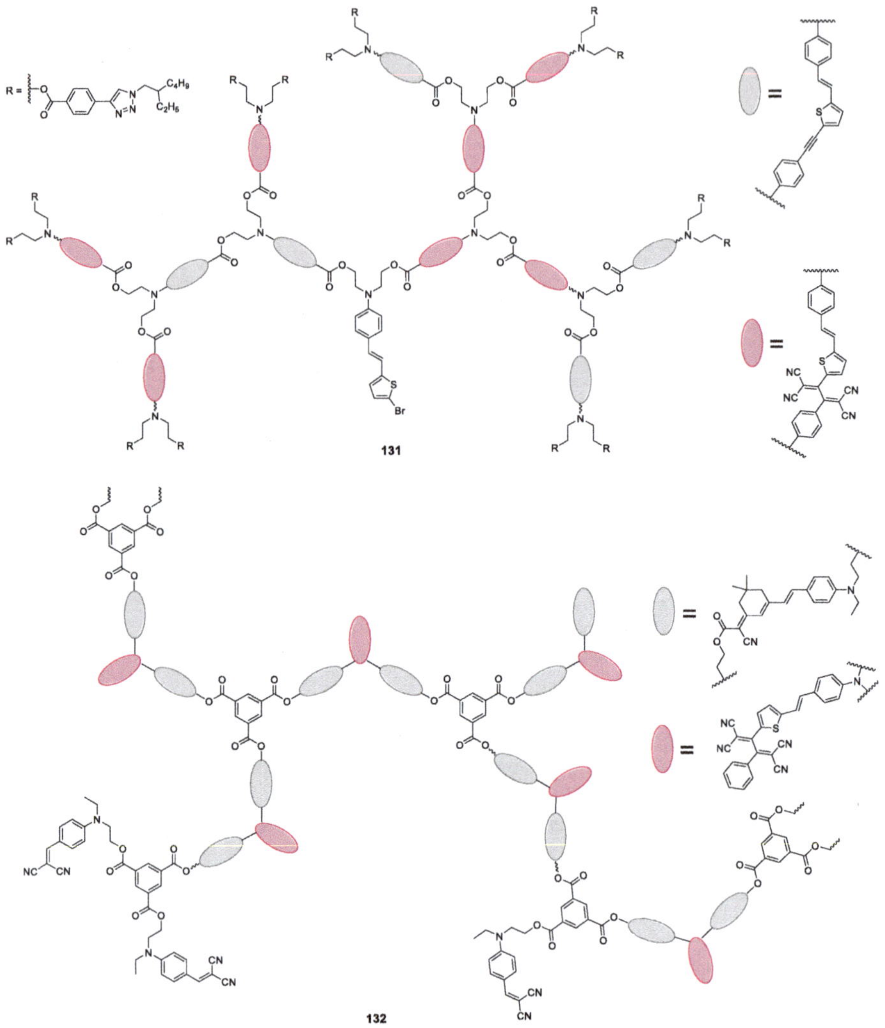

Fig. 18 TCBD-based hyperbranched polymers with high E-O activities

[134]. Among these, TCBD-based star molecule **133** achieved a 1.3% higher PCE compared to its alkyne precursor, due to its extended absorption into the near-infrared region, resulting from strong intramolecular CT interactions [135]. Donor-acceptor dyes, such as **9a** and **134–136**, synthesized through reactions between aniline-substituted alkynes and TCNE or TCNQ, were employed as photosensitizers [136]. It was suggested that the triphenylamine-substituted derivatives **135** and **136** demonstrated higher current densities (J_{sc}), while the TCNQ adducts **134** and **135** yielded higher open-circuit voltages (V_{oc}). Push-pull chromophores **137** and **138**, which contain carboxylic acid groups, were tested using TiO_2 or SnO_2 electrodes.

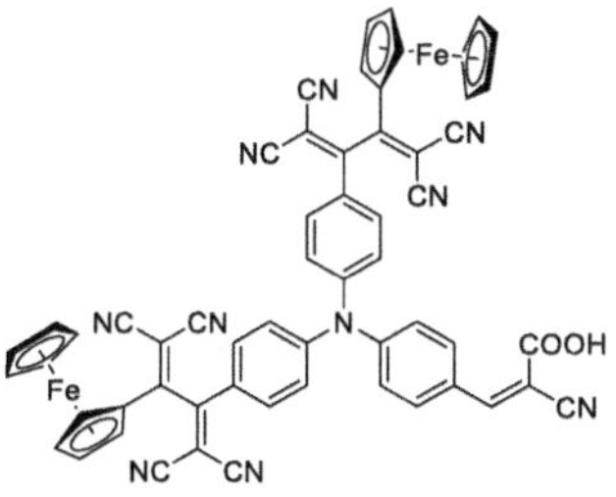

133

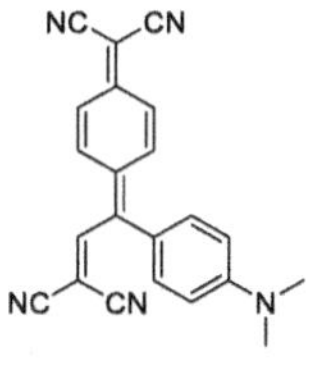

134

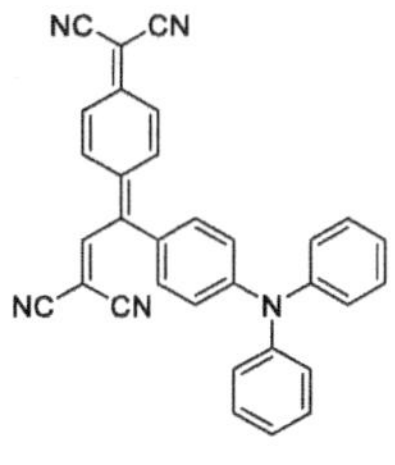

135

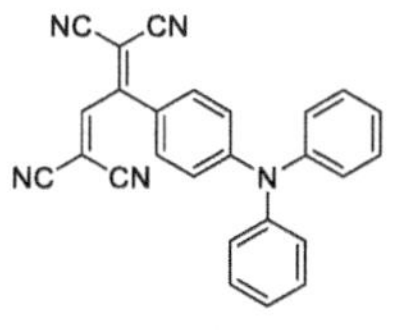

136

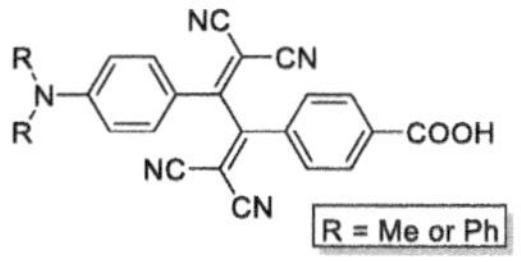

137

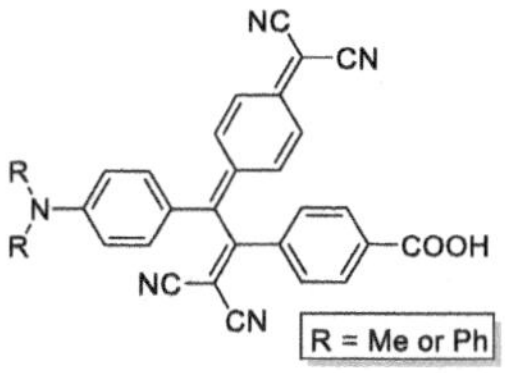

138

Fig. 19 [2 + 2] CA-RE dye sensitizers for DSSCs

The incident photon-to-current conversion efficiency (IPCE) spectra showed superior photocurrent generation when these dyes were used in SnO_2-based devices [137].

Most studies on CA-RE products in organic photovoltaics have focused on their performance in bulk heterojunction solar cells (BHJSCs). In this system, the active layer is composed of an intermixed electron donor and acceptor, creating a solid-state phase separation with interfaces optimized for charge generation and transport. Donors are either electron-rich polymer or small molecule, while acceptors are mainly divided into fullerene or non-fullerene acceptors. Due to their strong intramolecular CT interactions, CA-RE products show great promise as either donor or acceptor components in BHJSCs (Fig. 20). For example, a simple push-pull chromophore **139** was evaluated as an n-type semiconductor in BHJSCs, where efficient energy transfer was observed. [138] While these push-pull chromophores demonstrated extended absorption into the near-infrared region, this extension sometimes resulted in reduced absorption in the visible region, leading to a decrease in short-circuit current density (J_{sc}). Incorporating additional donor or acceptor units into the push-pull molecules can effectively overcome this issue. Therefore, chromophores **140–148** were synthesized via the [2 + 2] CA-RE reaction and used to fabricate photovoltaic devices through a solution process [139]. The results indicated that modifications to donor or acceptor units enabled precise optimization of PCE values. Interestingly, asymmetric V-shaped derivative **149** with one TCBD unit exhibited superior light-harvesting capabilities, which enhanced photocurrent density and overall device performance, achieving a maximum PCE of 7.58% [140]. Furthermore, diketopyrrolopyrrole-based push-pull chromophores **150–152** were utilized as acceptors in BHJSC devices, and the PCE value reached 9.29% [141]. Block

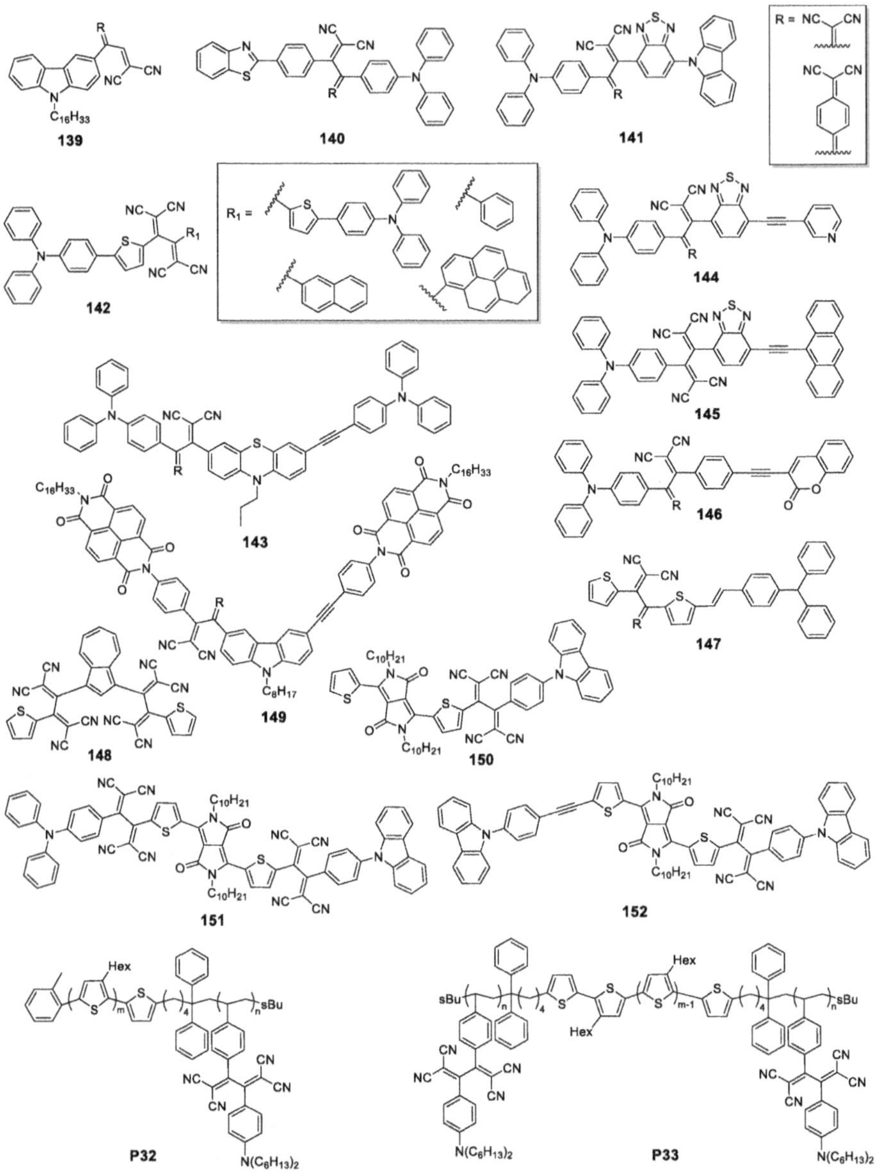

Fig. 20 [2 + 2] CA-RE donor-acceptor molecules and polymers for bulk heterojunction solar cells

copolymers synthesized through [2 + 2] CA-RE postfunctionalization were employed as an additive for BHJSC. The AB diblock copolymer **P32** and ABA triblock copolymer **P33** were prepared, and the incorporation of TCBD moieties led to the optimization of energy levels for nanoscale phase separation and improved

CS. Notably, BHJSC devices based on **P33** exhibited a 6.6% enhancement in the PCE value compared to the device without any additives [142].

The [2 + 2] CA-RE postfunctionalization method was also employed to modify the HOMO level of hole-transporting polymeric materials for perovskite solar cells. The introduction of TCBD moieties into a polyamine precursor lowered the HOMO energy level by 0.07 eV, leading to efficient charge transport and increased V_{oc} value due to improved energy-level alignment [143].

3.3 p-Dopants for Organic Semiconductors

The energy levels of push-pull chromophores can be effectively modulated by the [2 + 2] CA-RE reaction [144]. The redox properties of CA-RE products are evaluated using cyclic voltammetry (CV) and rotating disk voltammetry (RDV). These push-pull chromophores typically exhibited oxidation and reduction steps attributed to the donor and acceptor units, respectively. It has been reported that the first reduction potential ($E_{red,1}$) is a reliable indicator of electron-accepting power. A stronger acceptor is characterized by a more anodically shifted $E_{red,1}$ value. Generally, the presence of highly electron-deficient moieties enhances the electron-accepting ability. For instance, CA-RE products derived from F_4TCNQ exhibited a significant anodic shift in $E_{red,1}$ values. Compound **153** demonstrated an $E_{red,1}$ of 0.00 V (in CH_2Cl_2 with 0.1 M Bu_4NPF_6 versus Fc/Fc^+), highlighting its strong electron-accepting capability [59]. When $E_{red,1}$ values compete with that of the benchmark acceptor F_4TCNQ, they can be classified as a "super acceptor." The introduction of additional cyano groups, which are strong electron-withdrawing substituents, led to the design of **154**, which showed an even more anodically shifted $E_{red,1}$ value of +0.16 V, confirming its enhanced electron-accepting ability. However, F_4TCNQ adducts are susceptible to decomposition upon contact with glass surfaces, highlighting the need to explore alternative strategies for enhancing electron-accepting properties without relying on F_4TCNQ. One effective approach is the removal of donor moieties, as donor-acceptor interactions usually shift $E_{red,1}$ values cathodically. For example, TCBD **155** without any donor substituents displayed an $E_{red,1}$ value of −0.16 V, demonstrating the effectiveness of this strategy [145].

Octacyano[4]dendralene derivatives with varying alkyl substituents on the anilino unit were synthesized by the bis-addition of TCNE to symmetric 1,3-diyne precursors. The diisopropylamino-substituted derivative **156** exhibited a more anodic $E_{red,1}$ value of −0.08 V [146]. Compound **156** was utilized as a p-dopant for organic thermoelectric device applications, leveraging its strong electron-accepting property and excellent thermal stability. The device was fabricated by doping **156** into poly(4-ethylenedioxythiophene) (PEDOT). The highest electrical conductivity of 0.11 S cm^{-1} and an optimized thermoelectric power factor value of 3.68×10^{-4} μW m^{-1} K^{-2} were achieved at a doping ratio of 18 mol% [147] (Fig. 21).

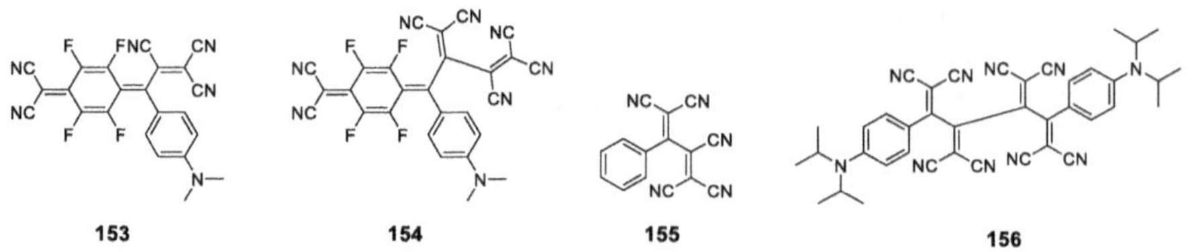

Fig. 21 "Super acceptors" obtained from the [2 + 2] CA-RE reaction

3.4 Colorimetric Sensors

The application of CA-RE products as ion sensors has been extensively studied. As already described, the coordination of Ag^+ with cyano groups resulted in a distinct color change and a bathochromic shift in the absorption spectra. This property highlighted the potential use of cyano-rich polymers, synthesized via CA-RE postfunctionalization, as colorimetric ion sensors [148]. Building on the antibacterial properties of Ag^+, antibacterial cross-linked films were produced by [2 + 2] CA-RE polymerization and subsequently coordinated with Ag^+ [149]. These Ag^+-loaded films demonstrated antibacterial activity against *E. coli*, and a correlation between the Ag^+ content and antibacterial performance was observed. Additionally, other metal ions, such as Hg^{2+} and Sn^{2+}, were found to coordinate with cyano groups in star molecules **157**. This coordination induced changes in the absorption and emission spectra, including shifts in wavelength and variations in peak intensity, indicating potential applications in the colorimetric detection of these metal ions [100d].

In addition to the metal ions identified through interactions with cyano groups, TCNQ adduct **158** has been shown to function as a colorimetric sensor for thiols [150]. The reaction between thiols and the acceptor unit reduced the electron-accepting ability and decreased the efficiency of intramolecular CT. Therefore, compound **158** exhibited chemodosimetric recognition of HS^- ions, characterized by a significant decrease in absorption at 477 nm and visible discoloration.

CA-RE products could also serve as colorimetric sensors for sulfide ions when sulfonamide groups were incorporated [101a]. The TCBD compound **159** exhibited a selective increase in absorption intensity in response to thiols, compared to other competing amino acids in an acetonitrile-aqueous solution, and a color change from yellow-green to nearly colorless was observed (Fig. 22).

3.5 Other Applications

A chiral spiro[pyrrol-benzopyran] scaffold with high asymmetry was synthesized using a [2 + 2] CA-RE cascade reaction, and its potential application in chiroptical sensing was explored. [151] The TCNE-adducted triphenylamine-based conjugated polymer **P34** exhibited efficient and specific detection of Hg^{2+}, a toxic pollutant,

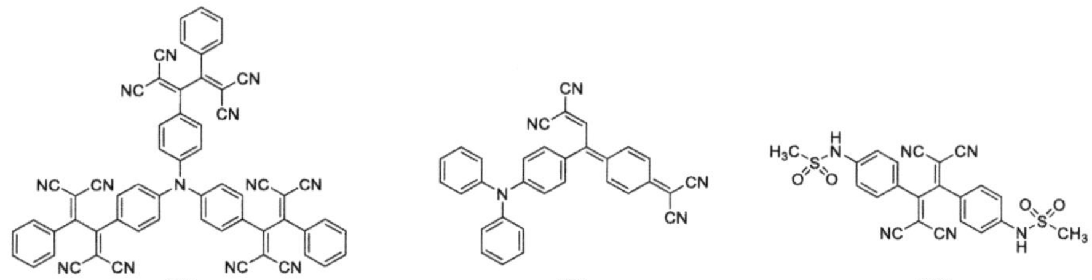

Fig. 22 CA-RE products as colorimetric sensors

through a distinct turn-on fluorescence response, achieving a detection limit as low as approximately 6.9 nM [100c]. In addition, the aggregation-induced emission (AIE) properties of TCBD derivatives were investigated. [152] Recent advancements include the successful conversion of TCBD derivatives from solution to the solid state by incorporating polystyrene (80 wt% relative to the chromophore) as a matrix. This approach facilitated the formation of white-light-emitting nanofibers via electrospinning [153].

Covalent organic frameworks (COFs) prepared using CA-RE products exhibited an extended absorption into the near-infrared region, granting them exceptional photothermal properties [154]. Additionally, TCNQ adduct **160** demonstrated a high photothermal conversion efficiency of 91.3% and effective near-infrared phototoxicity, showing excellent *in vivo* tumor growth inhibition in photothermal therapy [155b]. Fullerene-based TCNQ adducts were also explored for photothermal therapy due to their superior reactive oxygen species (ROS) generation efficiency [155a]. Moreover, functional CA-RE products have shown promising applications in drug delivery and nitric oxide (NO)-mediated disease therapies, underscoring the immense potential of the [2 + 2] CA-RE reaction in medical applications [156].

CA-RE products have also been reported for their applications in organic transistors. A functional organic photosensitizer was synthesized and used in organic phototransistors for detecting broadband spectra (250–700 nm) and implementing efficient CT [157]. Michinobu et al. investigated the thin-film transistor performance of carbazole-based platinum polyyne polymers **P35** and **P36**, synthesized via [2 + 2] CA-RE postfunctionalization [158]. It was observed that **P35** exhibited a shallower HOMO level compared to **P36**, enabling more efficient hole injection. As a result, **P35** demonstrated a higher hole mobility than **P36** (Fig. 23).

4 Conclusion

The [2 + 2] CA-RE reaction between donor-activated alkynes and electron-deficient alkenes is a powerful, versatile, and efficient method for constructing nonplanar push-pull chromophores. Under optimized conditions for selecting activated alkynes and acceptors, this transformation qualifies as a true click reaction, offering the

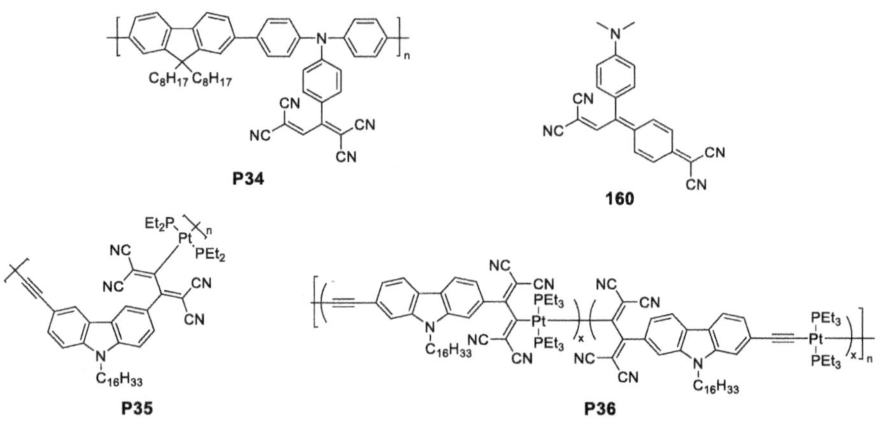

Fig. 23 CA-RE products in other applications

advantages of high yields, mild conditions, catalyst-free process, and no side reactions. The resulting chromophores exhibit strong intramolecular CT interactions, extended optical absorption into near-infrared region, and notable redox behavior. Additionally, various novel reactions, such as cascade and competing reactions, have been explored, expanding the synthetic versatility of this methodology. The opto-electronic properties of these chromophores can be precisely tuned by modifying the donor and acceptor moieties. As a click transformation, the [2 + 2] CA-RE reaction has also shown significant potential in step-growth polymerization and polymer postfunctionalization. The molecules and polymers obtained have been extensively investigated for diverse applications, including nonlinear optics, solar cells, p-dopants, colorimetric sensors, and more, showcasing the broad utility and adaptability of this reaction.

References

1. Bruce MI, Rodgers JR, Snow MR, Swincer AG (1981) J Chem Soc Chem Commun:271–272
2. (a) Bruce MI, Hambley TW, Rodgers JR, Snow MR, Swincer AG (1982) J Organomet Chem 226, C1–C4; (b) Bruce MI, Duffy DN, Liddell MJ, Snow MR, Tiekink ERT (1987) J Organomet Chem 335:365–378
3. (a) Onitsuka K, Takahashi S (1995) Synthesis and structure of s-cis-and s-trans-μ-butadiene-2,3-diyldiplatinum complexes by the reaction of μ-ethynediyldiplatinum complexes with tetracyanoethylene. J Chem Soc Chem Commun 2095–2096; (b) Onitsuka K, Ose N, Ozawa F, Takahashi S (1999) Reactions of acetylene-bridged diplatinum complexes with tetracyanoethylene. J Organomet Chem 578:169–177
4. (a) Wu X, Wu J, Liu Y, Jen AK-Y (1999) Highly efficient, thermally and chemically stable second order nonlinear optical chromophores containing a 2-phenyl-tetracyanobutadienyl acceptor. J Am Chem Soc 121:472–473; (b) Cai C, Liakatsa I, Wong M-S, Bçsch M, Bosshard C, Günter P, Concilio S, Tirelli N, Suter UW (1999) Org Lett 1:1847–1849

5. Morioka Y, Yoshizawa N, Nishida J-i, Yamashita Y (2004) Chem Lett 33:1190–1191
6. Michinobu T, May JC, Lim JH, Boudon C, Gisselbrecht J-P, Seiler P, Gross M, Biaggio I, Diederich F (2005) A new class of organic donor–acceptor molecules with large third-order optical nonlinearities. Chem Commun:737–739
7. Michinobu T, Boudon C, Gisselbrecht J-P, Seiler P, Frank B, Moonen NNP, Gross M, Diederich F (2006) Chem Eur J 12:1889–1905
8. Kolb HC, Finn MG, Sharpless KB (2001) Angew Chem Int Ed 40:2004–2021; (2001) Angew Chem 113:2056–2075
9. (a) Kivala M, Diederich F (2008) Conjugation and optoelectronic properties of acetylenic scaffolds and charge-transfer chromophores. Pure Appl Chem 80:411–427; (b) Kivala M, Diederich F (2009) Acetylene-derived strong organic acceptors for planar and nonplanar push–pull chromophores. Acc Chem Res 42, 235–248; (c) Kato S-I, Diederich F (2010) Chem Commun 46:1994–2006
10. Hansen JKS, Tortzen CG, Sørensen PG, Nielsen MB (2023) Chem Eur J 29:e202202833
11. Anastas P, Eghbali N (2010) Green chemistry: principles and practice. Chem Soc Rev 39:301–312
12. Neethu KM, Nag K, Dar AH, Bajaj A, Gopal SA, Gowri V, Nagpure M, Sartaliya S, Sharma R, Solanki AK, Ali ME, Muthukrishnan A, Jayamurugan G (2023) Org Biomol Chem 21:2922
13. Afrose SP, Ghosh C, Das D (2021) Substrate induced generation of transient self-assembled catalytic systems. Chem Sci 12:14674–14685
14. Donckele EJ, Finke AD, Ruhlmann L, Boudon C, Trapp N, Diederich F (2015) The [2 + 2] cycloaddition–retroelectrocyclization and [4 + 2] hetero-diels–alder reactions of 2-(dicyanomethylene)indan-1,3-dione with electron-rich alkynes: influence of Lewis acids on reactivity. Org Lett 17:3506–3509
15. Lehnherr D, Adam M, Murray AH, McDonald R, Hampel F, Tykwinski RR (2017) Synthesis, physical properties, and chemistry of donor–acceptor-substituted pentacenes. Can J Chem 95: 303–314
16. Donckele EJ, Reekie TA, Trapp N, Diederich F (2016) penta-2,4-dien-1-ones by formal [3+2] cycloaddition–rearrangement of electron-deficient diethyl 2-(dicyanomethylene)malonate with alk-ynes. Eur J Org Chem:716–724
17. Reekie TA, Donckele EJ, Ruhlmann L, Boudon C, Trapp N, Diederich F (2015) Ester-substituted electron-poor alkenes for cycloaddition–retroelectrocyclization (CA–RE) and related reactions. Eur J Org Chem:7264–7275
18. (a) Reisinger CM, Fuentes PR, Lampart S, Schweizer WB, Diederich F (2011) Chem Eur J 17: 12906–12911; (b) Xiao W, Li F, Liu X, Cao W, Feng X (2023) Catalytic asymmetric synthesis of axially and centrally chiral 1,2-dihydrobenzofuro[3,2-b]pyridines through a [2 + 2] cycloaddition/retroelectrocyclization/re-cycloaddition cascade. Org Lett 25:8005–8009
19. Yamada M, Schweizer WB, Schoenebeck F, Diederich F (2010) Unprecedented thermal rearrangement of push–pull-chromophore–[60]fullerene conjugates: formation of chiral 1,2,9,12-tetrakis-adducts. Chem Commun 46:5334–5336
20. Kerisit N, Finke AD, Trapp N, Leroux YR, Guillemin J-C, Trolez Y, Diederich F (2015) New reactivity of 6,6-bis-donor-substituted pentafulvenes: one-step synthesis of highly substituted [3]cumulene and dihydropentalene. Tetrahedron 71:4393–4399
21. Shoji T, Nagai D, Tanaka M, Araki T, Ohta A, Sekiguchi R, Ito S, Mori S, Okujima T (2017) Chem Eur J 23:5126–5136
22. Jayamurugan G, Finke AD, Gisselbrecht J-P, Boudon C, Schweizer WB, Diederich F (2014) One-pot access to push–pull oligoenes by sequential [2 + 2] cycloaddition–retroelectrocyclization reactions. J Org Chem 79:426–431
23. (a) Reutenauer P, Kivala M, Jarowski PD, Boudon C, Gisselbrecht J-P, Gross M, Diederich F (2007) New strong organic acceptors by cycloaddition of TCNE and TCNQ to donor-substituted cyanoalkynes. Chem Commun 4898–4900; (b) Jayamurugan G, Gisselbrecht J-P, Boudon C, Schoenebeck F, Schweizer WB, Bernet B, Diederich F (2011) Expanding

the chemical space for push-pull chromophores by non-concerted [2+2] and [4+2] cycloadditions: access to a highly functionalised 6,6-dicyanopentafulvene with an intense, low-energy charge-transfer band. Chem Commun 47:4520–4522; (c) Jayamurugan G, Dumele O, Gisselbrecht J-P, Boudon C, Schweizer WB, Bernet B, Diederich F (2013) Expanding the chemical structure space of opto-electronic molecular materials: unprecedented push–pull chromophores by reaction of a donor-substituted tetracyanofulvene with electron-rich alkynes. J Am Chem Soc 135:3599–3606

24. (a) Michinobu T, Diederich F (2018) The [2+2] cycloaddition-retroelectrocyclization (CA-RE) click reaction: facile access to molecular and polymeric push-pull chromophores. Angew Chem Int Ed 57:3552–3577; (b) Yamada M (2024) Perspectives on push–pull chromophores derived from click-type [2 + 2] cycloaddition–retroelectrocyclization reactions of electron-rich alkynes and electron-deficient alkenes. Beilstein J Org Chem 20:125–154
25. (a) Hein JE, Fokin VV (2010) Copper-catalyzed azide–alkyne cycloaddition (CuAAC) and beyond: new reactivity of copper(i) acetylides. Chem Soc Rev 39:1302–1315; (b) Haldón E, Nicasio MC, Pérez PJ (2015) Copper-catalysed azide–alkyne cycloadditions (CuAAC): an update. Org Biomol Chem 13:9528–9550
26. (a) Jewett JC, Bertozzi CR (2010) Cu-free click cycloaddition reactions in chemical biology. Chem Soc Rev 39:1272–1279; (b) Escorihuela J, Marcelis ATM, Zuilhof H (2015) Adv Mater Interface 2:1500135; (c) Liao Y, Wu B, Tang R, Lin F, Tan Y (2022) Prog Chem 34:2134–2145
27. (a) Lallana E, Fernandez-Megia E, Riguera R (2009) Surpassing the use of copper in the click functionalization of polymeric nanostructures: a strain-promoted approach. J Am Chem Soc 131:5748–5750; (b) Kennedy DC, McKay CS, Legault MCB, Danielson DC, Blake JA, Pegoraro AF, Stolow A, Mester Z, Pezacki JP (2011) J Am Chem Soc 133:17993–18001
28. Lowe AB (2010) Thiol-ene "click" reactions and recent applications in polymer and materials synthesis. Polym Chem 1:17–36
29. Tasdelen MA (2011) Diels–Alder "click" reactions: recent applications in polymer and material science. Polym Chem 2:2133–2145
30. Beels MT, Fleischman MS, Biaggio I, Breiten B, Jordan M, Diederich F (2012) Opt Mater Express 2:294–303
31. (a) Frank BB, Blanco BC, Jakob S, Ferroni F, Pieraccini S, Ferrarini A, Boudon C, Gisselbrecht J-P, Seiler P, Spada GP, Diederich F (2009) Chem Eur J 15:9005–9016; (b) Shoji T, Maruyama M, Maruyama A, Nagai D, Tanaka M, Sekiguchi R, Ito S, Okujima T (2017) Synthesis and electrochemical properties of azulene-substituted tetracyanobutadiene and dicyanoquinodimethane chromophores connected with naphthalene cores. Heterocycles 95:353–369
32. Dar AH, Gowri V, Gopal A, Muthukrishnan A, Bajaj A, Sartaliya S, Selim A, Ali ME, Jayamurugan G (2019) Designing of push–pull chromophores with tunable electronic and luminescent properties using urea as the electron donor. J Org Chem 84:8941–8947
33. (a) Tang X, Liu W, Wu J, Lee C-S, You J, Wang P (2010) Synthesis, crystal structures, and photophysical properties of triphenylamine-based multicyano derivatives. J.Org. Chem. 75: 7273–7278; (b) Gautam P, Maragani R, Misra R (2014) Tuning the HOMO–LUMO gap of donor-substituted benzothiazoles. Tetrahedron Lett 55:6827–6830; (c) Misra R, Gautam P (2014) Tuning of the HOMO–LUMO gap of donor-substituted symmetrical and unsymmetrical benzothiadiazoles. Org Biomol Chem 12:5448–5457; (d) Heyer E, Ziessel R (2015) Panchromatic push–pull dyes of elongated form from triphenylamine, diketopyrrolopyrrole, and tetracyanobutadiene modules. Synlett 26:2109–2116; (e) Patil Y, Misra R (2017) Tetracyanobutadiene bridged ferrocene and triphenylamine functionalized pyrazabole dimers. J Organomet Chem 840:23–29; (f) Rout Y, Gautam P, Misra R (2017) Unsymmetrical and symmetrical push–pull phenothiazines. J Org Chem 82:6840–6845; (g) Rao PS, Brixi S, Shaikh DB, Kobaisi MA, Lessard BH, Bhosale SV, Bhosale SV (2021) Eur J Org Chem:2615–2624
34. Bijesh S, Misra R (2018) Asian J.Org. Chem. 7:1882–1892

35. Kato S-i, Noguchi H, Jin S, Nakamura Y (2016) Synthesis and electronic, optical, and electrochemical properties of a series of tetracyanobutadiene-substituted carbazoles. Asian J Org Chem 5:246–256
36. Erden K, Dengiz C (2022) 3-alkynylindoles as building blocks for the synthesis of electronically tunable indole-based push–pull chromophores. J Org Chem 87:4385–4399
37. Huang S, Ma J, Yi Y, Li M, Cai P, Wu N (2022) Synthesis of orthogonal push–pull chromophoresviaclick reaction of arylynamines. Org Biomol Chem 20:4081–4085
38. Kogawa C, Fujiwara A, Sekiguchi R, Shoji T, Kawakami J, Okazaki M, Ito S (2018) Synthesis and photophysical properties of azuleno[1',2':4,5]pyrrolo[2,1-b]quinazoline-6,14-diones: azulene analogs of tryptanthrin. Tetrahedron 74:7018–7029
39. Okuno T, Iwahasih H (2013) Acta Crystallogr E Struct Rep Online 69:o665
40. (a) Poddar M, Misra R (2017) NIR-absorbing donor–acceptor based 1,1,4,4-tetracyanobuta-1,3-diene (TCBD)- and cyclohexa-2,5-diene-1,4-ylidene-expanded TCBD-substituted ferrocenyl phenothiazines. Chem Asian J 12:2908–2915; (b) Poddar M, Jang Y, Misra R, D'Souz F (2020) Chem Eur J 26:6869–6878; (c) Yadav IS, Jang Y, Rout Y, Thomas MB, Misra R, D'Souza F (2022) Chem Eur J 28:e202200348; (d) Yadav IS, Misra R (2022) Phenothiazine and phenothiazine-5,5-dioxide-based push–pull derivatives: synthesis, photophysical, electrochemical and computational studies. New J Chem 46:15999–16006
41. Pappenfus TM, Schneiderman DK, Casado J, Navarrete JTL, Delgado MCR, Zotti G, Vercelli B, Lovander MD, Hinkle LM, Bohnsack JN, Mann KR (2011) Oligothiophene tetracyanobutadienes: alternative donor—acceptor architectures for molecular and polymeric materials. Chem Mater 23:823–831
42. Li X, Wang Y, Jia C, Kim H, Son Y-A (2019) Mol Cryst Liq Cryst 689:83–91
43. Kato S-i, Kivala M, Schweizer WB, Boudon C, Gisselbrecht J-P, Diederich F (2009) Chem Eur J 15:8687–8691
44. García R, Herranz MÁ, Torres MR, Bouit P-A, Delgado JL, Calbo J, Viruela PM, Ortí E, Martín N (2012) Tuning the electronic properties of nonplanar exTTF-based push–pull chromophores by aryl substitution. J Org Chem 77:10707–10717
45. Shoji T, Ito S, Toyota K, Yasunami M, Morita N (2008) Synthesis, properties, and redox behavior of mono-, bis-, and tris[1,1,4,4,-tetracyano-2-(1-azulenyl)-3-butadienyl] chromophores binding with benzene and thiophene cores. Chem Eur J 14:8398–8408; (a) Shoji T, Ito S, Toyota K, Iwamoto T, Yasunami M, Morita N (2009) Reactions between 1-ethynylazulenes and 7,7,8,8-tetracyanoquinodimethane (TCNQ): preparation, properties, and redox behavior of novel azulene-substituted redox-active chromophores. Eur J Org Chem:4316–4324; (b) Shoji T, Maruyama M, Ito S, Morita N (2012) Bull Chem Soc Jpn 85:761–773; (c) Shoji T, Maruyama M, Shimomura E, Maruyaam A, Ito S, Okujima T, Toyota K, Morita N (2013) J Org Chem 78:12513–12524; (d) Shoji T, Shimomura E, Maruyama M, Maruyama A, Ito S, Okujima T, Toyota K, Morita N (2013) Synthesis and properties of azulene-substituted donor–acceptor chromophores connected by arylamine cores. Eur J Org Chem:7785–7799; (e) Shoji T, Maruyama M, Maruyama A, Ito S, Okujima T, Toyota K (2014) Chem Eur J 20:11903–11912; (f) Shoji T, Maruyama A, Shimomura E, Nagai D, Ito S, Okujima T, Toyota K, Synthesis, properties, and redox behavior of tris(1-azulenyltetracyanobutadiene) and tris[1-azulenylbis(tetracyanobutadiene)] chromophores connected to a 1,3,5-tri(1-azulenyl)benzene core;. Eur J Org Chem 2015:1979–1990; (g) Shoji T, Maruyama A, Tanaka M, Nagai D, Shimomura E, Fujimori K, Ito S, Okujima T, Toyota K, Yasunami M (2016) ChemSelect 1:49–56; (h) Shoji T, Ito S (2017) Azulene-based donor–acceptor systems: synthesis, optical, and electrochemical properties. Chem Eur J 23: 16696–16709; (i) Shoji T, Miura K, Araki T, Maruyama A, Ohta A, Sekiguchi R, Ito S, Okujima T (2018) Synthesis of 2-methyl-1-azulenyl tetracyanobutadienes and dicyanoquinodimethanes: substituent effect of 2-methyl moiety on the azulene ring toward the optical and electrochemical properties. J Org Chem 83:6690–6705
46. Shoji T, Sugiyama S, Takeuchi M, Ohta A, Sekiguchi R, Ito S, Yatsu T, Okujima T, Yasunam M (2019) J Org Chem 84:1257–1275

47. (a) Shoji T, Higashi J, Ito S, Okujima T, Yasunami M, Morita N (2011) Chem Eur J 17:5116–5129; (b) Shoji T, Higashi J, Ito S, Okujima T, Yasunami M, Morita N (2012) Synthesis of donor–acceptor chromophores by the [2 + 2] cycloaddition of arylethynyl-2H-cyclohepta[b] furan-2-ones with 7,7,8,8-tetracyanoquinodimethane. Org Biomol Chem 10:2431–2438
48. (a) Niu S, Ulrich G, Retailleau P, Ziessel R (2011) Regioselective Synthesis of 5-Monostyryl and 2-Tetracyanobutadiene BODIPY Dyes. Org Lett 13:4996–4999; (b) Niu S, Ulrich G, Retailleau P, Ziessel R (2011) BODIPY-bridged push–pull chromophores: optical and electrochemical properties. Tetrahedron Lett 52:4848–4853; (c) G. Ulrich, Barsella, A. Boeglin, S. Niu, R. Ziessel, Chem Phys Chem 2014, 15, 2693–2700; (d) Dhokale B, Jadhav T, Mobin SM, Misra R (2016) Tetracyanobutadiene functionalized ferrocenyl BODIPY dyes. Dalton Trans 45:1476–1483
49. (a) Winterfeld KA, Lavarda G, Guilleme J, Sekita M, Guldi DM, Torres T, Bottari G (2017) Subphthalocyanines axially substituted with a tetracyanobuta-1,3-diene–aniline moiety: synthesis, structure, and physicochemical properties. J Am Chem Soc 139:5520–5529; (b) Broløs L, Kilde MD, Nielsen MB (2020) Novel synthetic strategy towards subphthalocyanine-functionalized acetylenic scaffoldsviavarious dibromo-enynes. Org Biomol Chem 18:6077–6085
50. Philippe C, Melan J, Barsella A, Vives T, Leroux YR, Guen FR-L, Lemiègre L, Jacquemin D, Gauthier S, Trolez Y (2023) A comprehensive study of tetracyanobutadiene push-pull chromophores derived from γ-pyranylidene. Tetrahedron Chem 5:100036
51. Koszelewski D, Nowak-Król A, Gryko DT (2012) Selective cycloaddition of tetracyanoethene (TCNE) and 7,7,8,8-tetracyano-p-quinodimethane (TCNQ) to affordmeso-substituted phenylethynyl porphyrins. Chem Asian J 7:1887–1894
52. (a) Sasaki S-i, Mizutani K, Kunieda M, Tamiaki H (2013) Tetrahedron 69:9772–9778; (b) Tamiaki H, Mizutani K, Sasaki S-i, Tatebe T (2016) Tetrahedron 72:6626–6633
53. Perrin FG, Kiefer G, Jeanbourquin L, Racine S, Perrotta D, Waser J, Scopelliti R, Severin K (2015) Angew Chem Int Ed 54:13393–13396; (2015) Angew Chem 127:13591–13594
54. Philippe C, Bui AT, Beau M, Bloux H, Riobé F, Mongin O, Roisnel T, Cordier M, Paul F, Lemiègre L, Trolez Y (2022) Synthesis and photophysical properties of 1,1,4,4-tetracyanobutadienes derived from ynamides bearing fluorophores**. Chem Eur J 28: e202200025
55. (a) Betou M, Kerisit N, Meledje E, Leroux YR, Katan C, Halet JF, Guillemin J-C, Trolez Y (2014) Chem Eur J 20:9553–9557; (b) Betou M, Durand RJ, Sallustrau A, Gousset C, Coz EL, Leroux YR, Toupet L, Trzop E, Roisnel T, Trolez Y (2017) Chem Asian J 12:1338–1346
56. Pigulski B, Misiak K, Męcik P, Szafert S (2023) Cycloaddition–retro-electrocyclization click reaction of amine end-capped oligoynes with tetracyanoethylene. Chem Eur J 29:e202302725
57. (a) Mochida T, Yamazaki S (2002) Mono- and diferrocenyl complexes with electron-accepting moieties formed by the reaction of ferrocenylalkynes with tetracyanoethylene. J Chem Soc Dalton Trans:3559–3564; (b) Bruce MI, Low PJ, Hartl F, Humphrey PA, de Montigny F, Jevric M, Lapinte C, Perkins GJ, Roberts RL, Skelton BW, White AH (2005) Syntheses, structures, some reactions, and electrochemical oxidation of ferrocenylethynyl complexes of iron, ruthenium, and osmium. Organometallics 24:5241–5255; (c) Jordan M, Kivala M, Boudon C, Gisselbrecht J-P, Schweizer WB, Seiler P, Diederich F (2011) Switching the Regioselectivity in Cycloaddition-Retro-Electrocyclizations between Donor-Activated Alkynes and the Electron-Accepting Olefins TCNE and TCNQ. Chem Asian J 6:396–401; (d) Shoji T, Ito S, Okujima T, Morita N (2011) Synthesis and [2+2] cycloaddition with tetracyanoethylene of ene–diyne scaffolds bearing ferrocenes at the periphery. Eur J Org Chem 2011:5134–5140; (e) Misra R, Gautam P, Mobin SM (2013) Aryl-substituted unsymmetrical benzothiadiazoles: synthesis, structure, and properties. J Org Chem 78:12440–12452; (f) Shoji T, Ito S, Okujima T, Morita N (2013) Synthesis of 2-azulenyl-1,1,4,4-tetracyano-3-ferrocenyl-1,3-butadienes by [2+2] cycloaddition of (ferrocenylethynyl)azulenes with tetracyanoethylene. Chem Eur J 19:5721–5730; (g) Maragani R, Misra R (2014) Ferrocenyl substituted triphenylamine based donor–acceptor molecular systems; synthesis,

photophysical, electrochemical, and theoretical studies. Tetrahedron 70:3390–3399; (h) Misra R, Sharma R, Mobin SM (2014) Star shaped ferrocenyl truxenes: synthesis, structure and properties. Dalton Trans 43:6891–6896; (i) Dhokale B, Jadhav T, Mobin SM, Misra R (2015) Synergistic effect of donors on tetracyanobutadine (TCBD) substituted ferrocenyl pyrenes. RSC Adv 5:57692–57699; (j) Kraube N, Kielmann M, Ma J, Butenschçn H (2015) Eur J Org Chem:2622–2631; (k) Misra R, Gautam P, Maragain R (2015) Tetrahedron Lett 56: 1664–1666; (l) Shoji T, Kamata N, Maruyama A, Ito S, Okujima T (2015) Synthesis, properties, and redox behavior of ferrocenyl-1,1,4,4-tetracyano-1,3-butadienes connected by arylamine and azobenzene spacers. Bull Chem Soc Jpn 88:1338–1346; (m) Shoji T, Maruyama A, Yaku C, Kamata N, Ito S, Okujima T, Toyota K (2015) Synthesis, properties, and redox behavior of 1,1,4,4-tetracyano-2-ferrocenyl-1,3-butadienes connected by aryl, biaryl, and teraryl spacers. Chem Eur J 21:402–409; (n) Patil Y, Jadhav T, Dhokale B, Misra R (2016) Tuning of the HOMO–LUMO Gap of Symmetrical and Unsymmetrical Ferrocenyl-Substituted Diketopyrrolopyrroles. Eur J Org Chem 2016:733–738; (o) Kivrak A, Zobi C, Torlak Y, Çamlısoy Y, Kuş M, Kivrak H (2018) Appl Organomet Chem 32:e4512; (p) Shinde J, Thomas MB, Poddar M, Misra R, D'Souza F (2021) Does location of BF2-chelated dipyrromethene (BODIPY) ring functionalization affect spectral and electron transfer properties? Studies on α-, β-, and meso-functionalized BODIPY-derived donor–acceptor dyads and triads. J Phys Chem C 125:23911–23921; (q) Korb M, Low PJ (2022) Structural characterization of ((9-fluorenylidene) (ferrocenyl)methyl)palladium iodide as the catalytic intermediate in the synthesis of 9-(ferrocenyl (ferrocenylethynyl)methylene)-9H-fluorene. J Coord Chem 75:1548–1566; (r) Popli C, Misra R, Butenschön H (2024) J Organomet Chem 1006:122998

58. Mamane V, Ledoux-Rak I, Deveau S, Zyss J, Riant O (2003) Synthesis:455–467
59. Soto E, Helmink KL, Chin CP, Ferguson M, Peters SJ, Szczepura LF (2022) Rhenium selenide clusters containing alkynyl ligands: unexpected reactivity of σ-bound phenylacetylide. Organometallics 41:2688–2697
60. (a) Kivala M, Boudon C, Gisselbrecht J-P, Seiler P, Gross M, Diederich F (2007) A novel reaction of 7,7,8,8-tetracyanoquinodimethane (TCNQ): charge-transfer chromophores by [2 + 2] cycloaddition with alkynes. Chem Commun:4731–4733; (b) Fesser P, Iacovita C, Wäckerlin C, Vijayaraghavan S, Ballav N, Howes K, Gisselbrecht J-P, Crobu M, Boudon C, Stöhr M, Jung TA, Diederich F (2011) Chem Eur J 17:5246–5250
61. Kivala M, Boudon C, Gisselbrecht J-P, Enko B, Seiler P, Müller IB, Langer N, Jarowski PD, Gescheidt G, Diederich F (2009) Organic super-acceptors with efficient intramolecular charge-transfer interactions by [2+2] cycloadditions of TCNE, TCNQ, and F4-TCNQ to donor-substituted cyanoalkynes. Chem Eur J 15:4111–4123
62. Washino Y, Murata K, Ashizawa M, Kawauchi S, Michinobu T (2011) Creation of persistent charge-transfer interactions in TCNQ polyester. Polym J 43:364–369
63. (a) Jarowski PD, Wu Y-L, Boudon C, Gisselbrecht J-P, Gross M, Schweizer WB, Diederich F (2009) New donor–acceptor chromophores by formal [2+2] cycloaddition of donor-substituted alkynes to dicyanovinyl derivatives. Org Biomol Chem 7:1312–1322; (b) Wu Y-L, Jarowski PD, Schweizer WB, Diederich F (2010) Chem Eur J 16:202–211; (c) Rijkers DTS, de Prada López F, Liskamp RMJ, Diederich F (2011) A convenient [2+2] cycloaddition–cycloreversion reaction for the synthesis of 1,1-dicyanobuta-1,3-diene-scaffolded peptides as new imaging chromophores. Tetrahedron Lett 52:6963–6967; (d) Silvestri F, Jordan M, Howes K, Kivala M, Rivera-Fuentes P, Boudon C, Gisselbrecht J-P, Schweizer WB, Seiler P, Chiu M, Diederich F (2011) Regular acyclic and macrocyclic [AB] oligomers by formation of push–pull chromophores in the chain-growth step. Chem Eur J 17:6088–6097
64. Finke AD, Diederich F (2015) 6,6-dicyanopentafulvenes: teaching an old dog new tricks. Chem Rec 15:19–30
65. Finke AD, Dumele O, Zalibera M, Confortin D, Cias P, Jayamurugan G, Gisselbrecht J-P, Boudon C, Schweizer WB, Gescheidt G, Diederich F (2012) 6,6-dicyanopentafulvenes:

electronic structure and regioselectivity in [2 + 2] cycloaddition–retroelectrocyclization reactions. J Am Chem Soc 134:18139–18146
66. Galán E, Andreu R, Garín J, Orduna J, Villacampa B, Diosdado BE (2012) Cycloaddition reactions of polyenic donor–π-acceptor systems with an electron-rich alkyne: access to new chromophores with second-order optical nonlinearities. Org Biomol Chem 10:8684–8691
67. Chiu M, Jaun B, Beels MTR, Biaggio I, Gisselbrecht J-P, Boudon C, Schweizer WB, Kivala M, Diederich F (2012) N,N′-dicyanoquinone diimide-derived donor–acceptor chromophores: conformational analysis and optoelectronic properties. Org Lett 14:54–57
68. (a) Kato S-i, Beels MTR, Porta PL, Schweizer WB, Boudon C, Gisselbrecht J-P, Biaggio I, Diederich F (2010) Angew Chem Int Ed 49:6207–6211; (2010) Angew Chem 122:6343–6347; (b) Dengiz C, Dumele O, Kato S-i, Zalibera M, Cias P, Schweizer WB, Boudon C, Gisselbrecht J-P, Gescheidt G, Diederich F (2014) Chem Eur J 20:1279–1286
69. (a) Shoji T, Higashi J, Ito S, Yasunami M, Morita N (2011) Heterocycles 83:2271–2274; (b) Shoji T, Maruyama M, Shimomura E, Maruyama A, Ito S, Yasunami M, Higashi J, Toyota K, Morita N (2014) Heterocycles 88:319–329
70. Dengiz C, Prange C, Gawel P, Trapp N, Ruhlmann L, Boudon C, Diederich F (2016) Push–pull chromophores by reaction of 2,3,5,6-tetrahalo-1,4-benzoquinones with 4-(N, N-dialkylanilino)acetylenes. Tetrahedron 72:1213–1224
71. Reekie TA, Donckele EJ, Manenti G, Püntener S, Trapp N, Diederich F (2016) A three-step synthesis of tetrasubstituted NH-pyrroles. Org Lett 18:2252–2255
72. Chiu M, Tchitchanov BH, Zimmerli D, Sanhueza IA, Schoenebeck F, Trapp N, Schweizer WB, Diederich F (2015) Angew Chem Int Ed 54:349–354; (2015) Angew Chem 127:356–361
73. M. Yamada, P. Rivera-Fuentes, W. B. Schweizer, F. Diederich, Angew Chem Int Ed 2010, 49, 3532–3535; Angew Chem 2010, 122, 3611–3615
74. (a) Alonso-Gómez JL, Schanen P, Rivera-Fuentes P, Seiler P, Diederich F (2008) Chem Eur J 14:10564–10568; (b) Alonso-Gómez JL, Petrovic AG, Harada N, Rivera-Fuentes P, Berova N, Diederich F (2009) Chem Eur J 15:8396–8400; (c) Frank BB, Kivala M, Blanco BC, Breiten B, Schweizer WB, Laporta PR, Biaggio I, Jahnke E, Tykwinski RR, Boudon C, Gisselrbecht J-P, Diederich F (2010) Eur J Org Chem:2487–2503
75. Ozcelik A, Pereira-Cameselle R, Peña-Gallego Á, Alonso-Gómez JL (2022) A tetracyanobutadiene spirobifluorene: synthesis, enantiomeric resolution and chiroptical properties. Eur J Org Chem:e202101333
76. Lavarda G, Bhattacharjee N, Brancato G, Torres T, Bottari G (2020) Angew Chem Int Ed 59: 21224–21229; (2020) Angew Chem 132:21410–21415
77. (a) Wu Y-L, Tancini F, Schweizer WB, Paunescu D, Boudon C, Gisselbrecht J-P, Jarowski PD, Dalcanale E, Diederich F (2012) Proacetylenic reactivity of a push–pull buta-1,2,3-triene: new chromophores and supramolecular systems. Chem Asian J 7:1185–1190; (b) Gawel P, Wu YL, Finke AD, Trapp N, Zalibera M, Boudon C, Gisselbrecht J-P, Schwerizer WB, Gescheidt G, Diederich F (2015) Chem Eur J 21:6215–6225
78. (a) Gawel P, Dengiz C, Finke AD, Trapp N, Boudon C, Gisselbrecht J-P, Diederich F (2014) Angew Chem Int Ed 53:4341–4345; (2014) Angew Chem 126:4430–4434; (b) Kerisit N, Gawel P, Levandowski B, Yang Y-F, García-López V, Trapp N, Ruhlmann L, Boudon C, Houk KN, Diederich F (2018) Chem Eur J 24:159–168
79. (a) Congreve DN, Lee J, Thompson NJ, Hontz E, Yost SR, Reusswig PD, Bahlke ME, Reineke S, Voorhis TV, Baldo MA (2013) Science 340:334–337; (b) Margulies EA, Wu Y-L, Gawel P, Miller SA, Shoer LE, Schaller RD, Diederich F, Wasielewski MR (2015) Angew Chem Int Ed 54:8679–8683; (2015) Angew Chem 127:8803–8807; (c) Margulies EA, Kerisit N, Gawel P, Mauck CM, Ma L, Miller CE, Young RM, Trapp N, Wu Y-L, Diederich F, Wasielewski MR (2017) Substituent effects on singlet exciton fission in polycrystalline thin films of cyano-substituted diaryltetracenes. J Phys Chem C 121:21262–21271
80. Januszewski JA, Hampel F, Neiss C, Gçrling A, Tykwinski RR (2014) Angew Chem Int Ed 53:3743–3747; (2014) Angew Chem 126:3818–3822

81. (a) Kivala M, Boudon C, Gisselbrecht J-P, Seiler P, Gross M, Diederich F (2007) Angew Chem Int Ed 46:6357–6360; (2007) Angew Chem 119:6473–6477; (b) Kivala M, Stanoeva T, Michinobu T, Frank B, Gescheidt G, Diederich F (2008) Chem Eur J 14:7638–7647; (c) Wu Y-L, Stuparu MC, Boudon C, Gisselbrecht J-P, Schweizer WB, Baldridge KK, Siegel JS, Diederich F (2012) Structural, optical, and electrochemical properties of three-dimensional push–pull corannulenes. J Org Chem 77:11014–11026; (d) Misra R, Maragani R, Gautam P, Mobin SM (2014) Tetracyanoethylene substituted triphenylamine analogues. Tetrahedron Lett 55:7102–7105; (e) Dengiz C, Breiten B, Gisselbrecht J-P, Boudon C, Trapp N, Schweizer WB, Diederich F (2015) Synthesis and optoelectronic properties of janus-dendrimer-type multivalent donor–acceptor systems. J Org Chem 80:882–896; (f) Sharma R, Maragani R, Misra R (2018) C3-symmetric star shaped donor–acceptor truxenes: synthesis and photophysical, electrochemical and computational studies. New J Chem 42:882–890; (g) Sharma R, Thomas MB, Misra R, D'Souza F (2019) Angew Chem Int Ed 58:4350–4355; (2019) Angew Chem 131:4394–4399; (h) Winterfeld KA, Lavarda G, Guilleme J, Guldi DM, Torres T, Bottari G (2019) Subphthalocyanine–tetracyanobuta-1,3-diene–aniline conjugates: stereoisomerism and photophysical properties. Chem Sci 10:10997–11005; (i) Ripoche N, Betou M, Philippe C, Trolez Y, Mongin O, Dudek M, Pokladek Z, Matczyszyn K, Samoc M, Sahnoune H, Halet J-F, Roisnel T, Toupet L, Cordier M, Moxey GJ, Humphrey MG, Paul F (2021) Two-photon absorption properties of multipolar triarylamino/tosylamido 1,1,4,4-tetracyanobutadienes. Phys Chem Chem Phys 23:22283–22297; (j) Yadav IS, Alsaleh AZ, Martin B, Misra R, D'Souza F (2022) Star-shaped triphenylamine–tetracyanobutadiene–phenothiazine push–pull systems: role of terminal phenothiazine in improving charge transfer. J Phys Chem C 126:13300–13310; (k) Philippe C, Coste M, Bretonnière Y, Lemiègre L, Ulrich S, Trolez Y (2022) Quadruple functionalization of a tetraphenylethylene aromatic scaffold with ynamides or tetracyanobutadienes: synthesis and optical properties. Eur J Org Chem 2022:e202200049
82. (a) Zhou W, Xu J, Zheng H, Liu H, Li Y, Zhu D (2008) Charge transfer chromophore-stopped [2] rotaxane through [2 + 2] cycloaddition. J Org Chem 73:7702–7709; (b) Zhou W, Xu J, Zheng H, Yin X, Zuo Z, Liu H, Li Y (2009) Distinct nanostructures from a molecular shuttle: effects of shuttling movement on nanostructural morphologies. Adv Funct Mater 19:141–149; (c) Zhou W, Zhang S, Li G, Zhao Y, Shi Z, Liu H, Li Y (2009) Fluorescent alteration on a bistable molecular shuttle. ChemPhysChem 10:2066–2072; (d) Yang W, Li Y, Liu H, Chi L, Li Y (2012) Design and assembly of rotaxane-based molecular switches and machines. Small 8:504–516; (e) Li Y, Liu T, Liu H, Tian M-Z, Li Y (2014) Self-assembly of intramolecular charge-transfer compounds into functional molecular systems. Acc Chem Res 47:1186–1198; (f) Trolez Y, Finke AD, Silvestri F, Monti F, Ventura B, Boudon C, Gisselbrecht J-P, Schweizer WB, Sauvage J-P, Armaroli N, Diederich F (2018) Unconventional synthesis of a cuirotaxane with a superacceptor stopper: ultrafast excited-state dynamics and near-infrared luminescence. Chem Eur J 24:10422–10433
83. Cheng S, Li K, Hu J, He J, Zeller M, Xu Z (2020) Building conjugated donor–acceptor cross-links into metal–organic frameworks for photo- and electroactivity. ACS Appl Mater Interfaces 12:19201–19209
84. Leith GA, Rice AM, Yarbrough BJ, Berseneva AA, Ly RT, Buck III CN, Chusov D, Brandt AJ, Chen DA, Lamm BW, Stefik M, Stephenson KS, Smith MD, Vannucci AK, Pellechia PJ, Garashchuk S, Shustova NB (2020) Angew Chem Int Ed 59:6000–6006; (2020) Angew Chem 132:6056–6062
85. Hopf H, Kreutzer M, Jones PG (1991) Angew Chem Int Ed Engl 30:1127–1128; (1991) Angew Chem 103:1148–1149
86. Shoji T, Ito S, Okujima T, Morita N (2012) Synthesis of push–pull chromophores by the sequential [2 + 2] cycloaddition of 1-azulenylbutadiynes with tetracyanoethylene and tetrathiafulvalene. Org Biomol Chem 10:8308–8313
87. (a) Tancini F, Monti F, Howes K, Belbakra A, Listorti A, Schweizer WB, Reutenauer P, Alonso-Gómez J-L, Chiorboli C, Urner LM, Gisselbrecht J-P, Boudon C, Armaroli N,

Diederich F (2014) Cyanobuta-1,3-dienes as novel electron acceptors for photoactive multicomponent systems. Chem Eur J 20:202–216; (b) Urner LM, Sekita M, Trapp N, Schweizer WB, Wçrle M, Gisselbrecht J-P, Boudon C, Guldi DM, Diederich F (2015) Eur J Org Chem:91–108; (c) Monti F, Venturini A, Nenov A, Tancini F, Finke AD, Diederich F, Armaroli N (2015) Anilino-substituted multicyanobuta-1,3-diene electron acceptors: TICT molecules with accessible conical intersections. J Phys Chem A 119:10677–10683; (d) Sekita M, Ballesteros B, Diederich F, Guldi DM, Bottari G, Torres T (2016) Angew Chem Int Ed 55:5560–5564; (2016) Angew Chem 128:5650–5654
88. (a) Yamada M, Tancini F, Sekita M, Guldi DM, Boudon C, Gisselbrecht J-P, Alberti MN, Schweizer WB, Diederich F (2014) Fuller nanotub carbon nanostructures. 22:99–127; (b) Gautam P, Misra R, Thomas MB, D'Souza F (2017) Chem Eur J 23:9192–9200
89. Reekie TA, Sekita M, Urner LM, Bauroth S, Ruhlmann L, Gisselbrecht J-P, Boudon C, Trapp N, Clark T, Guldi DM, Diederich F (2017) Porphyrin donor and tunable push–pull acceptor conjugates—experimental investigation of marcus theory. Chem Eur J 23:6357–6369
90. Yadav IS, Alsaleh AZ, Misra R, D'Souza F (2021) Chem Sci 12:1109–1120
91. (a) Patil Y, Misra R (2019) Rational molecular design towards NIR absorption: efficient diketopyrrolopyrrole derivatives for organic solar cells and photothermal therapy. J Mater Chem C 7:13020–13031; (b) Jang Y, Rout Y, Misra R, D'Souza F (2021) Symmetric and asymmetric push–pull conjugates: significance of pull group strength on charge transfer and separation. J Phys Chem B 125:4067–4075
92. Yadav I, Sharma JK, Sankar M, D'Souza F (2023) Chem Eur J 29:e202301341
93. Winterfeld KA, Lavarda G, Yoshida K, Bayerlein MJ, Kise K, Tanaka T, Osuka A, Guldi DM, Torres T, Bottari G (2020) Synthesis and optical features of axially and peripherally substituted subporphyrins. a paradigmatic example of charge transfer versus exciplex states. J Am Chem Soc 142:7920–7929
94. Guragain M, Pinjari D, Misra R, D'Souza F (2023) Chem Eur J 29:e202302665
95. (a) Li L, Wang D, Wang L, Ramella D, Wang H, Gao H, Zhang J, Xing Y, Li B, Yang Z, Cao H, He W (2018) The photoacoustic effect of near-infrared absorbing porphyrin derivatives prepared via click chemistry. Dyes Pigments 148:501–507; (b) Zhao Z, Wang D, Gao H, Yang Z, Cao H, He W (2019) Photoacoustic effect of near-infrared absorbing fullerene derivatives with click moieties. Dyes Pigments 164:182–187
96. Bui AT, Philippe C, Beau M, Richy N, Cordier M, Roisnel T, Lemiègre L, Mongin O, Paul F, Trolez Y (2020) Synthesis, characterization and unusual near-infrared luminescence of 1,1,4,4-tetracyanobutadiene derivatives. Chem Commun 56:3571–3574
97. (a) Aviram A, Ratner MA (1974) Molecular rectifiers. Chem Phys Lett 29:277–283; (b) Jayamurugan G, Gowri V, Hernández D, Martin S, González-Orive A, Dengiz C, Dumele O, Pérez-Murano F, Gisselbrecht J-P, Boudon C, Schweizer WB, Breiten B, Finke AD, Jeschke G, Bernet B, Ruhlmann L, Cea P, Diederich F (2016) Design and synthesis of aviram–ratner-type dyads and rectification studies in langmuir–blodgett (LB) films. Chem Eur J 22:10539–10547
98. (a) Michinobu T (2011) Adapting semiconducting polymer doping techniques to create new types of click postfunctionalization. Chem Soc Rev 40:2306–2316; (b) Fujita H, Michinobu T, Tokita M, Ueda M, Higashihara T (2012) Synthesis and postfunctionalization of rod–coil diblock and coil–rod–coil triblock copolymers composed of poly(3-hexylthiophene) and poly (4-(4′-N,N-dihexylaminophenylethynyl)styrene) segments. Macromolecules 45:9643–9656; (c) Michinobu T (2015) Click functionalization of aromatic polymers for organic electronic device applications. Macromol Chem Phys 216:1387–1395
99. (a) Li Y, Ashizawa M, Uchida S, Michinobu T (2011) A novel polymeric chemosensor: dual colorimetric detection of metal ions through click synthesis. Macromol Rapid Commun 32: 1804–1808; (b) Li Y, Ashizawa M, Uchida S, Michinobu T (2012) Colorimetric sensing of cations and anions by clicked polystyrenes bearing side chain donor–acceptor chromophores. Polym Chem 3:1996–2005

100. (a) Li Y, Michinobu T (2010) Sequential double click reactions: a highly efficient post-functionalization method for optoelectronic polymers. Polym Chem 1:72–74; (b) Li Y, Tsuboi K, Michinobu T (2010) Double click synthesis and second-order nonlinearities of polystyrenes bearing donor–acceptor chromophores. Macromolecules 43:5277–5286; (c) Li Y, Tsuboi K, Michinobu T, Ishida Y, Hirai T, Hayakawa T, Kakimoto M-a (2010) J Photopolym Sci Technol 23:337–342; (d) Song S, Ko Y-G, Lee H, Wi D, Ree BJ, Li Y, Michinobu T, Ree M (2015) High-performance triazole-containing brush polymers via azide–alkyne click chemistry: a new functional polymer platform for electrical memory devices. NPG Asia Mater 7:e228
101. (a) Tang X, Liu W, Wu J, Zhao W, Zhang H, Wang P (2011) A colorimetric chemosensor for fast detection of thiols based on intramolecular charge transfer. Tetrahedron Lett 52:5136–5139; (b) Michinobu T, Li Y, Hyakutake T (2013) Polymeric ion sensors with multiple detection modes achieved by a new type of click chemistry reaction. Phys Chem Chem Phys 15:2623–2631; (c) Shi W, Ma F, Hui Y, Mi H, Tian Y, Lei Y, Xie Z (2014) TCNE-decorated triphenylamine-based conjugated polymer: click synthesis and efficient turn-on fluorescent probing for Hg2+. Dyes Pigments 104:1–7; (d) Grigoras M, Ivan T, Vacareanu L, Catargiu AM, Tigoianu R (2014) Synthesis and optoelectronic characterization of some triphenylamine-based compounds containing strong acceptor substituents. J Lumin 153:5–11; (e) Wang Y, Michinobu T (2017) Bull Chem Soc Jpn 90:1388–1400; (f) Li Y, Washino Y, Hyakutake T, Michinobu T (2017) Colorimetric ion sensors based on polystyrenes bearing side chain triazole and donor-acceptor chromophores. Anal Sci 33:599–604
102. Li Y, Michinobu T (2012) Multi-coloration of polyurethane derivatives through click postfunctionalization, electrochemical oxidation, and Ag+ ion complexation. J Mater Chem 22:9513–9521
103. Michinobu T (2008) Click-type reaction of aromatic polyamines for improvement of thermal and optoelectronic properties. J Am Chem Soc 130:14074–14075
104. (a) Michinobu T, Seo C, Noguchi K, Mori T (2012) Effects of click postfunctionalization on thermal stability and field effect transistor performances of aromatic polyamines. Polym Chem 3:1427–1435; (b) Ko Y-G, Kim DM, Kim K, Jung S, Wi D, Michinobu T, Ree M (2014) ACS Appl Mater Interfaces 6:8415–8425
105. Washino Y, Murata K, Michinobu T (2012) Postfunctionalization of aromatic polyamine by [2+2] cycloaddition of 7,7,8,8-tetracyanoquinodimethane with side chain alkynes. Polym Bull 69:137–147
106. (a) Yuan Y, Michinobu T (2011) J. Polym. Sci. Part A 49:225–233; (b) Yuan Y, Choi W, Nishide H, Michinobu T (2013) Sequential and click-type postfunctionalization of regioregular poly(3-hexylthiophene) for realization of n-doped multiplet state. Chem Sci 4: 345–350; (c) Huang W, Chen H (2013) Synthesis and characterization of a low-bandgap poly (arylene ethynylene) having donor–acceptor type chromophores in the side chain. Macromolecules 46:2032–2037; (d) Grigorasa M, Catargiua AM, Ivan T, Vacareanu L, Minaev B, Stromylo E (2015) Dyes Pigments 113:227–238
107. Michinobu T, Kumazawa H, Noguchi K, Shigehara K (2009) One-step synthesis of donor–acceptor type conjugated polymers from ferrocene-containing poly(aryleneethynylene)s. Macromolecules 42:5903–5905
108. (a) Michinobu T, Fujita H (2010) Postfunctionalization of alkyne-linked conjugated carbazole polymer by thermal addition reaction of tetracyanoethylene. Materials 3:4773–4783; (b) Fujita H, Tsuboi K, Michinobu T (2011) High-yielding alkyne-tetracyanoethylene addition reactions: a powerful tool for analyzing alkyne-linked conjugated polymer structures. Macromol Chem Phys 212:1758–1766; (c) Yuan Y, Michinobu T, Ashizawa M, Mori T (2011) J Polym Sci Part A 49:1013–1020; (d) Huang W (2014) Synthesis and characterization of a main-chain donor–acceptor type low-bandgap polymer by post-functionalization of a poly (arylene ethynylene). React Funct Polym 83:42–48
109. (a) Sonogashira K, Morimoto H, Takai Y, Takahashi S (1981) Symp Organomet Chem. 28: 64–66; (b) Yuan Y, Michinobu T (2012) Construction of donor–acceptor chromophores in

platinum polyyne polymer by [2 + 2] cycloaddition of organic acceptors. Macromol Chem Phys 213:2114–2119
110. Yuan Y, Michinobu T, Oguma J, Kato T, Miyake K (2013) Attempted inversion of semiconducting features of platinum polyyne polymers: a new approach for all-polymer solar cells. Macromol Chem Phys 214:1465–1472
111. Ohshita J, Kajihara T, Tanaka D, Ooyama Y (2014) Preparation of poly (disilanylenetetracyanobutadienyleneoligothienylene)s as new donor–acceptor type organosilicon polymers. J Organomet Chem 749:255–260
112. Öktem G, Sahre K, Voit B, Jordanb R, Kiriy A (2017) Polym Chem 8:2675–2685
113. Wang D, Michinobu T (2011) J Polym Sci A Polym Chem 49:72–81
114. (a) Washino Y, Michinobu T (2011) Application of alkyne-TCNQ addition reaction to polymerization. Macromol Rapid Commun 32:644–648; (b) Washino Y, Michinobu T (2012) J Photopolym Sci Technol 25:267–270
115. Washino Y, Michinobu T (2014) Sequence-regulated linear polymers with ion-sensing charge-transfer chromophores by the multiple click approach. Macromol Chem Phys 215: 1485–1490
116. (a) Washino Y, Tsuboi K, Michinobu T (2011) Simultaneous formation of donor-acceptor chromophores and cross-linking for electro-optic polymer materials. J Photopolym Sci Technol 24:305–309; (b) Washino Y, Michinobu T (2016) Cross-linking and postfunctionalization of polymer films by utilizing the orthogonal reactivity of 7,7,8,8-tetracyanoquinodimethane. Phys Chem Chem Phys 18:2288–2291
117. (a) Dehu C, Meyers F, Brédas JL (1993) Donor-acceptor diphenylacetylenes: geometric structure, electronic structure, and second-order nonlinear optical properties. J Am Chem Soc 115:6198–6206; (b) Tykwinski RR, Gubler U, Martin RE, Diederich F, Bosshard C, Günter PJ (1998) J Phys Chem B 102:4451–4465; (c) Breitena B, Biaggio I, Diederich F (2010) Chimia 64:409–413; (d) Biaggio I (2022) The appeal of small molecules for practical nonlinear optics. Chem Eur J 28:e202103168
118. Esembeson B, Scimeca ML, Michinobu T, Diederich F, Biaggio I (2008) A high-optical quality supramolecular assembly for third-order integrated nonlinear optics. Adv Mater 20: 4584–4587
119. (a) Koos C, Vorreau P, Vallaitis T, Dumon P, Bogaerts W, Baets R, Esembeson B, Biaggio I, Michinobu T, Diederich F, Freude W, Leuthold J (2009) Nat Photon 3:216–219; (b) Dallachiesa L, Biaggio I (2022) Electrically poled vapor-deposited organic glasses for integrated electro-optics. Opt Lett 47:1924–1927
120. (a) Zhao P, Wang D, Gao H, Zhang J, Xing Y, Yang Z, Cao H, He W (2016) Dyes Pigments 162:451–458; (b) Han P, Wang D, Gao H, Zhang J, Xing Y, Yang Z, Cao H, He W (2018) Third-order nonlinear optical properties of cyanine dyes with click chemistry modification. Dyes Pigments 149:8–15
121. Beels MT, Fleischman MS, Biaggio I, Breiten B, Jordan M, Diederich F (2012) Compact TCBD based molecules and supramolecular assemblies for third-order nonlinear optics. Opt Mater Express 2:294–303
122. (a) Mammadova F, Ozsinan S, Okutan M, Dengiz C (2020) Synthesis, characterization, and theoretical investigation of optical and nonlinear optical (NLO) properties of triazene-based push–pull chromophores. J Mol Struct 1220:128726; (b) Inyurt FC, Dengiz C (2024) ChemSelect 9:e202304394
123. Dengiz Ç (2021) Polycyclic aromatic hydrocarbon-substituted push–pull chromophores: an investigation of optoelectronic and nonlinear optical properties using experimental and theoretical approaches. Turk J Chem 45:1375–1390
124. (a) Miao Z, Han H, Wang D, Gao H, Gu J, Hu H (2016) Nonlinear optical and energy-level modulation of organic alkynes by click chemistry. Tetrahedron 72:4039–4046; (b) Pokladek Z, Ripoche N, Betou M, Trolez Y, Mongin O, Olesiak-Banska J, Matczyszyn K, Samoc M, Humphrey MG, Blanchard-Desce M, Paul F (2016) Chem Eur J 22:10155–10167; (c) Zhao Y, Liu X, Zhao X, Li Q, Zhao Y, Guo Z, He Z, Zhang H, Gao J,

Miao Z (2022) Preparation of symmetrical and asymmetrical multi-phenylene ring nonlinear optical materials with click chemical modifications and their properties. Tetrahedron 127: 132992; (d) Liu X, He Z, Li C, Miao Z, Wang D, Luan Y, Li Y, Zhao Y (2022) Click chemically modified azobenzene derivatives and their third-order nonlinear optical properties. Dyes Pigments 206:110595

125. Wang C, Fan C, Yuan C, Yang G, Li X, Ju C, Feng Y, Xu J (2017) Third- and high-order nonlinear optical properties of an intramolecular charge-transfer compound. RSC Adv 7: 4825–4829
126. Štefko M, Tzirakis MD, Breiten B, Ebert M-O, Dumele O, Schweizer WB, Gisselbrecht J-P, Boudon C, Beels MT, Biaggio I, Diederich F (2013) donor–acceptor (D–A)-substituted polyyne chromophores: modulation of their optoelectronic properties by varying the length of the acetylene spacer. Chem Eur J 19:12693–12704
127. (a) Zhang W-S, Wang D, Cao H, Yang H (2014) Energy level tunable pre-click functionalization of [60]fullerene for nonlinear optics. Tetrahedron 70:573–577; (b) Liu X, Wang D, Gao H, Yang Z, Xing Y, Cao H, He W, Wang H, Gu J, Hu H (2016) Click chemistry functionalization improving the wideband optical-limiting performance of fullerene derivatives. Phys Chem Chem Phys 18:7341–7348
128. (a) Mi Y, Liang P, Yang Z, Wang D, Cao H, He W, Yang H, Yu L (2016) Application of near-IR absorption porphyrin dyes derived from click chemistry as third-order nonlinear optical materials. ChemistryOpen 5:71–77; (b) Liu X, Wang D, Gao H, Yang Z, Xing Y, Cao H, He W, Wang H, Gu J, Hu H (2016) Nonlinear optical properties of symmetrical and asymmetrical porphyrin derivatives with click chemistry modification. Dyes Pigments 134: 155–163
129. (a) Jin Z, Wang D, Wang X, Liang P, Mi Y, Yang H (2013) Efficient modification of pyrene-derivative featuring third-order nonlinear optics via the click post-functionalization. Tetrahedron Lett 54:4859–4864; (b) Mi Y, Liang P, Jin Z, Wang D, Yang Z (2013) Synthesis and third-order nonlinear optical properties of triphenylene derivatives modified by click chemistry. ChemPhysChem 14:4102–4108; (c) Zhao Y, Yu Y, Song W, Yao R, Gao J, Zhang H, Zhao Y, Zhao Y, Wang D (2023) Click reactions to prepare symmetrical and asymmetrical and broad-band near-infrared absorbing dyes. New J Chem 47:11040–11050
130. Wang D, Zhang R, Gao H, Wang X, Wang H, Yang Z, He W, Cao H, Gu J, Hu H, Yang H (2016) Energy-level tuning of poly(p-phenylenebutadiynylene) derivatives by click chemistry-type postfunctionalization of side-chain alkynes. React Funct Polym 105:114–121
131. (a) Dalton LR, Steier WH, Robinson BH, Zhang C, Ren A, Garner S, Chen A, Londergan T, Irwin L, Carlson B, Fifield L, Phelan G, Kincaid C, Amend J, Jen AK-Y (1999) J Mater Chem 9:1905–1920; (b) Ma H, Chen B, Sassa T, Dalton LR, Jen AK-Y (2001) Highly efficient and thermally stable nonlinear optical dendrimer for electrooptics. J Am Chem Soc 123:986–987; (c) Luo J, Ma H, Haller M, Jen AK-Y, Barto RR (2002) Large electro-optic activity and low optical loss derived from a highly fluorinated dendritic nonlinear optical chromophore. Chem Commun:888–889; (d) Ma H, Liu S, Luo J, Suresh S, Liu L, Kang SH, Haller M, Sassa T, Dalton LR, Jen AK-Y (2002) Adv Funct Mater 12:565–574; (e) Pereverzev YV, Prezhdo OV, Dalton LR (2003) Structural origin of the enhanced electro-optic response of dendrimeric systems. Chem Phys Lett 373:207–212
132. (a) Zang X, Liu H, Li Q, Li Z, Li Z (2020) Polym Chem 11:5493–5499; (b) Li Z, Zang X, Li Y, Li Q, Li Z (2022) Macromol Rapid Commun 43:2200179
133. Zhang S, Yang X, Numata Y, Han L (2013) Highly efficient dye-sensitized solar cells: progress and future challenges. Energy Environ Sci 6:1443–1464
134. (a) Kim BG, Zhen CG, Jeong EJ, Kieffer J, Kim J (2012) Organic dye design tools for efficient photocurrent generation in dye-sensitized solar cells: exciton binding energy and electron acceptors. Adv Funct Mater 22:1606–1612; (b) Wu Y, Zhu W (2013) Organic sensitizers from D–π–A to D–A–π–A: effect of the internal electron-withdrawing units on molecular absorption, energy levels and photovoltaic performances. Chem Soc Rev 42:2039–2058; (c) Liang M, Chen J (2013) Arylamine organic dyes for dye-sensitized solar cells. Chem Soc Rev 42:3453–3488
135. Misra R, Maragani R, Patelb KR, Sharma GD (2014) RSC Adv 4:34904–34911

136. (a) Michinobu T, Satoh N, Cai J, Lia Y, Han L (2014) J Mater Chem C 2:3367–3372; (b) Pakravesh F, Izadyar M, Arkan F (2021) Effect of electron donor and acceptor on the photovoltaic properties of organic dyes for efficient dye-sensitized solar cells. Physica B 609: 412815
137. Yuan Y, Michinobu T, Satoh N, Ashizawa M, Han L (2015) J Nanosci Nanotechnol 15:5856–5866
138. Michinobu T, Yamada N, Washino Y, Nakayama K (2016) Novel Design of Carbazole-Based Donor–Acceptor Molecules for Fullerene-Free Organic Photovoltaic Devices. J Nanosci Nanotechnol 16:8662–8669
139. (a) Gautam P, Misra R, Siddiqui SA, Sharma GD (2015) Donor–acceptor–π–acceptor based charge transfer chromophore as electron donors for solution processed small molecule organic bulk heterojunction solar cells. Org Electron 19:76–82; (b) Gautam P, Misra R, Koukaras EN, Sharma A, Sharma GD (2015) Donor–acceptor–acceptor–donor small molecules for solution processed bulk heterojunction solar cells. Org Electron 27:72–83; (c) Gautam P, Misra R, Siddiqui SA, Sharma GD (2015) ACS Appl Mater Interfaces 7:10283–10292; (d) Gautam P, Misra R, Sharma GD (2016) Dicyanoquinodimethane-substituted benzothiadiazole for efficient small-molecule solar cells. Phys Chem Chem Phys 18:7235–7241; (e) Rao PS, Gupta A, Bhosale SV, Bilic A, Xiang W, Evans RA, Bhosale SV (2017) Dyes Pigments 146:502–511; (f) Raheem AA, Kamaraj S, Sannasi V, Praveen C (2018) New D–π-A push–pull chromophores as low band gap molecular semiconductors for organic small molecule solar cell applications. Org Chem Front 5:777–787; (g) Rout Y, Misra R, Singhal R, Biswasc S, Sharma GD (2018) Phys Chem Chem Phys 20:6321–6329; (h) Marqués PS, Castán JMA, Raul BAL, Londi G, Ramirez I, Pshenichnikov MS, Beljonne D, Walzer K, Blais M, Allain M, Cabanetos C, Blanchard P (2020) Chem Eur J 26:16422–16433; (i) Raheem AA, Kumar C, Shanmugam R, Murugan P, Praveen C (2021) J Mater Chem C 9:4562–4575; (j) Raheem AA, Murugan P, Shanmugam R, Praveen C (2021) Azulene bridged π-distorted chromophores: the influence of structural symmetry on optoelectrochemical and photovoltaic parameters. ChemPlusChem 86:1451–1460
140. Raoa PS, More VG, Jangale AD, Bhosale SV, Bhosale RS, Puyad AL, Chen J-Y, Li J-L, Bhosale SV, Gupta A, Sharma GD (2019) Dyes Pigments 171:107677
141. (a) Yamagata T, Kuwabara J, Kanbara T (2012) Synthesis and characterization of dioxopyrrolopyrrole derivatives having electron-withdrawing groups. Eur J Org Chem 2012: 5282–5290; (b) Patil Y, Jadhav T, Dhokale B, Misra R (2016) Asian JOrg Chem 5:1008–1014; (c) Patil Y, Misra R, Keshtov ML, Sharma GD (2016) 1,1,4,4-tetracyanobuta-1,3-diene substituted diketopyrrolopyrroles: an acceptor for solution processable organic bulk heterojunction solar cells. J Phys Chem C 120:6324–6335; (d) Patil Y, Misra R, Keshtovb ML, Sharma GD (2017) J Mater Chem A 5:3311–3319; (e) Bucher L, Desbois N, Harvey PD, Gros CP, Misra R, Sharma GD (2018) ACS Appl Energy Mater 1:3359–3368; (f) Seco CR, Vidal-Ferran A, Misra R, Sharma GD, Palomares E (2018) ACS Appl Energy Mater 1:4203–4210
142. Fujita H, Michinobu T, Fukuta S, Koganezawa T, Higashihara T (2016) Sequentially different AB Diblock and ABA triblock copolymers as P3HT:PCBM interfacial compatibilizers for bulk-heterojunction photovoltaics. ACS Appl Mater Interfaces 8:5484–5492
143. Morita K, Michinobu T (2019) Energy level tuning of aromatic polyamines by [2 + 2] cycloaddition-retroelectrocyclization for the optimization of device performances. Synth Met 257:116179
144. Wang D, Zhang W, Xing Y, Gao H, Wang X, Zhao Y, Yang H (2013) Energy-level modulation of organic alkynes by click chemistry. Tetrahedron 69:895–901
145. Lacy AR, Vogt A, Boudon C, Gisselbrecht J-P, Schweizer WB, Diederich F (2013) Post-cycloaddition–retroelectrocyclization transformations of polycyanobutadienes. Eur J Org Chem:869–879
146. (a) Breiten B, Wu Y-L, Jarowski PD, Gisselbrect J-P, Boudon C, Griesser M, Onitsch C, Gescheidt G, Schweizer WB, Langer N, Lennartz C, Diederich F (2011) Chem Sci 2:88–93;

(b) Breiten B, Jordan M, Taura D, Zalibera M, Griesser M, Confortin D, Boudon C, Gisselbrecht J-P, Schweizer WB, Gescheidt G, Diederich F (2013) Donor-substituted octacyano[4]dendralenes: investigation of π-electron delocalization in their radical ions. J Org Chem 78:1760–1767

147. Tu K-H, Wang Y, Kiyota Y, Iwahashi T, Ouchi Y, Mori T, Michinobu T (2020) A cyano-rich small molecule dopant for organic thermoelectrics. Org Electron 87:105978
148. Fujita H, Michinobu T (2018) Covalent layer-by-layer thin films with charge-transfer chromophores: side chain engineering for efficient Ag+ion recognition in aqueous solutions. Soft Matter 14:9055–9060
149. Fujita H, Nihei N, Bito M, Michinobu T (2018) Antibacterial polymeric films fabricated by [2+2] cycloaddition–retroelectrocyclization and Ag+ ion coordination. Macromol Biosci 18: 1800336
150. Pham QNN, Silpcharu K, Vchirawongkwin V, Sukwattanasinitt M, Rashatasakhon P (2022) 2,3-Diaryl-1,1,4,4-tetracyanobutadienes as Colorimetric Sensors for Sulfide Ion in Aqueous Media. Synlett 33:1335–1340
151. Tan J, Wang C, Lao HK, Wang W, Feng G, Yuan D, Wu C, Zhang X (2019) Chem Commun 55:7438–7441
152. (a) Xu J, Liu X, Lv J, Zhu M, Huang C, Zhou W, Yin X, Liu H, Li Y, Ye J (2008) Morphology transition and aggregation-induced emission of an intramolecular charge-transfer compound. Langmuir 24:4231–4237; (b) Chaudhary J, Laskar IR, Roy RK (2022) ACS Appl Electron Mater 4:3724–3738; (c) Roger M, Bretonnière Y, Trolez Y, Vacher A, Arbouch I, Cornil J, Félix G, Winter JD, Richeter S, Clément S, Gerbier P (2023) Int J Mol Sci 24:8715
153. Dar AH, Gowri V, Mishra RK, Khan R, Jayamurugan G (2022) Nanotechnology-assisted, single-chromophore-based white-light-emitting organic materials with bioimaging properties. Langmuir 38:430–438
154. (a) Tang X, Chen Z, Xu Q, Su Y, Xu H, Horike S, Zhang H, Li Y, Gu C (2022) CCS Chem 4: 2842–2853; (b) Lin Z, Zhong Y-H, Zhong L, Ye X, Chung L-H, Hu X, Xu Z, Yu L, He J (2023) Minimalist design for solar energy conversion: revamping the π-grid of an organic framework into open-shell superabsorbers. JACS Au 3:1711–1722
155. (a) Shi H, Gu R, Xu W, Huang H, Xue L, Wang W, Zhang Y, Si W, Dong X (2019) ACS Appl Mater Interfaces 11:44970–44977; (b) Zhong Y-H, Huang G-F, Zhao S-Y, Chung L-H, Zhang H-T, Zheng J-H, Yan Y-L, Ni W-X, He J (2024) Easy but efficient: facile approach to molecule with theoretically justified donor–acceptor structure for effective photothermal conversion and intravenous photothermal therapy. Adv Sci 11:2309068
156. (a) Liu W, Li B, Gao H, Wang D, Wang L, Yang Z, Cao H, He W, Wang H, Zhang J, Xing Y (2019) NIR absorbing thiophene derivatives with photoacoustic and photothermal effect used in drug release. Dyes Pigments 162:331–338; (b) Roy HS, M NK, Rajput S, Sadhukhan S, Gowri V, Dar AH, Monga M, Salaria N, Guha R, Chattopadhyay N, Jayamurugan G, Ghosh D (2023) Chem Eur J 29:e202301748
157. Lim D-H, Kang M, Jang S-Y, Hwang K, Kim I-B, Jung E, Jo Y-R, Kim Y-J, Kim J, Choi H, Kim T-W, Mathur S, Kim B-J, Kim D-Y (2020) Unsymmetrical small molecules for broadband photoresponse and efficient charge transport in organic phototransistors. ACS Appl Mater Interfaces 12:25066–25074
158. (a) Michinobu T (2019) Development of N-type semiconducting polymers for transistor applications. J Photopolym Sci Technol 32:563–570; (b) He W, Washino Y, Murata K, Nozaki N, Matsumoto H, Michinobu T (2022) [2+2] Cycloaddition-retroelectrocyclization reactivity and thin film transistor performances of carbazole-based platinum polyyne polymers. Mater Chem Phys 281:125861

Top Heterocycl Chem (2026) 63: 123–160
https://doi.org/10.1007/7081_2025_98

Published online: 30 December 2025

The Thiol-ene "Click" Reaction for Synthesis of Biomolecular Thioether and Thioester Derivatives

Pierre Milbeo and Chiara Zanato

Contents

1 Introduction 124
1.1 Development of the Thiol-ene "Click" Reaction 125
1.2 Bioorthogonality of the "Photo-Click" Mechanism 126
2 Application to Peptide Science 127
2.1 Peptide and Protein Lipidation 128
2.2 Peptide and Protein Glycosylation 132
2.3 Peptide Stapling and Cyclization 135
3 Application to Carbohydrate Science 141
3.1 Thioglycoside Synthesis 142
3.2 Glycoside Bioconjugates 145
4 Application to Biomaterials: Biomedical Hydrogels 149
5 Conclusion 152
References 152

Abstract Over the past century, the radical thiol-ene reaction has demonstrated great value for the controlled synthesis of polymers and functionalized materials and for biomolecule conjugation. The free radical aspect offers many advantages, the three most prominent being water compatibility, a metal-free protocol (useful for the development of new therapeutic bioconjugates), and a bioorthogonal mode of action.

This chapter covers recent developments in the use of the free radical thiol-ene "click" reaction for the synthesis and modification of biomolecules, bioconjugates,

The original version of the chapter has been revised. The title captions' layout for Schemes 11 and 17 has been updated to display under the figure; In page 152, the line "Acknowledgments: I suggest including acknowledgements for funding here.", has been removed.

P. Milbeo (✉) and C. Zanato
BioCIS, UMR 8076, CY Cergy Paris Université, CNRS, Cergy-Pontoise, France

BioCIS, UMR 8076, Université Paris-Saclay, CNRS, Orsay, France
e-mail: pierre.milbeo@cyu.fr

and biomaterials. Applications in peptide and protein science are also discussed, including major contributions to the synthesis of glycopeptides and lipopeptides as well as peptide stapling and cyclization. The various synthetic strategies towards thiosugars, glycoconjugates, and glycoclusters are also highlighted with a focus on the regio- and stereoselectivity of the process. An overview of the thiol-ene-mediated cross-linking strategy for the development of hydrogel biomaterials is also included, which demonstrates the diversity and versatility of the reaction in a biomolecular context.

Keywords Thiol-ene · Biomolecules · Bioconjugates · Biomaterials

1 Introduction

The thiol-ene reaction is the direct formation of a carbon-sulfur bond via the coupling between a thiol and an olefin. The first occurrence of such reactivity was reported in the early 1900s [1], and it has since proven to be useful for the preparation of functionalized materials and polymers. More recently, the emergence of the concept of "click" chemistry, initially defined by Sharpless et al. [2], gave to the thiol-ene reaction a new dimension by revealing specific features allowing its application in biomolecule synthesis and modification [3]. Indeed, the thiol-ene reaction meets the defined criteria characterizing the click chemistry, namely high yields, lack of chemical waste production, compatibility with aqueous conditions (and, therefore, biocompatibility), and orthogonality to many synthetic processes. Considering the high value, polyfunctionality, and water solubility of biomolecules, "click" chemistry is particularly suited for their synthesis and functionalization as it does not require protective groups or organic solvents and it facilitates easier purification. More specifically, the thiol-ene "click" reaction (TEC), in contrast with the copper-catalyzed azide-alkyne cycloaddition (CuAAC), does not require any metal catalyst whose removal can be troublesome, and remaining traces can result in toxicity as far as biological applications are concerned.

Generally speaking, TEC refers to the radical-mediated and (often) photoinitiated coupling between a thiyl radical (a highly reactive species, characterized by a sulfur atom with an unpaired electron) and an alkene and forms the focus of this chapter. Michael acceptors such as electron-deficient alkenes are also considered to be substrates for thiol-ene coupling although this type of coupling proceeds via an ionic mechanism assisted by a Brønsted base [4]. The use of alkynes as acceptors in the so-called thiol-yne reaction has also been extensively reviewed as a tool for bioconjugation [5, 6] and synthesis of biomaterials and in polymer science, where the ability of the yne-bond to react with two equivalents of thiyl radical via a sequential thiol-yne/thiol-ene process allows access to branched functionalized thioethers [3, 7] (Scheme 1).

In this chapter, we propose to cover the following topics. Firstly, we will consider recent progress made in the development of radical-mediated bioorthogonal thiol-ene click reactions for the synthesis and functionalization of biomolecules. Then, after introducing the TEC reaction focusing on mechanistic aspects, selected recent applications in peptide science, glycoscience, and biomaterials will be presented

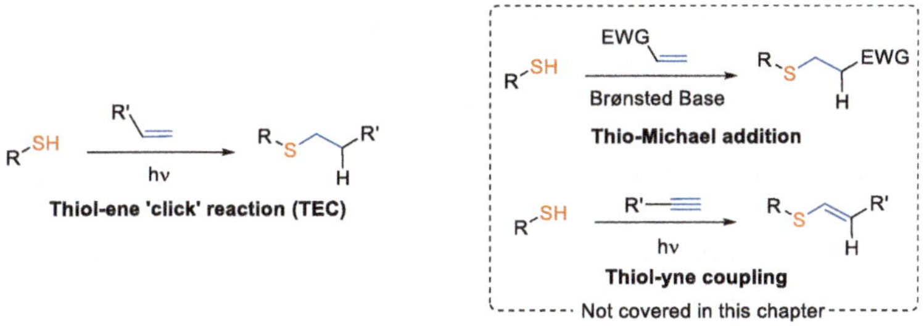

Scheme 1 The thiol-ene "click" reaction and its variations

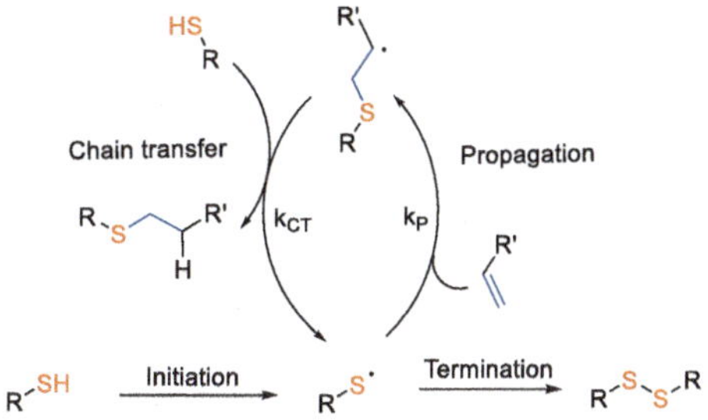

Scheme 2 Free radical chain mechanism of the thiol-ene reaction (TEC)

throughout this chapter. This document has no pretension to comprehensively review this field of research but instead to provide selected examples demonstrating the diversity and versatility of the TEC reaction in the development of biomolecules. Recent and older well-documented reviews on the topic can provide an exhaustive discussion on every section of this chapter [8–10].

1.1 Development of the Thiol-ene "Click" Reaction

Although the first report of a thiol addition to olefins goes back to Posner's work in 1905 [1], the free radical chain mechanism of thiol-ene-based couplings was proposed for the first time by Kharach et al. in 1938 [11]. The initiation of this mechanism relies on the homolytic dissociation of the S–H bond (bond dissociation energy around 87 kcal mol^{-1}) [12] via thermal decomposition of a radical initiator or via photoinitiation using UV irradiation. The formed thiyl radical propagates across the C=C bond to give a carbon-centered radical that abstracts a hydrogen atom from another equivalent of thiol and produces the thioether product along with a thiyl radical that allows further propagation. The termination step relies on a radical-radical coupling resulting in the formation of either a disulfide or a 1,2-dithioether, depending on the radical partners involved (Scheme 2).

The TEC was shown to be exothermic, with reaction enthalpies ranging from −10.5 kcal mol^{-1} for electron-rich alkenes to −22.6 kcal mol^{-1} for electron-poor double bonds [13]. Kinetically speaking, the balance between propagation and chain transfer rates (k_P/k_{CT}) determines whether the reaction is first or half-order, with the chain transfer often being the rate-limiting step. Recent advancements, including Bowman's first-order kinetic model [14] and Aviyente's quantum chemical studies [15], have clarified key reaction parameters and the influence of alkene and thiol functionalities. The reaction rate exhibits first-order dependence on thiol concentration when the k_P/k_{CT} ratio is high, making chain transfer the rate-limiting step. This scenario is typical for alkyl thiols with less easily abstractable hydrogen atoms, which slow down chain transfer. When k_P and k_{CT} are similar, the reaction shows half-order dependence on both alkene and thiol concentrations. However, when the k_P/k_{CT} ratio is low, especially with less reactive alkene functionalities, the reaction rate becomes alkene dependent, with propagation as the rate-limiting step.

The TEC reaction, as with any other free radical addition, has an anti-Markovnikov regioselectivity resulting from the stabilization of the carbon-radical intermediate on the most substituted carbon of the olefin [16]. In terms of stereochemical outcome, the ability of the carbon-centered radical to retain its configuration across the chain transfer step is controlling the stereoselectivity of the thiol coupling. Therefore, stereoselectivity is rather challenging to achieve and often requires substrates with highly restricted free rotation, e.g., cyclic olefins. The most significant examples of stereoselective TEC reactions were developed on glycals and other unsaturated glycoconjugates [17].

The light-mediated thiol-ene reaction has gained popularity in recent decades due to its simple reaction conditions and wide-ranging applications in biomaterials, polymer chemistry, pharmaceuticals, and drug synthesis. In the late 1970s, the process demonstrated its first industrial applications in materials science including relief printing plates (Letterflex process) or conformal coating on electronics via the work developed at W.R. Grace chemical company on thiol-ene photocuring processes [18]. Norland further advanced the technology in the late 1970s with thiol-ene-based optical and electronic adhesives, and by 2004, thiol-ene chemistry had become a vital tool for designing polymer materials for applications such as glasses, optical materials, and adhesives [13].

1.2 Bioorthogonality of the "Photo-Click" Mechanism

Bioorthogonality is the ability of a chemical reaction to take place in a biological environment without affecting or being affected by usual biochemical processes. Originally theorized by Carolyn Bertozzi's research group, the concept of bioorthogonal chemistry has built bridges between synthetic chemistry and biology, allowing organic synthesis to be performed in living organisms and cells [19]. To be considered bioorthogonal, a chemical reaction must comply with different requirements including inertness to any nucleophiles, electrophiles, reductants, or oxidants found in the biological environments and compatibility with aqueous media and physiological pH. Most importantly, the bioorthogonal reaction must involve functional groups that are not present in living systems. In this aspect, the TEC reaction cannot be considered strictly bioorthogonal as thiols are widely present in biomolecules, e.g., cysteine, glutathione, dithiothreitol, and bovine serum albumin. Although underrepresented compared to other natural amino acids, cysteine occurrence in living organisms is still non-negligible (~2.26% in mammals and 0.5% in bacteria) [20].

Thus, the thiol-ene reaction is widely used in biological systems, particularly for posttranslational functionalization of cysteine side chains in peptides and proteins [21]. Although the mechanism of the TEC has been well known for almost a century, the application of TEC reaction in a biomolecular context has emerged much later, driven by its ability to be performed in aqueous media. In addition, the radical, photoactivated mechanism of the reaction proved to be orthogonal to the mostly ionic reactivities of biomolecules, facilitating in situ precise modifications and diversifying the toolbox of the biochemists for the development of new therapeutics.

2 Application to Peptide Science

Due to their unique therapeutic properties, peptides and proteins are gaining attention as promising drug candidates. They exhibit several advantages over traditional small-molecule drugs, offering high specificity, strong target binding affinity, reduced off-target effects, and biodegradability into nontoxic amino acids [22]. However, their clinical application is hindered by several limitations that must be addressed to optimize their efficacy [23]. One of the primary challenges is proteolytic instability. Peptides are highly susceptible to enzymatic degradation, leading to a short half-life and necessitating frequent administration. Additionally, poor oral bioavailability restricts their use in convenient oral formulations, as they degrade in the gastrointestinal tract and exhibit limited intestinal absorption. Another major drawback is the limited membrane permeability of peptides, which prevents efficient intracellular delivery and reduces their effectiveness against specific therapeutic targets.

To overcome these limitations and enhance the therapeutic potential of peptides, site-selective peptide and protein modifications have been developed as a key strategy. Techniques such as peptide stapling or cyclization and conjugation to lipid or sugar moieties could significantly improve stability, bioavailability, and target specificity [24]. Moreover, peptide and protein modifications are highly valuable for studying various cell functions [25], as well as for the development of biotherapeutics [26]. Peptide and protein modifications also enable their tracking in vivo and in vitro through fluorescent tagging [27].

Thiol-ene chemistry offers distinct advantages over other bioorthogonal reactions [10, 28]. One of its primary benefits in peptide and protein applications is its selectivity for cysteine residues, eliminating the need for structural modifications to introduce a reactive anchoring site. Cysteine residues are mainly forming disulfide bonds with each other in peptides and proteins, therefore remaining "protected" and

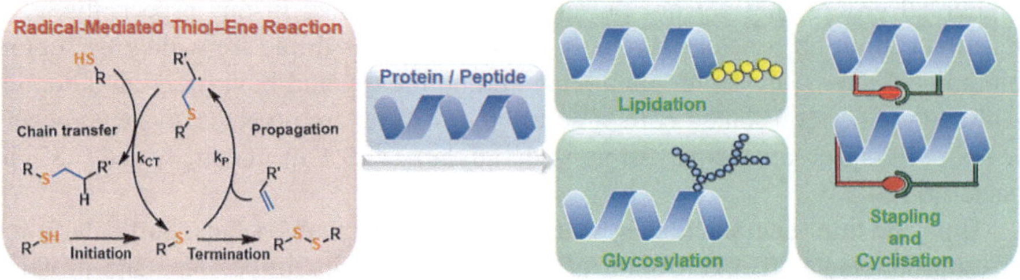

Fig. 1 Different applications of thiol-ene chemistry in protein and peptide science: lipidation, glycosylation, stapling, and cyclization

unreactive unless reduced by agents such as dithiothreitol (DTT), 2-mercaptoethanol (2-Me), glutathione (GSH), or tris(2-carboxyethyl)phosphine (TCEP). Moreover, compared to other nucleophilic residues such as lysine, serine, or tyrosine, cysteine's thiol group is significantly more reactive under physiological conditions (pH 6.5–7.5). Furthermore, the thioether linkage formed through the thiol-ene reaction is remarkably stable across diverse chemical environments, including various pH conditions. This stability contrasts with other commonly employed bioconjugation linkages, such as triazoles produced via copper-catalyzed azide-alkyne cycloaddition (CuAAC).

Proteins and peptides can undergo thiol-ene modification via two principal pathways [29]: (i) reaction of a preinstalled alkene, either introduced during protein synthesis or through posttranslational modification of a thiol-containing residue, or (ii) direct conjugation of an alkene-bearing molecule with an available cysteine residue. This flexibility has enabled the application of thiol-ene chemistry in mimicking posttranslational modifications, such as peptide lipidation, glycosylation, stapling, and cyclization (Fig. 1) [10, 28].

2.1 *Peptide and Protein Lipidation*

Protein lipidation is a co- or posttranslational modification (CTM/PTM) in which proteins undergo covalent attachment to various lipids, such as fatty acids, cholesterol, and isoprenoids [28]. This modification is crucial for regulating protein structure, localization, interactions, and function. Dysregulation of protein lipidation has been linked to numerous diseases, including cancer, neurological disorders, metabolic conditions, and infections [30]. Fatty acids typically vary in chain length from 8 to over 20 carbons, and their attachment can occur at the N-terminus or C-terminus or within the protein sequence. This modification is relatively stable and typically occurs at specific amino acids, including cysteine, serine, threonine, and lysine. Peptide lipidation is a powerful strategy to enhance the pharmacokinetic properties of peptides, improving their stability, membrane permeability, and biological activity. Lipopeptides have gained significant interest in medicinal chemistry

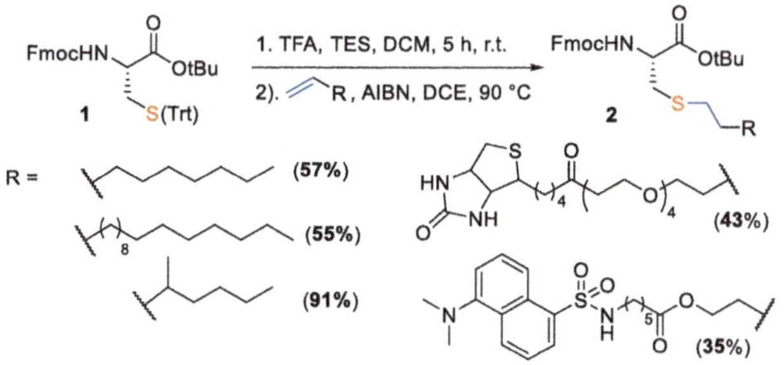

Scheme 3 Waldmann's experimental conditions for the synthesis of S-alkylated cysteines via thiol-ene reaction [21]

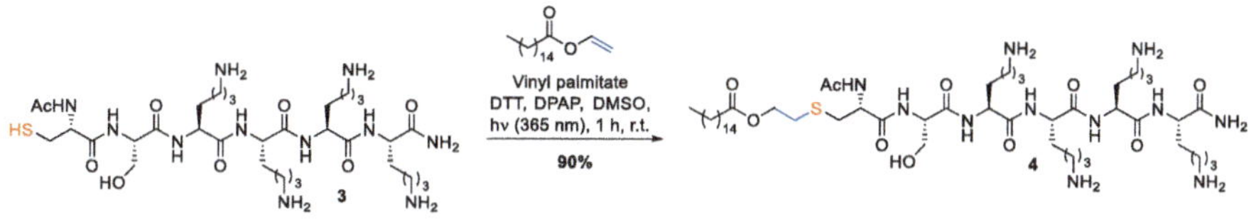

Scheme 4 Brimble's thiol-ene reaction conditions on AcNH-CSKKKK-NH_2 (**3**) at low temperature in the presence of DMPA as radical initiator [33]

due to their diverse biological activities, including antiviral, antimicrobial, antitumor, antifungal, and anti-inflammatory properties [31]. Produced mainly by fungi and bacteria, lipopeptides often contain D- and L-amino acids, enhancing stability against proteolytic enzymes and allowing oral or intravenous administration. Many are cyclized via ester or amide linkages, restricting their conformation. Their primary mode of action involves membrane disruption, making resistance development rare. As a result, lipopeptides are considered promising agents in combating antimicrobial resistance [32]. Among the diverse lipidation methods available, thiol-ene chemistry has gained significant attention due to its specificity, efficiency, and compatibility with various biomolecules.

Waldmann's group [21] pioneered the racemization-free synthesis of S-alkylated cysteine via thiol-ene reaction. The reaction was performed at 90°C in dichloroethane (DCE) in the presence of 1-hexadecene, cysteine methyl ester (**1**), and azobisisobutyronitrile (AIBN) as radical initiator. The lipidated cysteine was obtained with a 57% yield. Waldmann demonstrated the compatibility of his methodology with several modified alkenes, like biotinylated and fluorescent compounds (**2**, Scheme 3).

Later, the group of Brimble developed new reaction conditions that allowed performing the thiol-ene reaction at lower temperatures on more complex substrates (such as longer peptides as AcNH-CSK4-NH_2, **3**) without epimerization [33]. The optimized conditions involved the use of vinyl palmitate, 2,2-dimethoxy-2-phenylacetophenone (DPAP), irradiation with UV light (365 nm), and dithiothreitol (DTT) to reduce the undesired polymerization of the vinyl palmitate (Scheme 4). The palmitoylated peptide **4** was obtained in 90% yield.

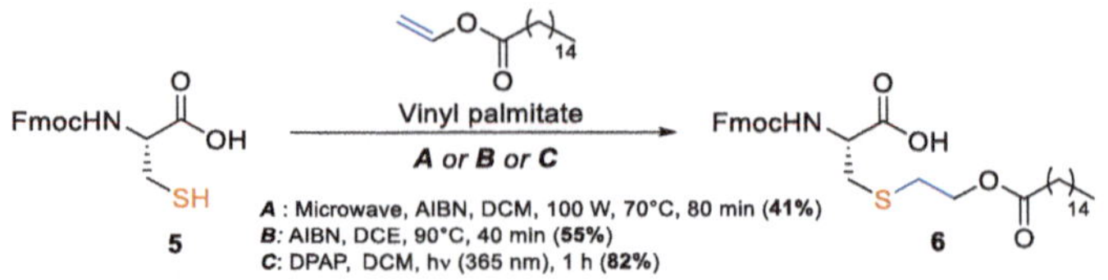

Scheme 5 CLipPA strategy for the synthesis of the Fmoc cysteine palmitate [34]

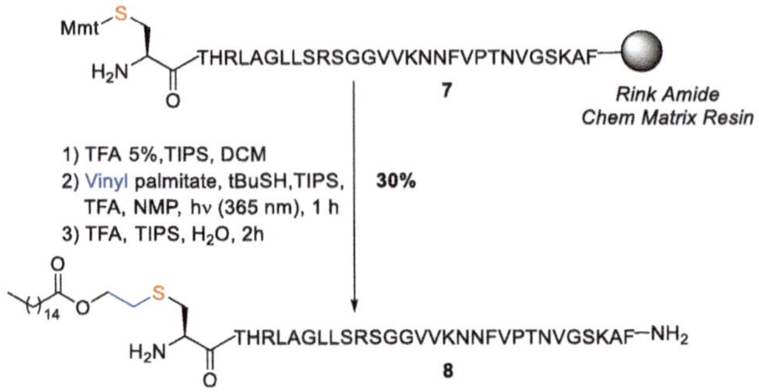

Scheme 6 Palmitoylation of the αCGRP$_{8–37}$ peptide using SP-CLipPA strategy [35]

Subsequently, in 2016, Brimble's group developed lipidated cysteine building blocks for direct use in solid-phase peptide synthesis (SPPS). Brimble coined the term "CLipPA" (cysteine lipidation on a peptide or amino acid) to describe this novel strategy [34]. Different N-protecting groups (Fmoc, Boc, Ac), radical initiators (DPAP, AIBN), and activation methods (thermal, UV light, microwave) were tested and screened to find the highest yielded thiol-ene reaction conditions. An example, starting from Fmoc-protected cysteine **5** to its palmitoylated version **6**, is reported in Scheme 5. In this case, UV irradiation in the presence of DPAP in DCM gave the best yields.

A solid-phase variant of CLipPA, the SP-CLipPA, was developed by the same research group a few years later in 2018 [35]. An example of the application of the SP-CLipPA is reported in Scheme 6. A CGRP receptor antagonist, employed in migraine treatment, was synthesized by the N-terminal lipidation of the αCGRP$_{8–37}$ peptide **7**. The reaction conditions included vinyl palmitate, 2,2-dimethoxy-2-phenylacetophenone (DPAP), triisopropylsilane (TIPS), tBuSH, and 5% of TFA (trifluoroacetic acid) under UV light (365 nm) in degassed N-methyl-2-pyrrolidone (NMP) at room temperature. The SP-CLipPA approach yielded mono-S-palmitoylated αCGRP$_{8–37}$ peptide 8 in 30% yield.

Later, Yang's team explored alternative thiol radical chain propagators in CLipPA and SP-CLipPA to replace volatile and odorous tBuSH [36]. The alternative molecule needed to be bulky enough to prevent S-alkylation and less volatile than tBuSH. Trityl thiol (TrtSH) and *tert*-nonyl mercaptan (*t*-nonylSH) were tested. TrtSH led to complex product mixtures, while *t*-nonylSH produced results comparable to tBuSH. Several reactions with *t*-nonylSH were performed, and all achieved high conversions, mostly exceeding 95% of yield.

In 2020, the Brimble group applied the CLipPA methodology to the lipidation of cyclic peptides obtained via native chemical ligation (NCL) [37]. The CLipPA method was employed to synthesize an analog of iturin A, a cyclic lipopeptide with antifungal properties. The SPPS generated the linear peptide thioester, which underwent NCL, yielding the corresponding cyclic peptide. Lipidation with vinyl palmitate, TIPS, *t*-nonylSH, DPAP, and TFA in NMP achieved 82% conversion. The strategy was extended to the synthesis of six other iturin A analogs with varying lipid modifications, showing high conversion rates (81–85%).

CLipPA methodology was applied in the following years by various research groups to synthesize bioactive molecules. Yim et al. applied CLipPA to obtain linear battacin analogs with different lipid chains using 2,2-dimethoxy-2-phenylacetophenone (DPAP), TIPS, and *t*-nonylSH in NMP. The lipopeptides were obtained with 1–32% yield and with a high degree of purity (>94%) [38]. The lipid chains were anchored to the N-terminal Cys or 3-mercaptopropionate (MPA) handles and tested for antimicrobial activity against Gram-negative and Gram-positive bacteria. MPA-linked peptides had equal or lower activity than Cys-linked versions. Anyway, this study highlights CLipPA's potential for generating bioactive peptide libraries.

Tong et al. used the CLipPA strategy to synthesize paenipeptin C analogs accessorized with different lipid moieties via the MPA handle [39]. Several analogs exhibited equal or enhanced activity against *E. coli* and *S. aureus* compared to paenipeptin C, further showcasing CLipPA's effectiveness in generating biologically active lipopeptides.

Yang et al. used CLipPA to synthesize lipidated connexin analogs to enhance cell membrane anchoring and to increase metabolic resistance [40]. Even in this case, peptides reacted at the SH of the Cys or the MPA handle, allowing the S-lipidation. While the half-life of the analogs remained unchanged, short lipid chains improved functional efficacy, further validating CLipPA for bioactive lipopeptide synthesis.

Yim and co-workers took advantage of CLipPA to synthesize truncated S-lipidated teixobactin analogs by capping the N-terminus with 3-mercaptopropionic acid [41]. Various lengths of lipid chains were attached, but no activity was observed against *S. aureus*. The activity loss was attributed to the increased hydrophilicity from the sulfur bridge.

In 2021, the Brimble group explored the thiol-ene-based lipidation of peptides as a strategy for developing new anti-hepatitis B virus (HBV) therapeutics [42]. Brimble's team synthesized S-lipidated analogs of Myrcludex B, a myristoylated peptide that inhibits HBV entry by targeting the NTCP receptor on hepatocytes. Using the CLipPA thiol-ene reaction (DPAP, TIPS, TFA, *t*-nonylSH, NMP, 360 nm, at 1 h), they generated a diverse library of lipidated peptides, demonstrating that heteroatoms in lipid anchors are well tolerated. Some S-lipidated analogs showed nanomolar IC50 values, confirming their potent antiviral activity. This work marks the first report of potent S-lipidated antiviral peptides, expanding the chemical space for HBV inhibitors.

In conclusion, the CLipPA methodology is one of the most notable approaches in the domain of lipopeptides. This method has been extensively developed to achieve site-specific modification with high yield and minimal side reactions [43, 44].

2.2 *Peptide and Protein Glycosylation*

Glycosylation is one of the most common and diverse protein modifications, influencing structure, stability, function, and cellular interactions. Over 50% of human proteins are glycosylated, impacting immune response, inflammation, and cell signaling. Aberrant glycosylation is linked to various diseases, highlighting its biological significance [45]. Over one-third of approved biopharmaceuticals are glycoproteins, highlighting their importance in medicine [46]. However, producing homogeneous glycoproteins to study their biology deeply or to develop a potential drug remains challenging due to microheterogeneity, as glycosylation occurs posttranslationally and is not genetically controlled. Traditional recombinant methods struggle to isolate pure glycoforms, necessitating efficient chemical synthesis.

In general, two main strategies exist to synthesize glycopeptides and glycoproteins chemically: (1) stepwise incorporation of glycosylated amino acids during peptide synthesis or (2) post-assembly glycan attachment [47, 48]. Due to the complexity of O- and N-linked glycosylation, alternative synthetic approaches using C- and S-linked glycans have been explored during the last year. S-linked glycopeptides, more stable than natural O- and N-linked counterparts, offer therapeutic potential. In this context, the thiol-ene reaction plays a central role as a robust, chemo selective method for S-glycosylation. This radical reaction provides a biocompatible peptide and protein modification tool, advancing glycobiology and glycol-medicinal research.

In 2001, Klaffke developed a straightforward photochemical approach for modifying N-allyl glycoside backbones (**9a–d**) using the thiol-ene reaction to obtain glycosylated peptides **10a–d** (Scheme 7) [49]. Under UV irradiation (254 nm) in methanol and inert atmosphere, cysteamine hydrochloride ($HS\text{-}CH_2\text{-}CH_2\text{-}NH_2\cdot HCl$) was grafted onto glycosides, yielding amino-thioether oligosaccharides in good yields (70–80%). These modified sugars were subsequently utilized to synthesize neoglycopeptides (**11a–d**) through enzymatic cross-linking with Cbz-Gln-Gly via acyl-transfer catalysis.

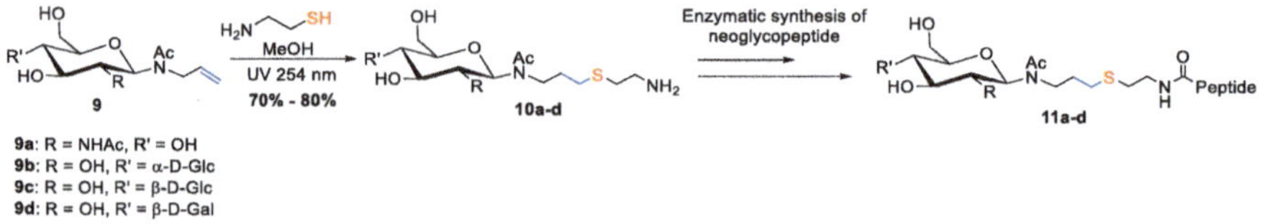

Scheme 7 Example of photoinduced thiol-ene reaction between cysteamine and an N-allyl glycoside [49]

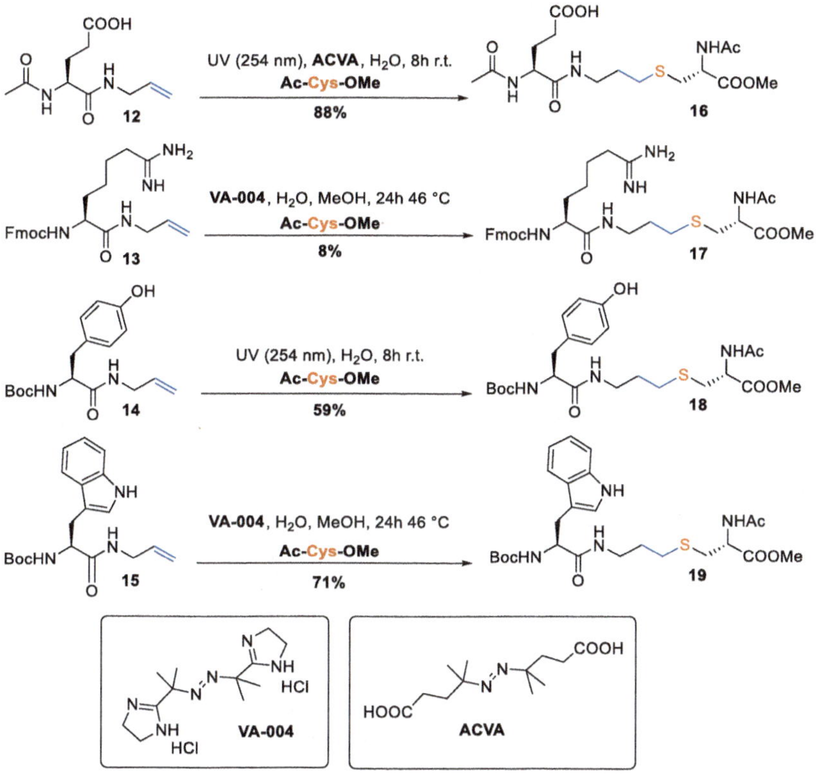

Scheme 8 Photoinduced thiol-ene reaction between cysteamine and an N-allyl glycoside [50]

In 2007, Kunz et al. explored the photoinduced thiol-ene reaction for synthesizing immune-compatible antitumor vaccines [50]. The group first examined the radical-mediated formation of thioether linkages between Ac-Cys-OMe and various vinyl-containing amino acids (**12–15**) under UV irradiation, with or without radical initiators (4,40-azobis(4-cyanopentanoic acid) (ACPA); VA-044, 2,20-azobis [2-(2-imidazolin-2-yl)propane]dihydrochloride) providing thioesters **16–19** (Scheme 8). Then, this optimized strategy was applied to conjugate synthetic MUC1 tumor-associated glycopeptide antigens to bovine serum albumin (BSA), demonstrating the technique's compatibility with biomolecules.

In 2009, the Dondoni and Davis groups simultaneously published a photoinduced strategy for glycoconjugation of peptides and proteins (Scheme 9) [51, 52].

Dondoni's approach involved hydrothiolation between peptides or proteins containing sulfhydryl groups (**20**) and an alkenyl C-glycoside (**21**). The reaction, providing conjugates **22**, was facilitated by UV irradiation at 360 nm and DPAP as the radical initiator [51]. In more detail, Dondoni first synthesized different S-linked galactosyl amino acids and dipeptides from peracetylated allyl C-galactosides and N-Fmoc cysteine derivatives. The reactions were performed under inert conditions,

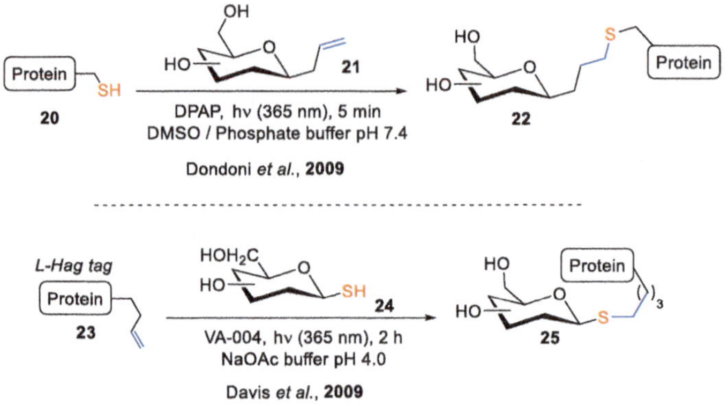

Scheme 9 Glycoconjugation strategy developed by Dondoni's team [51] and Davis' team [52]

yielding high-efficiency single diastereomers (95%). Then, they performed S-glycosylation on glutathione to validate its applicability, revealing that solvent choice significantly impacted yields (81% of yield was achieved in a DMF/H_2O mixture compared to 15% in DMF alone). Further, glycosylation of a nonapeptide (TALNCNDSL) with allyl α-C-galactoside in a phosphate buffer/DMSO medium (pH 7.4) led to complete conversion to the corresponding S-linked C-glycopeptide. Finally, the method was also applied to bovine serum albumin (BSA), where the photoinduced coupling of an α-D-galactoside resulted in the attachment of three sugar residues, likely due to disulfide bond reduction during irradiation. Dondoni's team also explored the "reverse strategy" (like Davis's approach, Scheme 9), performing thiol-ene coupling between O-acetyl-protected glycosyl thiols and N-protected C-allyl or C-vinyl glycines under UV light (365 nm) with DPAP, yielding various S-glycosyl amino acids in 53–90% yield.

Davis's team utilized L-homoallylglycine (L-Hag)-tagged proteins (**23**) and glycosyl thiols (**24**) to form glycoconjugates **25** through thiol-ene addition (Scheme 9) [52]. The reaction was carried out using VA-044 as an initiator. Additionally, Davis demonstrated the feasibility of glycoconjugation at multiple sites within a viruslike particle (VLP).

More recently, in 2023, the group of Scanlan [53] developed a thiol-ene-mediated peptide bioconjugation approach optimized for green solvents and continuous flow conditions. Scanlan's team demonstrated the feasibility of radical-based peptide modifications using deep eutectic solvents (DESs), bio-based solvents, and fully aqueous systems, significantly reducing the need for hazardous organic solvents. By employing the water-soluble photoinitiator Irgacure 2959, Scanlan's team successfully glycosylated biologically relevant peptides like the CREKA peptide **26**, under mild conditions in the presence of alkene-equipped sugar like **27** to obtain the conjugate like **28** (Scheme 10). The study highlights the potential of continuous flow thiol-ene chemistry for sustainable peptide modifications, enhancing efficiency while minimizing environmental impact.

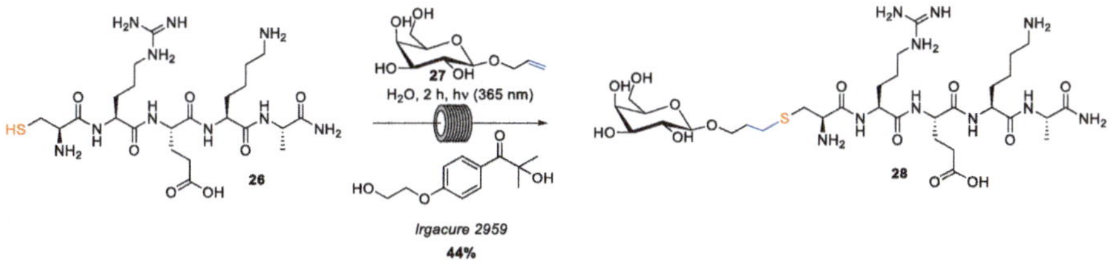

Scheme 10 Scanlan thiol-ene reaction for the glycosylation of CREKA peptide 26 under continuous flow conditions [53]

2.3 *Peptide Stapling and Cyclization*

As stated above, peptides hold great promise for therapeutic development but face challenges such as poor stability, low bioavailability, and limited cell permeability. Their rapid degradation by proteases and poor absorption due to low lipophilicity hinder oral delivery, the preferred route for drug administration [54]. Chemical modifications, such as lipidations, have been explored to enhance peptide stability, cell penetration, and therapeutic potency to overcome these limitations. Another practical approach is to rigidify the peptide scaffold. This rigidification introduces conformational constraints that improve the resistance of the peptide to degradation, enhancing its drug efficacy. The α-helix, a key structural motif in peptides and proteins, has been widely used to design stabilized peptides with improved resistance to degradation [55]. Stapled α-helical peptides offer advantages such as better cell permeability, oral absorption, and lower production costs, making them promising drug candidates. Various strategies have been developed to stabilize peptide helices [56–58], including covalent cross-linking of side chains, enhancing hydrogen bonding and electrostatic interactions, or incorporating helix-inducing nonnatural amino acids like aminoisobutyric acid (Aib) to improve protease resistance. Backbone rigidification via cross-links or hydrogen-bond surrogates has become a key technique in peptide drug design, offering new possibilities for developing next-generation therapeutics.

Another approach to stabilization of peptides involves the formation of cyclic peptides. Peptide cyclization is a posttranslational modification that naturally occurs in biological systems. Cyclic peptides can be categorized into four structural types based on their ring-closure mechanisms: (i) head-to-tail cyclic peptides (formed via a covalent bond between the peptide's C- and N-terminus); (ii) head-to-side chain cyclic peptides (generated by linking a side chain functional group to the N-terminus); (iii) tail-to-side chain cyclic peptides (formed by a covalent connection between a side chain functional group and the C-terminus); and (iv) side chain-to-side chain cyclic peptides (characterized by the presence of two covalently linked side chain functionalities). Cyclic peptides are abundant in nature and exhibit diverse biological activities, making them valuable drug candidates [59]. Similarly to stapling, the cyclization of linear peptides enhances receptor binding by restricting

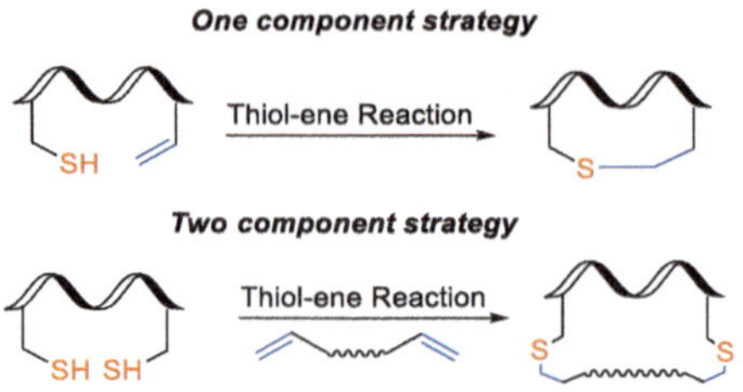

Scheme 11 Simplified representation of the one- and two-component stapling/cyclization approaches using thiol-ene reaction

conformational flexibility, thereby improving peptide stability against enzymatic degradation, leading to longer half-lives in plasma and improved oral bioavailability. Moreover, the rigidified structure minimizes the peptide's polar surface area, increasing the likelihood of internal hydrogen bond formation, which can enhance cell membrane permeability. As a result, significant research efforts have been dedicated to developing new and efficient chemical strategies for peptide cyclization [60]. By leveraging their inherent stability, target specificity, and favorable pharmacokinetic properties, researchers continue to explore cyclic peptides as potential treatments for various diseases.

There are many different chemical biocompatible methodologies for stapling or cyclizing a peptide [61]. Among them, TEC has emerged as a valuable tool for both peptide stapling and cyclization. Thiol-ene represents a strong and versatile approach for stabilizing α-helical structures in unprotected, expressed, or synthetic peptides. It enables selective cysteine modification using readily available alkene-containing building blocks, avoiding the need for costly or complex unnatural amino acids. The stapling/cyclization can be divided into one- and two-component systems (Scheme 11). One-component stapling/cyclization methods typically involve a cysteine and an activated long aliphatic alkene side chain attached to the peptide scaffold. In contrast, two-component stapling methods involve the reaction of two native Cys residues with dienes of varying lengths (such as a double-thiol-ene reaction).

One-component peptide thiol-ene stapling/cyclization methods rely on at least one modified or nonnatural amino acid alongside a complementary SH group on another amino acid, which may be synthetic or naturally occurring (cysteine). This approach enables controlled ring closure by incorporating all the necessary modifications during linear peptide synthesis, without requiring additional linkers. Possible unwanted intermolecular reactions represent the main drawback of this strategy. This problem can often be mitigated by using high dilutions or performing the stapling/cyclization on SPPS, taking advantage of the pseudo-dilution effects. Below are some examples of applying thiol-ene reaction using the one-component approach.

In 2010, Aimetti and co-workers pioneered an on-resin peptide macrocyclization strategy using the thiol-ene reaction [62]. The thiol-ene donor partner consisted of the SH of an unmodified natural cysteine. The alkene moiety was introduced in the peptide chain, modifying a lysine's side chain with either an allyloxycarbonyl (Alloc) or a norbornene group (**29** and **30**, Scheme 12).

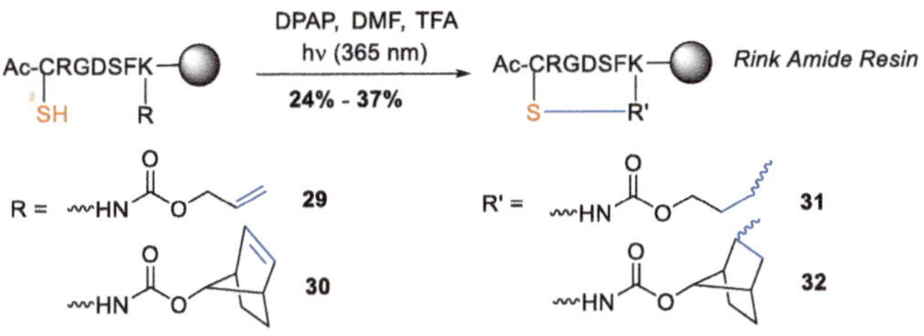

Scheme 12 Example of photoinduced thiol-ene reaction between cysteamine and an N-allyl glycoside [62]

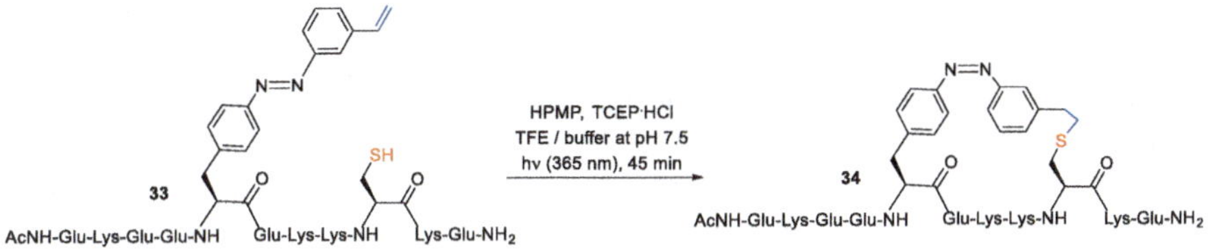

Scheme 13 Thiol-ene stapling on a photoswitchable peptide reported by Hoppmann [64, 65]

These linear peptides were then subjected to UV-induced thiol-ene coupling in the presence of the photoinitiator DPAP in DMF, TFA, and UV light. The approach yielded cyclic peptides **31** and **32** at 24% and 37% efficiency, respectively. Notably, norbornene-containing peptides cyclized rapidly (within 20 min) due to the increased reactivity of the strained bicyclic system. NMR analysis of **32** indicated the formation of *endo* and *exo* conformations. While thermal initiation was explored, it resulted in lower yields, and solution-phase reactions proved less efficient due to the need for additional purification steps. The resulting cyclic Arg-Gly-Asp (RGD) peptides demonstrated enhanced biological activity compared to their linear counterparts.

Further refinement of a similar thiol-ene macrocyclization was carried out in 2018 by the group of LeValley [63]. In the linear peptide CGRGDdvK(alloc)DRK(Az), the alloc function of the side chain lysine was cyclized with the SH of the cysteine. The peptide was equipped with an azide (attached to the terminal lysine) to perform another click reaction using CuAAC. The thiol-ene cyclization performed on SPPS provided low yields, while a solution-phase reaction using lithium phenyl-2,4,6-trimethylbenzoylphosphinate (LAP) as a soluble photoinitiator in water yielded approximately 70% of the cyclic peptide while preserving the azide functionality, which was subsequently used for hydrogel and fluorophore conjugation.

Hoppmann et al. introduced a novel photoswitchable peptide cross-linking method using a photoswitchable click amino acid (PSCaa), which contained an alkene handle on an azobenzene moiety (**33**, Scheme 13) [64, 65]. The azobenzene moiety, submitted to the UV irradiation, isomerized from *trans* to *cis* and then, in the presence of the photoinitiator 2-hydroxy-1-[4-(2-hydroxyethoxy)phenyl]-2-methyl-1-propanone (HPMP), reacted through a thiol-ene mechanism to give the cyclic

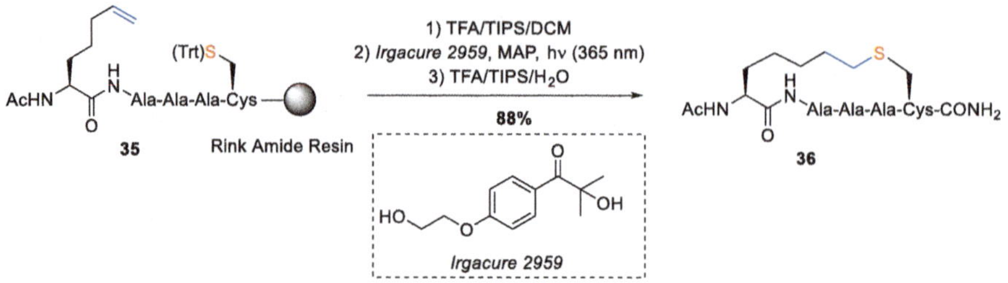

Scheme 14 Synthesis of cyclic peptide 36 developed by Li's team [66]

peptide **34**. The addition of 2,2,2-trifluoroethanol (TFE) kept the peptide in an α-helical conformation, while in an aqueous buffered solution, no preferred conformation was observed. The reducing agent Tris(2-carboxyethyl)phosphine hydrochloride (TCEP) was also employed to minimize disulfide bond formation.

In 2016, another notable example of the application of thiol-ene reaction on peptide cyclization was demonstrated by Li's group, who developed an on-resin macrocyclization between a cysteine and an alkene-functionalized side chains, ranging from four to six carbons in length [66]. After optimizing the reaction conditions, they performed the reaction on different pentapeptides, varying the composition of the amino acids (Arg, Glu, Gly, Lys, Pro, Trp, and Tyr). An example of Li's methodology, starting from peptide **35**, is shown in Scheme 14. First, the Trt (trytil)-protecting group of the cysteine was selectively removed by treating the rink amide resin with trifluoroacetic acid and triisopropylsilane. Then, the cyclization was performed in the presence of the radical initiator Irgacure 2,959 and 4-methoxyacetophenone (MAP) as additive under 365 nm UV irradiation giving the cyclic peptide **36** (Ac-(cyclo-1,5)-[mS5AAAC]-$CONH_2$) with 88% conversion. To demonstrate biological relevance, Li's group synthesized an *i*, *i* + 4 stapled analog of an 11-mer linear peptide targeting the estrogen receptor co-activator binding site. Fluorescence polarization assays revealed that the cyclic peptide displayed stronger binding affinity than its linear counterpart and outperformed the native ligand.

Overall, these one-component thiol-ene macrocyclization strategies highlight the versatility of this chemistry for cyclic peptide synthesis, whether performed on solid support or in solution. The ability to install either simple or functionally enhanced linkers without requiring multiple unnatural amino acids makes this method appealing. However, challenges remain regarding selectivity when multiple cysteine residues are present within a peptide. Further research into selective functionalization or protecting group strategies could address this limitation, expanding the utility of thiol-ene chemistry in peptide design.

Two-component peptide stapling involves the reaction of a peptide with an external linker containing two functional groups that complement reactive sites on the peptide chain. Typically, amino acids at the stapling positions are substituted with residues that feature appropriate reactive side chains, with cysteine being a common choice due to the reactivity of its thiol group. This approach enables thiol-based radical formation, facilitating attachment of the external staple. However, a

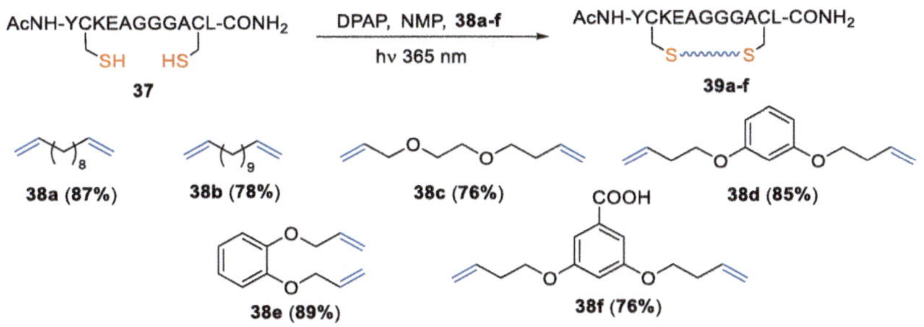

Scheme 15 Thiol-ene stapling conditions for macrocyclization of unmodified peptide sequences [67]

primary challenge in two-component systems is the risk of forming side products, where the peptide reacts with two separate linker molecules instead of creating the intended intramolecular staple. Careful positioning of the cysteines on the peptide scaffold and structural preorganization of the peptide chain has been employed to avoid intramolecular cyclization over intermolecular.

The first reported application of thiol-ene chemistry for two-component peptide stapling was demonstrated by Chou and colleagues, who utilized diene-based staples to link two cysteine residues in peptides [67]. Notably, this study showed that thiol-ene stapling could be used for macrocyclization of unmodified peptide sequences. The stapling was performed with various dienes and peptides, and the optimized conditions were the following: DMF or N-methyl-2-pyrrolidone (NMP) as a solvent, and UV light irradiation (365 nm) and DPAP as the photoinitiator. Scheme 15 shows the thiol-ene reaction on the linear peptide **37** with different dienes (**38a–f**). Cyclic compounds **39a–f** were obtained in good-excellent yields (76–89%).

To evaluate the impact of thiol-ene stapling on peptide function, Chou's team synthesized from native unprotected peptide an *i*, *i* + 4-stapled Axin mimetic, previously developed via ring-closing metathesis (RCM) by the Verdine group [68]. Circular dichroism (CD) experiments revealed similar α-helical content between the RCM-stapled Axin analog synthesized by the Verdine team and the thioether-stapled peptides. Chou's team also synthesized an *i*, *i* + 7-stapled p53 mimetic, previously obtained via RCM [69]. Even in this case, the dithioether staple variant, obtained using 1,8-nonadiene as linker, displayed comparable α-helicity to its RCM-stapled counterpart. Furthermore, enzyme-linked immunosorbent assay (ELISA) experiments assessing the inhibition of p53-MDM2 interactions confirmed that the dithioether-stapled peptide mimic of the p53 was as effective as the RCM-stapled versions, while the unstapled peptide lacked inhibitory activity. Importantly, this study highlighted that the thiol-ene stapling method is compatible with unprotected peptides and does not require nonnatural amino acids. Further development of this approach was carried out by Wang et al. [70] who adapted the thiol-ene stapling methodology for aqueous conditions by employing the water-soluble radical initiator VA-044 in conjunction with diallylurea-based water-soluble staples. When performed in water at pH 4, this method achieved 53% conversion to the stapled product.

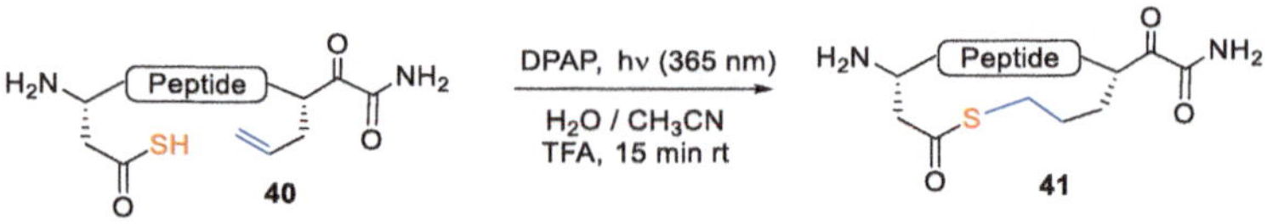

Scheme 16 Scanlan's photochemical acyl thiol-ene reaction [73]

More recently, the Brimble group [71] explored the use of divinyl diester staples in thiol-ene peptide stapling, offering an alternative to traditional hydrocarbon staples with significantly enhanced hydrophilicity. Brimble developed a stapled Axin analog based on CLipPA strategy (DPAP, TIPO, *t*-nonylSH, TFA in NMP). The i, $i + 4$ stapling was confirmed via mass analysis, and CD spectroscopy revealed an α-helical content of 89%, comparable to the 88% reported for Chou's dithioether-stapled peptide.

In 2023, Scanlan and Nielsen presented a high-throughput, nanomole-scale photochemical thiol-ene reaction system that enables rapid late-stage diversification of peptide macrocycles [72]. The thiol-ene reaction was leveraged in an automated high-throughput workflow using 1,536-well plates. A photoreactor facilitates efficient photoinitiated thiol-ene reactions with high conversions. The process employs the photoinitiator 2,2-dimethoxy-2-phenylacetophenone (DPAP), which generates volatile byproducts easily removed under vacuum, enhancing product purity. Through systematic optimization, this study demonstrates the broad applicability of thiol-ene chemistry to peptide macrocycles bearing various functional groups, including amines and carboxylic acids. This work established a scalable and efficient approach to late-stage peptide modification, advancing the application of high-throughput photochemistry in the synthesis of bioactive macrocycles. The method's compatibility with automated liquid handling and acoustic droplet ejection further underscores its potential for accelerating lead compound optimization in pharmaceutical research.

In 2024, the group of Scanlan reported a photochemical acyl thiol-ene reaction as a novel approach for the rapid cyclization of unprotected peptides containing both a thioacid and an alkene, leading to the formation of peptide thiolactones [73]. The method offers a metal-free, radical-mediated alternative to traditional peptide cyclization strategies, such as macrolactamization and thiolactone formation via activated leaving groups. The optimized reaction, using 2,2-dimethoxy-2-phenylacetophenone (DPAP) as a photoinitiator, proceeds under mild conditions with high chemoselectivity, achieving cyclization in 15 min (from **40** to **41**, Scheme 16). The study demonstrates broad substrate compatibility, including tolerance to functional groups like phenols, amines, and carboxylic acids, making it suitable for synthesizing biologically relevant thiolactone-containing peptides. This efficient methodology facilitates high-throughput library generation of cyclic peptides and offers a promising platform for drug discovery and peptide-based therapeutics.

Finally, in 2025, the group of McGouran [74] explored thiol-ene chemistry as a method for selective protein labeling, focusing on cysteine residues. Three activation methods – UV light, visible light, and redox initiation – were tested in different settings, including chemical models, recombinant proteins, and complex cell lysates. Results indicate that light-based activation, particularly with UV initiators like Irgacure 2959, is the most efficient approach and does not require an oxygen-free environment, making it more biocompatible. Redox activation using manganese acetate was also effective, though less efficient in complex mixtures. The study highlights the advantage of protein-protein interactions in enhancing reaction efficiency, as induced proximity between thiol and alkene significantly improves labeling selectivity. This work again demonstrates the broad applicability of thiol-ene coupling in protein profiling, offering a versatile tool for studying enzyme activity and biomolecular interactions.

These studies highlight the versatility of thiol-ene chemistry in two-component peptide stapling, enabling the incorporation of diverse linker functionalities, including ester, ether, and urea groups. Unlike conventional methods that rely on aliphatic staples, this approach influences both solubility and structural properties while maintaining compatibility with natural cysteine residues, eliminating the need for unnatural amino acids. The ability to operate under biocompatible conditions without harsh reagents marks a significant advancement in peptide and protein stapling, offering a powerful strategy for stabilizing α-helices, improving target binding, and enhancing peptide-based therapeutics.

3 Application to Carbohydrate Science

Carbohydrates have prominent roles in a large range of biological processes, being a major component of cell surfaces and therefore involved in both the structural integrity of organisms and the molecular recognition in various biological mechanisms [75]. Consequently, the development of glycomimetics stable to enzymatic degradation has been the focus of much research work in chemical biology and medicinal chemistry. In this context, thio-linked glycoconjugates are interesting synthetic targets offering this additional stability by replacement of the readily hydrolyzable native glycosidic bond with a thioether moiety [76, 77].

Various routes towards these valuable glycoconjugates have been developed capitalizing on the high nucleophilicity of the thiolate group [78], but the radical-mediated thiol-ene reaction remains one of the most efficient methods for both the synthesis of thioglycosides and the functionalization or bioconjugation of glycosides [8, 17, 79].

3.1 Thioglycoside Synthesis

The photoinitiated thiol-ene reaction has been applied to the construction of thiosugar derivatives via reactions involving different types of alkene-containing glycosides, i.e., enoses. *Endo*- and *exo*-glycals are among the most described enose substrates in thioglycoside synthesis. They have been extensively studied as thiol-ene acceptors and a systematic evaluation of the impact of their structure, as well as the temperature and the nature of the thiol counterpart, on the efficacy and the stereochemical outcome of the reaction have been performed throughout the years [17].

Initial work on this field was published in the early 1970s by Honma et al. [80] and Kushida et al. [81] (Scheme 16). Honma reported that the addition of thioacetic acid on glycal resulted in a complete regioselectivity at the C2 position; however, the diastereoselectivity was rather low (Scheme 17a). Kushida demonstrated that an acetoxy substitution on the C2 position of the glycal starting material resulted in a complete inversion of the regioselectivity of the thiol addition towards the anomeric thiosugar. Interestingly, complete alpha selectivity was obtained in this case (Scheme 17b).

More recent (2012) work in this field was mainly carried out by the group of Dondoni and the group of Borbás. The initial observation in terms of regio- and stereoselectivity was confirmed via the use of sugar thiols towards the synthesis of disaccharide. The use of glycals bearing no substituent on the alkene moiety resulted in exclusive C2 addition of sugar thiol upon photoinitiated thiol-ene reaction. Interestingly, the diastereoselectivity was reported to be highly dependent on the initial configuration of the C3 and C4 carbon centers of the glycal allowing to reach full diastereoselectivity in the best scenario (Scheme 18a) [82]. The use of substituted glycals with an acetoxy group in the C2 position allowed the formation of a new thioglycosidic bond with complete alpha selectivity. This stereoselectivity was confirmed with multiple thiol partners including thiopyranose, thiofuranose, 2-aminothiosugar, but also cysteinyl derivatives (Scheme 18b) [83]. More recently, the complete regio- and stereoselectivity of the TEC reaction on glycal was exploited to build thiooligosaccharide via the use of 1,4-dithiopyranosides and 4-thioglycals

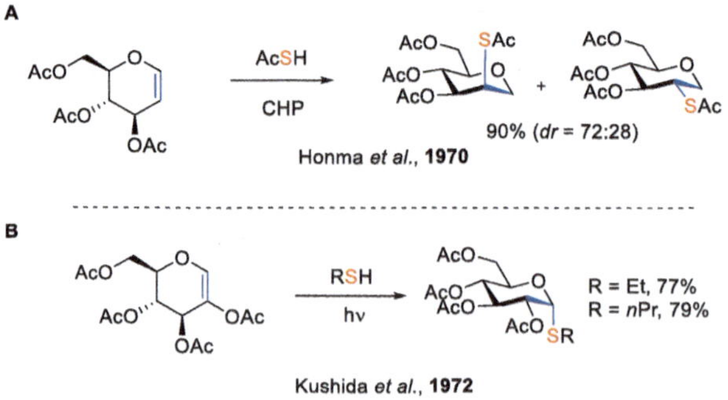

Scheme 17 Initial work on thiol-ene coupling using enose acceptors. (**a**) Honma's works [80] and (**b**) Kushida's works [81]

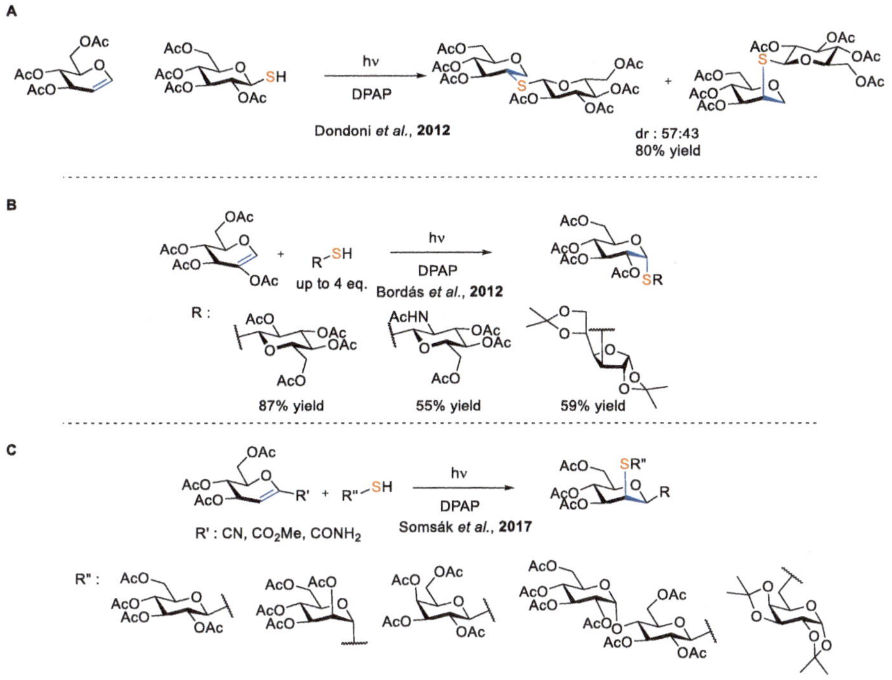

Scheme 18 The effect of glycal's substituents on the outcome of the reaction. (**a**) Unsubstituted glycals [82]; (**b**) 2-substituted glycals [83]; (**c**) 1-substituted glycals [85]

that are prepared very efficiently from the same starting material [84]. The thiol-ene approach, in this case, allows to simplify the protecting group strategy for a straightforward route towards the pentasaccharide with a complete selectivity for the α-S-linkage.

C1-substituted glycals have also been studied as acceptor for the free radical thiol-ene coupling. The addition revealed a complete regioselectivity on the less hindered C2 position as well as a complete diastereoselectivity towards the axial linkage via a preferred β-side approach of the thiol (Scheme 18c) [85]. Borbás and co-workers have also demonstrated the crucial effect of the temperature of the thiol-ene coupling on glycal acceptors [86, 87]. Cooling the reaction media resulted in an important increase of the conversion that has been related to the displacement of the propagation equilibrium towards the carboradical intermediate and therefore upon chain transfer towards the final product. It was also observed that low temperature would tend to decrease the premature formation of disulfide by quenching the thiyl radicals and thus allowed the use of less equivalent of thiol-ene donors (Scheme 19).

Other types of enoses have been used to exemplify this reactivity, expanding the scope of glycoside derivatives that can be accessed via the TEC reaction. The reaction of diverse thiols on 2,3-unsaturated glycosides was reported to proceed with high yield at low temperature. Interestingly, the regio- and stereoselectivity of

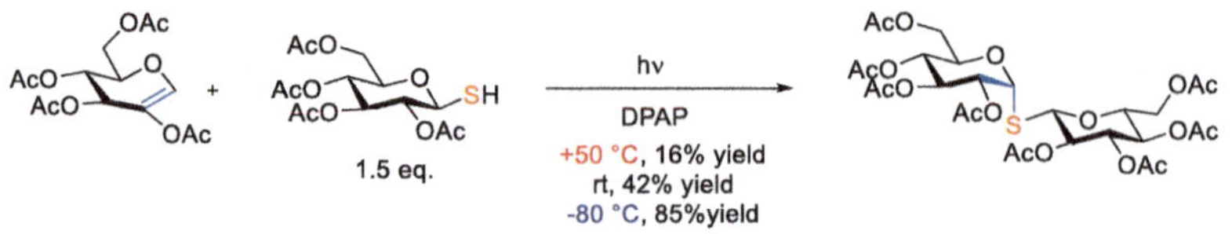

Scheme 19 The effect of the temperature of the thiol-ene coupling [86, 87]

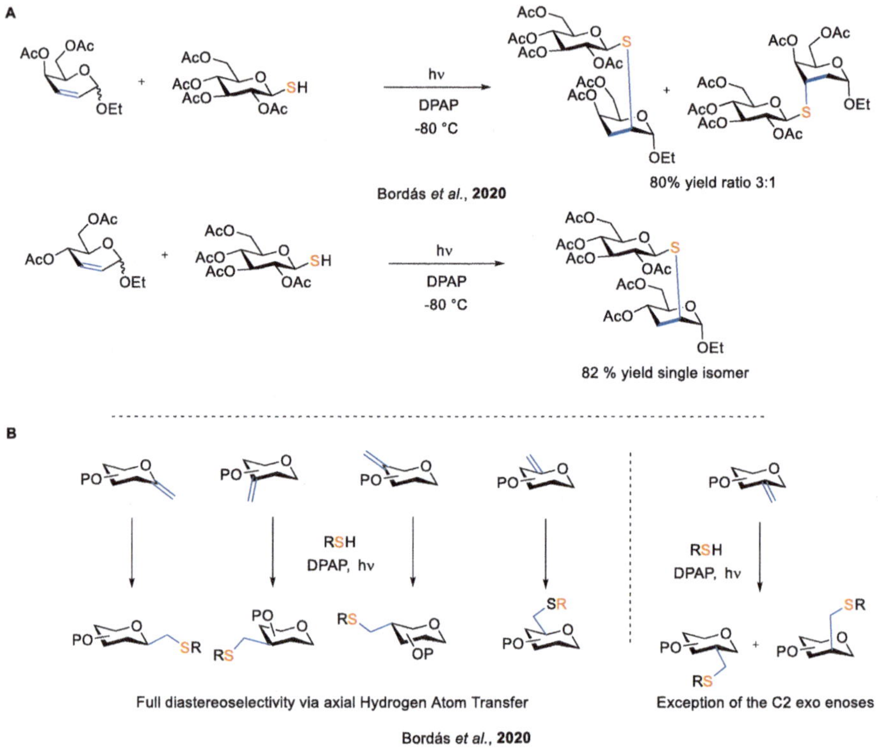

Scheme 20 Reactivity of enoses other than endo glycals. (**a**) 2,3-Endoenoses [83, 88] and (**b**) exoenoses [89]

the addition was highly dependent on the stereochemistry of the thiol-ene donor and thiol-ene acceptor glycosides involved in the coupling. General rules can be identified for the control of the selectivity of the TEC reaction in this specific case: the thiol attack is favored on the less substituted and less hindered position of the alkene, and the axial approach is predominant, resulting in high diastereoselectivities (Scheme 20a) [83, 88].

Exo enoses have also been studied as thiol-ene acceptors to create a carbon-sulfur bridge between two glycosidic residues and therefore enhance the resistance to enzymatic hydrolysis of the so-formed glycomimetics (Scheme 20b) [89–92]. In the specific example of *exo*-enose with the double bond in the pseudo-anomeric position (exo-glycals), the thiol-ene addition resulted in a complete selectivity for the β-configuration at the pseudo-anomeric position. It is assumed that the carbon-radical intermediate formed after the attack of the thiyl radical is stabilized by SOMO-O_n orbital overlap, thus favoring the axial S-H abstraction. Interestingly, this selectivity was reproduced with substrates bearing an exo double bond on the C3, C4, and C5 but the C2-substituted enose provided a mixture of epimers upon TEC reaction with diverse thiols [89]. This result was rationalized by the coexistence of two conformers of the carbon-radical intermediate: the chair conformer subjected to an axial S-H abstraction and providing the stereoisomer with an equatorial glycosidic connection, and the half-chair conformer subjected to an axial S-H abstraction either on the top or on the bottom face and therefore providing a 1:1 ratio of axial:equatorial glycosidic connection. Similar observations were made on exo cyclic alkene thiol-ene acceptors such as unsaturated nucleosides in which stereochemical outcome of the addition has been connected to the relative stability of the different conformers of the carbon-radical intermediate [93]. This last example of TEC reaction involving nucleosides was successfully applied to the development of novel compounds exhibiting cytostatic activities against cancer cell lines [94].

3.2 *Glycoside Bioconjugates*

The thiol-ene reaction has proved to be particularly suited to the synthesis of thiosaccharides with controlled regio- and stereoselectivity. Consequently, the free radical methodology was applied to the functionalization of glycosides in order to build bioconjugates with therapeutic activities. As previously stated, carbohydrates are involved in a large number of biological processes making their bioconjugates valuable candidates for the development of novel active molecules for therapeutics or diagnosis. Among those bioconjugates, glycopeptides and glycoproteins have already been discussed above. However, many other thiol-ene-built glycoconjugates have been studied.

Borbás and co-workers have exploited "click" chemistry in an optimal manner, associating the free radical thiol-ene ligation and the Huisgen cycloaddition to build hydrolytically stable S-linked mannobioside head ligands linked to C60 fullerene nanoparticles or calix[4]arenes. These so-formed "mannoclusters" revealed an important activity in the inhibition of mannose-binding lectins paving the way to the development of antiviral and antibacterial agents (Scheme 21) [95, 96].

The same group later developed two new S-linked sugar-nucleotide analogs, potential inhibitors of glycosyltransferases. The synthetic strategy relied entirely on thiol-ene ligations using either a 1,2-ethanedithiol linker or an allylamine linker to achieve molecular complexity in a fast and straightforward way [97]. The use of thiol-ene ligation for the incorporation of S-linked spacers is indeed a frequent strategy in the construction of glyco-bioconjugates. Klaffke and coworkers introduced for instance a cysteamine linker in order to build a series of spacer bridged carbohydrates ready to be conjugated to peptides under enzymatic catalysis [49]. On

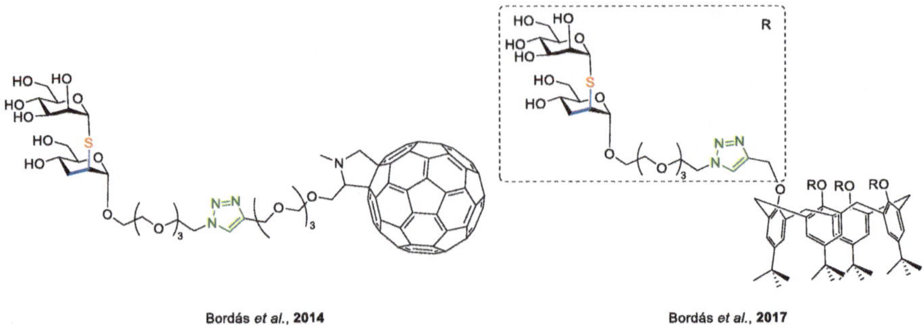

Scheme 21 Synergistic use of "click" chemistry for the synthesis of "mannoclusters" [95, 96]

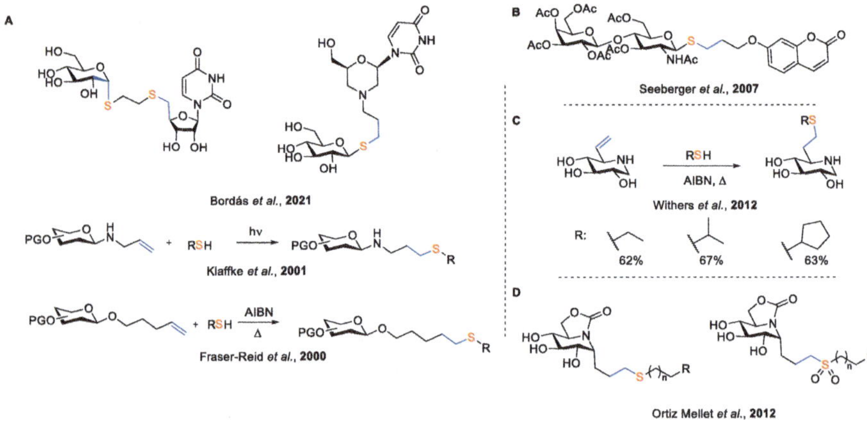

Scheme 22 Spanning the diversity of thiol-ene glycoconjugates. (**a**) Illustration of the use of thiol-ene spacers [49, 98]; (**b**) aglycon conjugates [99]; (**c**) tuning the lipophilicity of iminosugars [100]; (**d**) examples of glycolipids [101]

the same topic, the group of Fraser-Reid proposed a complete study of various spacers introduced via thiol-ene ligation to build neo-glycoconjugates seeking broad applications for the study of biological processes (Scheme 22a) [98].

Seeberger and colleagues utilized thiol-ene ligation to attach a 3-mercaptopropyl linker to the reducing end of carbohydrates using anomeric glycosyl thiols. They demonstrated efficient TEC conjugation of thioglucose and thioglucosamine with various allylic aglycones, followed by enzymatic glycosylation to produce disaccharide-linker-aglycon conjugates (Scheme 22b) [99]. Meanwhile, Withers and his team [100] applied TEC to a D-xylose-derived iminosugar, reacting it with different thiols to create a library of lipophilic iminosugars for Gaucher disease treatment. By modifying the thiol component, they adjusted the hydrophobicity of these biological chaperones (Scheme 21c).

Glycolipids have also been developed using the photoinitiated thiol-ene coupling. Ortiz Mellet and co-workers demonstrated that iminosugar-derived glycolipids can

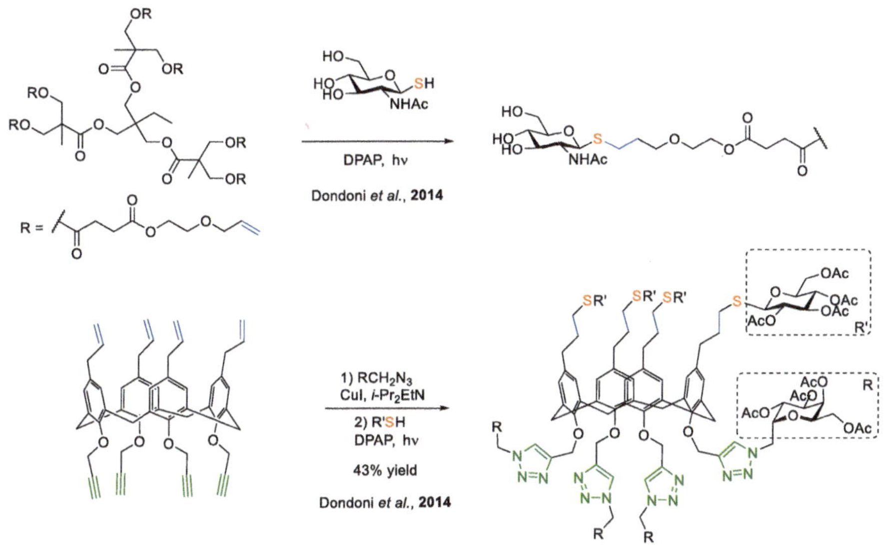

Scheme 23 Examples of glycodendrimers and glycoclusters by Dondoni and co-workers [104, 105]

be readily obtained via stereoselective C-allylation followed by "click" thiol-ene coupling of a variety of thiols with lipidic tails. The obtained glycoconjugates were successfully tested in the inhibition of glycosidases (Scheme 22d) [101].

Ever since the conceptualization of the multivalence effect in modern glycobiology in which a multivalent ligand displays a greater affinity as a whole than the sum of its part, chemists have been focusing on the development of new glycoclusters to study carbohydrate-protein interactions and elaborate novel inhibitors of these interactions [102]. Dendrimers and cyclodextrins have been two types of scaffolds for which the radical-mediated thiolation has been of great use for the development of glycoclusters. Dondoni and Mara reported the complete thioglycosylation of dendrimers bearing up to 48 alkene groups via an efficient photochemically induced thiol-ene reaction [103, 104]. The authors revealed that many of these glycodendrimers were good ligands of the wheat germ agglutinin showing relative potency per sugar units about 10^6 times higher than the monosaccharidic ligand. The same group reported the dual use of thiol-ene "click" reaction and Huisgen cycloaddition "click" reaction to build a glycocluster based on calix[4]arene clusters and differently functionalized on the upper and lower faces of the structure (Scheme 23) [105]. Later, the team of Mara applied the same thiol-ene conditions for the first ligation of imino-disaccharides on calix[4]arenes paving the way to the discovery of new glycosidase and glycosyltransferase inhibitors [106].

In the family of glycoclusters, cyclodextrin scaffolds present a better biocompatibility and a higher water solubility than the calix[4]arene scaffold. Stoddart and co-workers have successfully used the photoaddition of thiols to alkenes in order to build functionalized β-cyclodextrins with thiosugars. The authors reported the

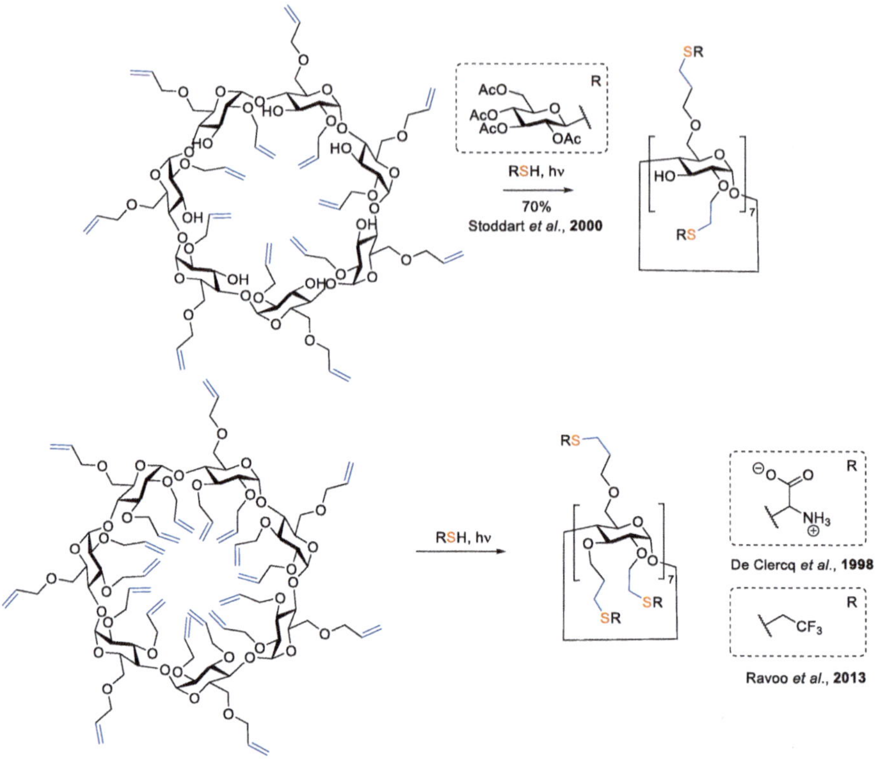

Scheme 24 Thiol-ene-mediated functionalization of cyclodextrins [107–110]

selective functionalization of the primary and the secondary faces of the scaffold as well as both faces together with excellent yields (Scheme 24) [107, 108]. Other groups have been taking advantage of the efficiency of the TEC reaction for the development of functionalized cyclodextrins. The use of cysteines as thiol-ene donors allowed De Clercq and co-workers to build polyzwitterionic structures with anti-HIV activity [109].

On another hand, Ravoo and co-workers proposed the synthesis of fluorinated cyclodextrins in quantitative yields with both 2,2,2-trifluoroethanethiol and perfluoroalkylthiols. This strategy allowed to tune the solubility of the structure and enable their use as contrast agent in ^{19}F-MRI (magnetic resonance imaging, Scheme 23) [110].

Altogether, in addition to the thiol-ene-mediated glycosylation of peptides and proteins that have been treated in the previous part of this chapter, it is clear that the TEC reaction is particularly suited to carbohydrate chemistry. In that field of research, the methodology not only proved to be an efficient manner to conjugate complex glycosyl structure but revealed a high regio- and stereoselectivity which is not a given for free radical processes.

4 Application to Biomaterials: Biomedical Hydrogels

Hydrogels are cross-linked polymer networks that are able to retain large amounts of water. They are mostly used for biomedical application such as tissue regeneration or drug carriers able to control the delivery of pharmaceutics [111]. Indeed, this class of biomaterials demonstrates a high efficiency in terms of cargo loading and biocompatibility. Most developed hydrogels are loaded with therapeutically active molecules by physical encapsulation as it prevents any structural modification of the bioactive cargo compared to covalent fixation. Light-driven processes offer precise control over the initiation, speed, and location of the hydrogel formation by controlling the application of light and therefore are particularly attractive for biomedical applications [9]. In this context, the radical-mediated thiol-ene ligation has unsurprisingly been identified as a valuable strategy to prepare hydrogels by covalent cross-linking of hydrophilic polymers containing thiols and alkenes (Scheme 25) [112].

A broad diversity of thiol-ene hydrogels was reported to date, mostly relying on the use of biopolymers or biocompatible polymers that are pre-functionalized with suitable functional groups. This initial step usually takes advantage of the chemical functionalities that are naturally present on biopolymers such as primary amines (chitosan, collagen), carboxylic acids (hyaluronic acid, heparin), or hydroxyl group (collagen, dextran, cyclodextrin) to introduce via chemical modifications the desired alkenes and thiols. As far as thiol functionalization is concerned, it was described via the coupling of a cysteine or a cysteamine residue [113, 114] or via the reaction of 2-iminothiolane on a primary amine [115], but also via the coupling of thiopropionic acid [116] or a thiourea on a primary hydroxyl group [117] (Scheme 26). Some thiols containing macromers were found to be commercially available, such as hyaluronic acids (HS-HA) [118], or biocompatible synthetic polymers such as thiolated poly (ethylene glycol) [119].

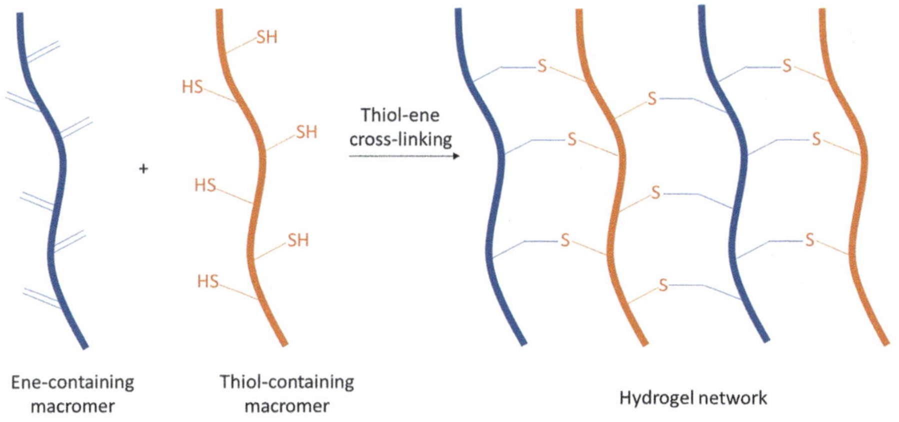

Scheme 25 Cross-linking strategy involving ene macromers and thiol macromers

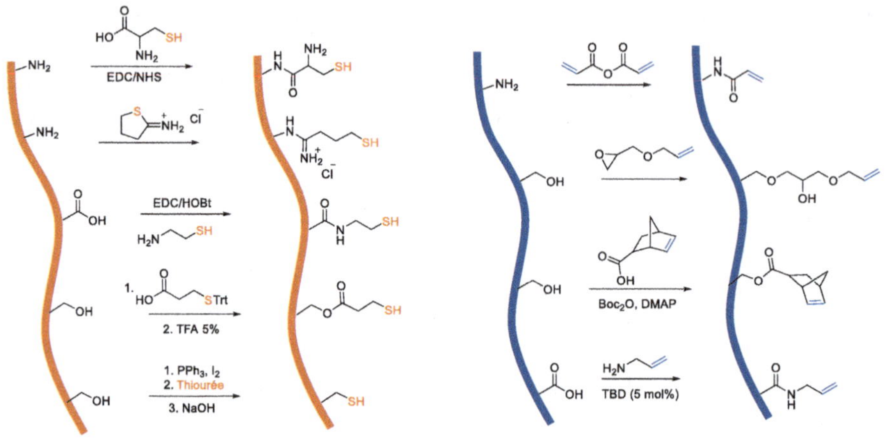

Scheme 26 Pre-functionalization strategies for thiol-containing macromers (orange) and alkene-containing macromers (blue)

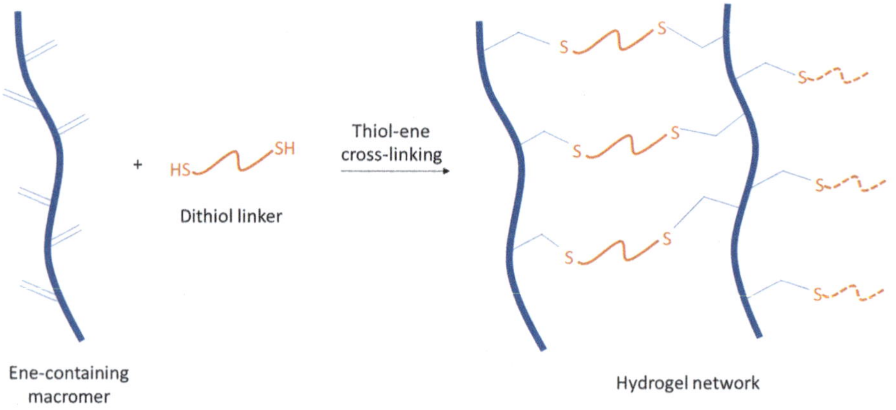

Scheme 27 Alternative cross-linking strategy involving dithiol spacer

Concerning the alkene functionality, it has been introduced as an acrylate [113, 118] or an allyl group [117, 120, 121] offering a reactive terminal double bond, but also as a norbornene [115, 122] which allowed to explore visible light-mediated hydrogel formation, improving the cytocompatibility of the process compared to the use of UV light (Scheme 26).

Another strategy of thiol-ene-mediated hydrogelation consists of the use of a simple dithiol cross-linker, such as dithiothreitol (DTT), to connect two alkene macromers with each other. It allows considerable simplification of the synthesis as only one macromer needs to be designed and prepared (Scheme 27) [123].

The thiol-ene reaction not only enables rapid and controlled in situ hydrogel formation but also allows for further covalent functionalization of the biomaterial to fine-tune its physical properties. This approach has been demonstrated in various studies, including the work of Burdick’s group on thiol-norbornene chemistry and a recent example by Hoogenboom and colleagues, who developed hydrogels based on a synthetic platform of poly(N-allyl acrylamide) [121, 124].

Altogether, from a synthetic perspective, the thiol-ene reaction offers an ideal solution for cross-linking biopolymers with limited functional modification requirements that can be achieved straightforwardly. The strategy also enables good cytocompatibility while proposing a controlled solution for encapsulation of a bioactive cargo [9].

Thiol-ene-based hydrogels have been used for very diverse biomedical applications throughout the years as highlighted in a very comprehensive manner in recent review articles [112, 122, 125]. The example reported by Hughes and co-workers in 2018 is particularly interesting as it offers a complete study from macromer preparation to hydrogel construction and clinical evaluation [113]. A photo-cross-linking strategy was developed to generate a gelatin-based hydrogel able to repair corneal wounds. Lysine side chains of gelatin were pre-functionalized with cysteine on the one hand and acrylate on the other hand to constitute the hydrogel precursors. Local UV irradiation allowed precise targeting of the wound location for in situ hydrogel formation with no detrimental effect on the cornea. The formed hydrogel demonstrated great transparency, high cell viability, and biocompatibility towards rabbit cornea and good biodegradability thanks to the presence of collagenase in tears. The authors claimed that this noninvasive strategy for corneal wound repair could become an attractive alternative to cornea implants.

More recently, a study from Li, Pu, and co-workers highlighted the development of a pH-responsive thiol-ene-based hydrogel allowing active modulation of drug release properties [120]. The authors developed a modified chitosan macromer bearing O-allyl groups for cross-linkage and preserving free primary amine group for pH response. Using thiolated PEG cross-linkers, the photoinitiated thiol-ene reaction quickly resulted in the hydrogel formation. The biomaterial exhibited pH-responsive behavior in terms of swelling properties. At basic pH, the authors observed a shrinking of the gel compared to acidic pH. Encapsulation of doxorubicin via in situ gel formation allowed monitoring the drug’s release in the function of the pH. A dramatic acceleration of the doxorubicin release was observed at basic pH, while the gel shrinks, compared to neutral or acidic behavior. Opposite behavior was observed replacing doxorubicin with bovine albumin serum, providing a drug carrier with active modulating opposite drug release behavior.

Nowadays, the hydrogel research area is experiencing an important breakthrough with the introduction of 3D printing technologies [126]. 3D bioprinting allows precise deposition of a biomaterial (bioink) in order to form finely tuned structures comparable to tissue. Considering the fast and reproducible nature of the thiol-ene photo-cross-linking, it was identified as a great strategy to formulate novel bioinks for 3D bioprinting potentially addressing identified challenges related to the mechanical and rheological properties of the material. Such bioprinted tissues offer significant potential for personalized medicine and biomedical applications and quite surely are the promise of a bright future for thiol-ene cross-linking technologies.

5 Conclusion

The photoinitiated radical thiol-ene "click" reaction offers the ability to selectively react a thiol and an alkene in the presence of mixtures of nucleophiles and electrophiles that can be found in natural media. Although initially applied to polymer science, recent years have experienced the fast development of thiol-ene-mediated strategies for the synthesis and modification of biomolecules.

The methodology enables high tolerance to aqueous conditions, preserving its "click" properties such as high speed, high yield, and high atom economy and avoiding the use of potentially toxic metal catalysts. Consequently, a broad list of bioconjugates has been developed based on the thiol-ene reaction as depicted by the representative sample that was detailed in this chapter. The methodology is applicable to the glycosylation and lipidation of peptides, the development of stapled peptidic architectures and cyclic peptides, the design of multivalent glycoconjugates, and the conception of biomaterials with controlled mechanical and rheological properties.

The photoinitiated character of the thiol-ene "click" reaction confers to the methodology an "on/off" ability allowing to perform ligation in situ and to precisely control the localization of the reaction. If technical issues have been associated with the photochemical nature of the reaction, especially for nonspecialist research teams, an in-depth understanding of the kinetics and thermodynamics of the process and accessibility of lab equipment related to photochemistry can contribute to the fast acceleration of the development of thiol-ene-mediated synthetic strategies. We can confidently expect further development in the future, especially using continuous flow technologies that have already proved to be very efficient for enabling more efficient and sustainable photochemical reactions [127].

References

1. Posner T (1905) Beiträge zur Kenntniss der ungesättigten Verbindungen. II. Ueber die Addition von Mercaptanen an ungesättigte Kohlenwasserstoffe. Berichte Dtsch Chem Ges 38:646–657. https://doi.org/10.1002/cber.190503801106
2. Kolb HC, Finn MG, Sharpless KB (2001) Click chemistry: diverse chemical function from a few good reactions. Angew Chem Int Ed 40:2004–2021. https://doi.org/10.1002/1521-3773(20010601)40:11<2004::AID-ANIE2004>3.0.CO;2-5
3. Hoyle CE, Lowe AB, Bowman CN (2010) Thiol-click chemistry: a multifaceted toolbox for small molecule and polymer synthesis. Chem Soc Rev 39:1355–1387. https://doi.org/10.1039/B901979K
4. Nair DP, Podgórski M, Chatani S, Gong T, Xi W, Fenoli CR, Bowman CN (2014) The thiol-Michael addition click reaction: a powerful and widely used tool in materials chemistry. Chem Mater 26:724–744. https://doi.org/10.1021/cm402180t
5. Dondoni A, Marra A (2013) Free-radical thiol-ene and thiol-yne couplings as click processes for glycoconjugation. In: Click chemistry in glycoscience. John Wiley & Sons, Ltd., pp 45–75
6. Massi A, Nanni D (2012) Thiol-yne coupling: revisiting old concepts as a breakthrough for up-to-date applications. Org Biomol Chem 10:3791–3807. https://doi.org/10.1039/C2OB25217A

7. Lowe AB, Hoyle CE, Bowman CN (2010) Thiol-yne click chemistry: a powerful and versatile methodology for materials synthesis. J Mater Chem 20:4745–4750. https://doi.org/10.1039/B917102A
8. Dondoni A, Marra A (2012) Recent applications of thiol-ene coupling as a click process for glycoconjugation. Chem Soc Rev 41:573–586. https://doi.org/10.1039/C1CS15157F
9. Kharkar PM, Rehmann MS, Skeens KM, Maverakis E, Kloxin AM (2016) Thiol-ene click hydrogels for therapeutic delivery. ACS Biomater Sci Eng 2:165–179. https://doi.org/10.1021/acsbiomaterials.5b00420
10. Nolan MD, Scanlan EM (2020) Applications of thiol-ene chemistry for peptide science. Front Chem 8:583272. https://doi.org/10.3389/fchem.2020.583272
11. Kharash MS, Read J, Mayo FR (1938) J Soc Chem Ind 57:752–754. https://doi.org/10.1002/jctb.5000573206
12. Dénès F, Pichowicz M, Povie G, Renaud P (2014) Thiyl radicals in organic synthesis. Chem Rev 114:2587–2693. https://doi.org/10.1021/cr400441m
13. Hoyle CE, Lee TY, Roper T (2004) Thiol-enes: chemistry of the past with promise for the future. J Polym Sci A Polym Chem 42:5301–5338. https://doi.org/10.1002/pola.20366
14. Cramer NB, Reddy SK, O'Brien AK, Bowman CN (2003) Thiol-ene photopolymerization mechanism and rate limiting step changes for various vinyl functional group chemistries. Macromolecules 36:7964–7969. https://doi.org/10.1021/ma034667s
15. Fındık V, Degirmenci I, Çatak Ş, Aviyente V (2019) Theoretical investigation of thiol-ene click reactions: a DFT perspective. Eur Polym J 110:211–220. https://doi.org/10.1016/j.eurpolymj.2018.11.030
16. Griesbaum K (1970) Problems and possibilities of the free-radical addition of thiols to unsaturated compounds. Angew Chem Int Ed Engl 9:273–287. https://doi.org/10.1002/anie.197002731
17. Borbás A (2020) Photoinitiated thiol-ene reactions of enoses: a powerful tool for stereoselective synthesis of glycomimetics with challenging glycosidic linkages. Chem Eur J 26:6090–6101. https://doi.org/10.1002/chem.201905408
18. Morgan CR, Magnotta F, Ketley AD (1977) Thiol/ene photocurable polymers. J Polym Sci Polym Chem Ed 15:627–645. https://doi.org/10.1002/pol.1977.170150311
19. Bird RE, Lemmel SA, Yu X, Zhou QA (2021) Bioorthogonal chemistry and its applications. Bioconjug Chem 32:2457–2479. https://doi.org/10.1021/acs.bioconjchem.1c00461
20. Miseta A, Csutora P (2000) Relationship between the occurrence of cysteine in proteins and the complexity of organisms. Mol Biol Evol 17:1232–1239. https://doi.org/10.1093/oxfordjournals.molbev.a026406
21. Triola G, Brunsveld L, Waldmann H (2008) Racemization-free synthesis of S-alkylated cysteines via thiol-ene reaction. J Org Chem 73:3646–3649. https://doi.org/10.1021/jo800198s
22. Wang L, Wang N, Zhang W, Cheng X, Yan Z, Shao G, Wang X, Wang R, Fu C (2022) Therapeutic peptides: current applications and future directions. Signal Transduct Target Ther 7:1–27. https://doi.org/10.1038/s41392-022-00904-4
23. Otvos L, Wade JD (2014) Current challenges in peptide-based drug discovery. Front Chem 2:00062. https://doi.org/10.3389/fchem.2014.00062
24. Lamers C (2022) Overcoming the shortcomings of peptide-based therapeutics. Future Drug Discov 4:FDD75. https://doi.org/10.4155/fdd-2022-0005
25. Griffin BA, Adams SR, Tsien RY (1998) Specific covalent labeling of recombinant protein molecules inside live cells. Science 281:269–272. https://doi.org/10.1126/science.281.5374.269
26. Zalipsky S (1995) Functionalized poly(ethylene glycols) for preparation of biologically relevant conjugates. Bioconjug Chem 6:150–165. https://doi.org/10.1021/bc00032a002
27. Rao J, Dragulescu-Andrasi A, Yao H (2007) Fluorescence imaging *in vivo*: recent advances. Curr Opin Biotechnol 18:17–25. https://doi.org/10.1016/j.copbio.2007.01.003

28. Ahangarpour M, Kavianinia I, Harris PWR, Brimble MA (2021) Photo-induced radical thiol-ene chemistry: a versatile toolbox for peptide-based drug design. Chem Soc Rev 50:898–944. https://doi.org/10.1039/D0CS00354A
29. Li Y, Yang M, Huang Y, Song X, Liu L, Chen PR (2012) Genetically encoded alkenyl-pyrrolysine analogues for thiol-ene reaction mediated site-specific protein labeling. Chem Sci 3:2766–2770. https://doi.org/10.1039/C2SC20433A
30. Tate EW, Soday L, de la Lastra AL, Wang M, Lin H (2024) Protein lipidation in cancer: mechanisms, dysregulation and emerging drug targets. Nat Rev Cancer 24:240–260. https://doi.org/10.1038/s41568-024-00666-x
31. Vicente-Garcia C, Colomer I (2023) Lipopeptides as tools in catalysis, supramolecular, materials and medicinal chemistry. Nat Rev Chem 7:710–731. https://doi.org/10.1038/s41570-023-00532-8
32. Mukherjee S, Dey R, Acharya Y, Haldar J (2023) Chapter 10: Antimicrobial lipopeptides: multifaceted designs to curb antimicrobial resistance. In: Dhara AK, Nayak AK, Chattopadhyay D (eds) Antibiotics – therapeutic spectrum and limitations. Academic Press, pp 203–232
33. Wright TH, Brooks AES, Didsbury AJ, Williams GM, Harris PWR, Dunbar PR, Brimble MA (2013) Direct peptide lipidation through thiol-ene coupling enables rapid synthesis and evaluation of self-adjuvanting vaccine candidates. Angew Chem Int Ed 52:10616–10619. https://doi.org/10.1002/anie.201305620
34. Yang S-H, Harris PWR, Williams GM, Brimble MA (2016) Lipidation of cysteine or cysteine-containing peptides using the thiol-ene reaction (CLipPA). Eur J Org Chem 2016:2608–2616. https://doi.org/10.1002/ejoc.201501375
35. Williams ET, PWR H, Jamaluddin MA, Loomes KM, Hay DL, Brimble MA (2018) Solid-phase thiol-ene lipidation of peptides for the synthesis of a potent CGRP receptor antagonist. Angew Chem Int Ed 57:11640–11643. https://doi.org/10.1002/anie.201805208
36. Yang S-H, Hermant YOJ, Harris PWR, Brimble MA (2020) Replacement of the acrid tert-butylthiol and an improved isolation protocol for cysteine lipidation on a peptide or amino acid (CLipPA). Eur J Org Chem 2020:944–947. https://doi.org/10.1002/ejoc.201901696
37. Yim VV, Kavianinia I, Cameron AJ, Harris PWR, Brimble MA (2020) Direct synthesis of cyclic lipopeptides using intramolecular native chemical ligation and thiol-ene CLipPA chemistry. Org Biomol Chem 18:2838–2844. https://doi.org/10.1039/D0OB00203H
38. Yim V, Kavianinia I, Knottenbelt MK, Ferguson SA, Cook GM, Swift S, Chakraborty A, Allison JR, Cameron AJ, Harris PWR, Brimble MA (2020) "CLipP"ing on lipids to generate antibacterial lipopeptides. Chem Sci 11:5759–5765. https://doi.org/10.1039/D0SC01814G
39. Tong JTW, Kavianinia I, Ferguson SA, Cook GM, Harris PWR, Brimble MA (2020) Synthesis of paenipeptin C′ analogues employing solution-phase CLipPA chemistry. Org Biomol Chem 18:4381–4385. https://doi.org/10.1039/D0OB00950D
40. Yang S-H, Clemett CA, Brimble MA, O'Carroll SJ, Harris PWR (2020) Synthesis and biological evaluation of S-lipidated lipopeptides of a connexin 43 channel inhibitory peptide. RSC Med Chem 11:1041–1047. https://doi.org/10.1039/D0MD00172D
41. Yim VV, Cameron AJ, Kavianinia I, Harris PWR, Brimble MA (2020) Thiol-ene enabled chemical synthesis of truncated s-lipidated teixobactin analogs. Front Chem 8:00568. https://doi.org/10.3389/fchem.2020.00568
42. Shepperson OA, Cameron AJ, Wang CJ, Harris PWR, Taylor JA, Brimble MA (2021) Thiol-ene enabled preparation of S-lipidated anti-HBV peptides. Org Biomol Chem 19:220–232. https://doi.org/10.1039/D0OB01997F
43. Hermant YO, Cameron AJ, Harris PWR, Brimble MA (2020) Synthesis of antimicrobial lipopeptides using the "CLipPA" thiol-ene reaction. Methods Mol Biol (Clifton, NJ) 2103: 263–274. https://doi.org/10.1007/978-1-0716-0227-0_18
44. Yim V, Hermant YO, Harris PWR, Brimble MA (2021) Synthesis of lipopeptides by CLipPA chemistry. In: Hussein WM, Stephenson RJ, Toth I (eds) Peptide conjugation: methods and protocols. Springer, New York, pp 253–263

45. He M, Zhou X, Wang X (2024) Glycosylation: mechanisms, biological functions and clinical implications. Signal Transduct Target Ther 9:1–33. https://doi.org/10.1038/s41392-024-01886-1
46. Seeberger PH, Freedberg DI, Cummings RD (2022) Glycans in biotechnology and the pharmaceutical industry. In: Varki A, Cummings RD, Esko JD, Stanley P, Hart GW, Aebi M, Mohnen D, Kinoshita T, Packer NH, Prestegard JH, Schnaar RL, Seeberger PH (eds) Essentials of glycobiology, 4th edn. Cold Spring Harbor Laboratory Press, Cold Spring Harbor
47. Doelman W, van Kasteren SI (2022) Synthesis of glycopeptides and glycopeptide conjugates. Org Biomol Chem 20:6487–6507. https://doi.org/10.1039/D2OB00829G
48. Zhuang H, Hao H, Qiu Y, Gan J, Li T (2023) Recent advances in glycopeptide and glycoprotein synthesis: a refined synthetic probe towards the biological world. Chin J Chem 41: 2010–2024. https://doi.org/10.1002/cjoc.202300035
49. Ramos D, Rollin P, Klaffke W (2001) Chemoenzymatic synthesis of neoglycopeptides: application to an α-gal-terminated neoglycopeptide. J Org Chem 66:2948–2956. https://doi.org/10.1021/jo001439c
50. Wittrock S, Becker T, Kunz H (2007) Synthetic vaccines of tumor-associated glycopeptide antigens by immune-compatible thioether linkage to bovine serum albumin. Angew Chem Int Ed 46:5226–5230. https://doi.org/10.1002/anie.200700964
51. Dondoni A, Massi A, Nanni P, Roda A (2009) A new ligation strategy for peptide and protein glycosylation: photoinduced thiol-ene coupling. Chem Eur J 15:11444–11449. https://doi.org/10.1002/chem.200901746
52. Floyd N, Vijayakrishnan B, Koeppe JR, Davis BG (2009) Thiyl glycosylation of olefinic proteins: S-linked glycoconjugate synthesis. Angew Chem Int Ed 48:7798–7802. https://doi.org/10.1002/anie.200903135
53. González IR, McLean JT, Ostrovitsa N, Fitzgerald S, Mezzetta A, Guazzelli L, O'Shea DF, Scanlan EM (2024) A thiol-ene mediated approach for peptide bioconjugation using 'green' solvents under continuous flow. Org Biomol Chem 22:2203–2210. https://doi.org/10.1039/D4OB00122B
54. Pereira AJ, de Campos LJ, Xing H, Conda-Sheridan M (2024) Peptide-based therapeutics: challenges and solutions. Med Chem Res 33:1275–1280. https://doi.org/10.1007/s00044-024-03269-1
55. Errington N, Iqbalsyah T, Doig AJ (2006) Structure and stability of the α-helix. In: Guerois R, de la Paz ML (eds) Protein design: methods and applications. Humana Press, Totowa, pp 3–26
56. Moiola M, Memeo MG, Quadrelli P (2019) Stapled peptides – a useful improvement for peptide-based drugs. Molecules 24:3654. https://doi.org/10.3390/molecules24203654
57. Chandramohan A, Josien H, Yuen TY, Duggal R, Spiegelberg D, Yan L, Juang Y-CA, Ge L, Aronica PG, Kaan HYK, Lim YH, Peier A, Sherborne B, Hochman J, Lin S, Biswas K, Nestor M, Verma CS, Lane DP, Sawyer TK, Garbaccio R, Henry B, Kannan S, Brown CJ, Johannes CW, Partridge AW (2024) Design-rules for stapled peptides with in vivo activity and their application to Mdm2/X antagonists. Nat Commun 15:489. https://doi.org/10.1038/s41467-023-43346-4
58. Li Y, Wu M, Fu Y, Xue J, Yuan F, Qu T, Rissanou AN, Wang Y, Li X, Hu H (2024) Therapeutic stapled peptides: efficacy and molecular targets. Pharmacol Res 203:107137. https://doi.org/10.1016/j.phrs.2024.107137
59. Ji X, Nielsen AL, Heinis C (2024) Cyclic peptides for drug development. Angew Chem Int Ed 63:e202308251. https://doi.org/10.1002/anie.202308251
60. Buchanan D, Mori S, Chadli A, Panda SS (2025) Natural cyclic peptides: synthetic strategies and biomedical applications. Biomedicine 13:240. https://doi.org/10.3390/biomedicines13010240
61. Zhan W, Duan H, Li C (2024) Recent advances in metal-free peptide stapling strategies. Chem Bio Eng 1:593–605. https://doi.org/10.1021/cbe.3c00123

62. Aimetti AA, Shoemaker RK, Lin C-C, Anseth KS (2010) On-resin peptide macrocyclization using thiol-ene click chemistry. Chem Commun 46:4061–4063. https://doi.org/10.1039/C001375G
63. LeValley PJ, Ovadia EM, Bresette CA, Sawicki LA, Maverakis E, Bai S, Kloxin AM (2018) Design of functionalized cyclic peptides through orthogonal click reactions for cell culture and targeting applications. Chem Commun 54:6923–6926. https://doi.org/10.1039/C8CC03218A
64. Hoppmann C, Schmieder P, Heinrich N, Beyermann M (2011) Photoswitchable click amino acids: light control of conformation and bioactivity. ChemBioChem 12:2555–2559. https://doi.org/10.1002/cbic.201100578
65. Hoppmann C, Kühne R, Beyermann M (2012) Intramolecular bridges formed by photoswitchable click amino acids. Beilstein J Org Chem 8:884–889. https://doi.org/10.3762/bjoc.8.100
66. Zhao B, Zhang Q, Li Z (2016) Constructing thioether-tethered cyclic peptides via on-resin intra-molecular thiol-ene reaction. J Pept Sci 22:540–544. https://doi.org/10.1002/psc.2902
67. Wang Y, Chou DH-C (2015) A thiol-ene coupling approach to native peptide stapling and macrocyclization. Angew Chem Int Ed 54:10931–10934. https://doi.org/10.1002/anie.201503975
68. Grossmann TN, Yeh JT-H, Bowman BR, Chu Q, Moellering RE, Verdine GL (2012) Inhibition of oncogenic Wnt signaling through direct targeting of β-catenin. Proc Natl Acad Sci 109:17942–17947. https://doi.org/10.1073/pnas.1208396109
69. Walensky LD, Kung AL, Escher I, Malia TJ, Barbuto S, Wright RD, Wagner G, Verdine GL, Korsmeyer SJ (2004) Activation of apoptosis in vivo by a hydrocarbon-stapled BH3 helix. Science 305:1466–1470. https://doi.org/10.1126/science.1099191
70. Wang Y, Bruno BJ, Cornillie S, Nogieira JM, Chen D, Cheatham IIITE, Lim CS, Chou DH-C (2017) Application of thiol-yne/thiol-ene reactions for peptide and protein macrocyclizations. Chem Eur J 23:7087–7092. https://doi.org/10.1002/chem.201700572
71. Paterson DL, Flanagan JU, Shepherd PR, Harris PWR, Brimble MA (2020) Variable-length Ester-based Staples for α-helical peptides by using a double thiol-ene reaction. Chem Eur J 26: 10826–10833. https://doi.org/10.1002/chem.202001478
72. Nolan MD, Schüttel M, Scanlan EM, Nielsen AL (2024) Nanomole-scale photochemical thiol-ene chemistry for high-throughput late-stage diversification of peptide macrocycles. Pept Sci 116:e24310. https://doi.org/10.1002/pep2.24310
73. Benny A, Scanlan EM (2024) Synthesis of macrocyclic thiolactone peptides via photochemical intramolecular radical acyl thiol-ene ligation. Chem Commun 60:7950–7953. https://doi.org/10.1039/D4CC02442G
74. Campani�o A, Baran M, Bowie AG, Longley DB, Harrison T, McGouran JF (2025) Chemical- and photo-activation of protein-protein thiol-ene coupling for protein profiling. Commun Chem 8:1–10. https://doi.org/10.1038/s42004-025-01412-6
75. Ernst B, Magnani JL (2009) From carbohydrate leads to glycomimetic drugs. Nat Rev Drug Discov 8:661–677. https://doi.org/10.1038/nrd2852
76. Driguez H (2001) Thiooligosaccharides as tools for structural biology. ChemBioChem 2:311–318. https://doi.org/10.1002/1439-7633(20010504)2:5<311::AID-CBIC311>3.0.CO;2-L
77. Lázár L, Csávás M, Tóth M, Somsák L, Borbás A (2015) Thio-click approach to the synthesis of stable glycomimetics. Chem Paper 69:889–895. https://doi.org/10.1515/chempap-2015-0085
78. Pachamuthu K, Schmidt RR (2006) Synthetic routes to thiooligosaccharides and thioglycopeptides. Chem Rev 106:160–187. https://doi.org/10.1021/cr040660c
79. McSweeney L, Dénès F, Scanlan EM (2016) Thiyl-radical reactions in carbo-hydrate chemistry: from thiosugars to glycoconjugate synthesis. Eur J Org Chem 2016:2080–2095. https://doi.org/10.1002/ejoc.201501543
80. Igarashi K, Honma T (1970) Addition reactions of glycals. IV. Free-radical addition of thiolacetic acid to D-glucal triacetate. J Org Chem 35:606–610. https://doi.org/10.1021/jo00828a012

81. Araki Y, Matsuura K, Ishido Y, Kushida K (1973) Synthetic studies of carbohydrate derivatives with photochemical reaction. VII. Photochemical addition of ethanethiol and 1-propanethiol to enoses. Chem Lett 2:383–386. https://doi.org/10.1246/cl.1973.383
82. Staderini S, Chambery A, Marra A, Dondoni A (2012) Free-radical hydrothiolation of glycals: a thiol-ene-based synthesis of *S*-disaccharides. Tetrahedron Lett 53:702–704. https://doi.org/10.1016/j.tetlet.2011.11.140
83. Lázár L, Csávás M, Herczeg M, Herczegh P, Borbás A (2012) Synthesis of S-linked glycoconjugates and S-disaccharides by thiol-ene coupling reaction of enoses. Org Lett 14: 4650–4653. https://doi.org/10.1021/ol302098u
84. Lázár L, Borbás A, Somsák L (2018) Synthesis of thiomaltooligosaccharides by a thio-click approach. Carbohydr Res 470:8–12. https://doi.org/10.1016/j.carres.2018.09.005
85. Lázár L, Juhász L, Batta G, Borbás A, Somsák L (2017) Unprecedented β-manno type thiodisaccharides with a C-glycosylic function by photoinitiated hydrothiolation of 1-C-substituted glycals. New J Chem 41:1284–1292. https://doi.org/10.1039/C6NJ03751H
86. Eszenyi D, Kelemen V, Balogh F, Bege M, Csávás M, Herczegh P, Borbás A (2018) Promotion of a reaction by cooling: stereoselective 1,2-cis-α-thioglycoconjugation by thiol-ene coupling at −80 °C. Chem Eur J 24:4532–4536. https://doi.org/10.1002/chem.201800668
87. Kelemen V, Bege M, Eszenyi D, Debreczeni N, Bényei A, Stürzer T, Herczegh P, Borbás A (2019) Stereoselective thioconjugation by photoinduced thiol-ene coupling reactions of hexo- and pentopyranosyl d- and l-glycals at low-temperature – reactivity and stereoselectivity study. Chem Eur J 25:14555–14571. https://doi.org/10.1002/chem.201903095
88. Kelemen V, Csávás M, Hotzi J, Herczeg M, Poonam RB, Herczegh P, Jain N, Borbás A (2020) Photoinitiated thiol-ene reactions of various 2,3-unsaturated O-, C- S- and N-glycosides – scope and limitations study. Chem Asian J 15:876–891. https://doi.org/10.1002/asia.201901560
89. Csávás M, Eszenyi D, Mező E, Lázár L, Debreczeni N, Tóth M, Somsák L, Borbás A (2020) Stereoselective synthesis of carbon-sulfur-bridged glycomimetics by photoinitiated thiol-ene coupling reactions. Int J Mol Sci 21:573. https://doi.org/10.3390/ijms21020573
90. József J, Juhász L, Illyés TZ, Csávás M, Borbás A, Somsák L (2015) Photoinitiated hydrothiolation of pyranoid *exo*-glycals: the d-*galacto* and d-*xylo* cases. Carbohydr Res 413: 63–69. https://doi.org/10.1016/j.carres.2015.05.008
91. József J, Juhász L, Somsák L (2019) Thio-click reaction of 2-deoxy-exo-glycals towards new glycomimetics: stereoselective synthesis of C-2-deoxy-D-glycopyranosyl compounds. New J Chem 43:5670–5686. https://doi.org/10.1039/C8NJ06138F
92. Fiore M, Marra A, Dondoni A (2009) Photoinduced thiol-ene coupling as a click ligation tool for thiodisaccharide synthesis. J Org Chem 74:4422–4425. https://doi.org/10.1021/jo900514w
93. Bege M, Bereczki I, Herczeg M, Kicsák M, Eszenyi D, Herczegh P, Borbás A (2017) A low-temperature, photoinduced thiol-ene click reaction: a mild and efficient method for the synthesis of sugar-modified nucleosides. Org Biomol Chem 15:9226–9233. https://doi.org/10.1039/C7OB02184D
94. Bege M, Kiss A, Kicsák M, Bereczki I, Baksa V, Király G, Szemán-Nagy G, Szigeti MZ, Herczegh P, Borbás A (2019) Synthesis and cytostatic effect of 3′-deoxy-3′-C-sulfanylmethyl nucleoside derivatives with d-xylo configuration. Molecules 24:2173. https://doi.org/10.3390/molecules24112173
95. Csávás M, Demeter T, Herczeg M, Timári I, Kövér KE, Herczegh P, Borbás A (2014) Rapid synthesis of self-assembling 1,2-thiomannobioside glycoconjugates as potential multivalent ligands of mannose-binding lectins. Tetrahedron Lett 55:6983–6986. https://doi.org/10.1016/j.tetlet.2014.10.104
96. Csávás M, Malinovská L, Perret F, Gyurkó M, Illyés ZT, Wimmerová M, Borbás A (2017) Tri- and tetravalent mannoclusters cross-link and aggregate BC2L-A lectin from *Burkholderia cenocepacia*. Carbohydr Res 437:1–8. https://doi.org/10.1016/j.carres.2016.11.008

97. Debreczeni N, Bege M, Borbás A (2021) Synthesis of potential glycosyl transferase inhibitors by thio-click reactions. Eur J Org Chem 2021:6743–6747. https://doi.org/10.1002/ejoc.202101220
98. Buskas T, Söderberg E, Konradsson P, Fraser-Reid B (2000) Use of n-pentenyl glycosides as precursors to various spacer functionalities. J Org Chem 65:958–963. https://doi.org/10.1021/jo9909554
99. Merbouh N, Wallner FK, Cociorva OM, Seeberger PH (2007) 3-Mercaptopropanol as a traceless linker for chemical and enzymatic synthesis of oligosaccharides. Org Lett 9:651–653. https://doi.org/10.1021/ol062922y
100. Goddard-Borger ED, Tropak MB, Yonekawa S, Tysoe C, Mahuran DJ, Withers SG (2012) Rapid assembly of a library of lipophilic iminosugars via the thiol-ene reaction yields promising pharmacological chaperones for the treatment of Gaucher disease. J Med Chem 55:2737–2745. https://doi.org/10.1021/jm201633y
101. Sánchez-Fernández EM, García-Moreno MI, García-Hernández R, Padrón JM, García Fernández JM, Gamarro F, Ortiz Mellet C (2019) Thiol-ene "click" synthesis and pharmacological evaluation of C-glycoside sp2-Iminosugar glycolipids. Molecules 24:2882. https://doi.org/10.3390/molecules24162882
102. Lundquist JJ, Toone EJ (2002) The cluster glycoside effect. Chem Rev 102:555–578. https://doi.org/10.1021/cr000418f
103. Ghirardello M, Öberg K, Staderini S, Renaudet O, Berthet N, Dumy P, Hed Y, Marra A, Malkoch M, Dondoni A (2014) Thiol-ene and thiol-yne-based synthesis of glycodendrimers as nanomolar inhibitors of wheat germ agglutinin. J Polym Sci A Polym Chem 52:2422–2433. https://doi.org/10.1002/pola.27262
104. Conte ML, Robb MJ, Hed Y, Marra A, Malkoch M, Hawker CJ, Dondoni A (2011) Exhaustive glycosylation, pegylation, and glutathionylation of a [G4]-ene48 dendrimer via photoinduced thiol-ene coupling. J Polym Sci A Polym Chem 49:4468–4475. https://doi.org/10.1002/pola.24888
105. Fiore M, Chambery A, Marra A, Dondoni A (2009) Single and dual glycoside clustering around calix[4]arene scaffolds via click thiol-ene coupling and azide–alkyne cycloaddition. Org Biomol Chem 7:3910–3913. https://doi.org/10.1039/B912686D
106. Zelli R, Dumy P, Marra A (2020) Metal-free synthesis of imino-disaccharides and calix-iminosugars by photoinduced radical thiol-ene coupling (TEC). Org Biomol Chem 18:2392–2397. https://doi.org/10.1039/D0OB00198H
107. Fulton DA, Stoddart JF (2000) An efficient synthesis of cyclodextrin-based carbohydrate cluster compounds. Org Lett 2:1113–1116. https://doi.org/10.1021/ol005668x
108. Fulton DA, Stoddart JF (2001) Synthesis of cyclodextrin-based carbohydrate clusters by photoaddition reactions. J Org Chem 66:8309–8319. https://doi.org/10.1021/jo010705z
109. Leydet A, Moullet C, Roque JP, Witvrouw M, Pannecouque C, Andrei G, Snoeck R, Neyts J, Schols D, De Clercq E (1998) Polyanion inhibitors of HIV and other viruses. 7. Polyanionic compounds and polyzwitterionic compounds derived from cyclodextrins as inhibitors of HIV transmission. J Med Chem 41:4927–4932. https://doi.org/10.1021/jm970661f
110. Becker MM, Zeng Z, Ravoo BJ (2013) Multivalent functionalization of cyclodextrins by photochemical thiol-ene addition reaction. Eur J Org Chem 2013:6831–6843. https://doi.org/10.1002/ejoc.201300880
111. Zöller K, To D, Bernkop-Schnürch A (2025) Biomedical applications of functional hydrogels: innovative developments, relevant clinical trials and advanced products. Biomaterials 312:122718. https://doi.org/10.1016/j.biomaterials.2024.122718
112. Degirmenci A, Sanyal R, Sanyal A (2024) Metal-free click-chemistry: a powerful tool for fabricating hydrogels for biomedical applications. Bioconjug Chem 35:433–452. https://doi.org/10.1021/acs.bioconjchem.4c00003
113. Li L, Lu C, Wang L, Chen M, White J, Hao X, McLean KM, Chen H, Hughes TC (2018) Gelatin-based photocurable hydrogels for corneal wound repair. ACS Appl Mater Interfaces 10:13283–13292. https://doi.org/10.1021/acsami.7b17054

114. Goh M, Kim Y, Gwon K, Min K, Hwang Y, Tae G (2017) *In situ* formation of injectable and porous heparin-based hydrogel. Carbohydr Polym 174:990–998. https://doi.org/10.1016/j.carbpol.2017.06.126
115. Holmes R, Yang X-B, Dunne A, Florea L, Wood D, Tronci G (2017) Thiol-ene photo-click collagen-PEG hydrogels: impact of water-soluble photoinitiators on cell viability, gelation kinetics and rheological properties. Polymers 9:226. https://doi.org/10.3390/polym9060226
116. Soiberman U, Kambhampati SP, Wu T, Mishra MK, Oh Y, Sharma R, Wang J, Al Towerki AE, Yiu S, Stark WJ, Kannan RM (2017) Subconjunctival injectable dendrimer-dexamethasone gel for the treatment of corneal inflammation. Biomaterials 125:38–53. https://doi.org/10.1016/j.biomaterials.2017.02.016
117. Arslan M, Gevrek TN, Sanyal R, Sanyal A (2015) Fabrication of poly(ethylene glycol)-based cyclodextrin containing hydrogels via thiol-ene click reaction. Eur Polym J 62:426–434. https://doi.org/10.1016/j.eurpolymj.2014.08.018
118. Lee HJ, Fernandes-Cunha GM, Myung D (2018) *In situ*-forming hyaluronic acid hydrogel through visible light-induced thiol-ene reaction. React Funct Polym 131:29–35. https://doi.org/10.1016/j.reactfunctpolym.2018.06.010
119. Chung S, Lee H, Kim H-S, Kim M-G, Lee LP, Lee JY (2016) Transdermal thiol–acrylate polyethylene glycol hydrogel synthesis using near infrared light. Nanoscale 8:14213–14221. https://doi.org/10.1039/C6NR01956K
120. Ding H, Li B, Jiang Y, Liu G, Pu S, Feng Y, Jia D, Zhou Y (2021) pH-responsive UV crosslinkable chitosan hydrogel via "thiol-ene" click chemistry for active modulating opposite drug release behaviors. Carbohydr Polym 251:117101. https://doi.org/10.1016/j.carbpol.2020.117101
121. Verjans J, Sedlačík T, Jerca VV, Bernhard Y, Van Guyse JFR, Hoogenboom R (2023) Poly (N-allyl acrylamide) as a reactive platform toward functional hydrogels. ACS Macro Lett 12:79–85. https://doi.org/10.1021/acsmacrolett.2c00650
122. Lin C-C, Ki CS, Shih H (2015) Thiol–norbornene photoclick hydrogels for tissue engineering applications. J Appl Polym Sci 132. https://doi.org/10.1002/app.41563
123. Mũnoz Z, Shih H, Lin C-C (2014) Gelatin hydrogels formed by orthogonal thiol–norbornene photochemistry for cell encapsulation. Biomater Sci 2:1063–1072. https://doi.org/10.1039/C4BM00070F
124. Gramlich WM, Kim IL, Burdick JA (2013) Synthesis and orthogonal photopatterning of hyaluronic acid hydrogels with thiol-norbornene chemistry. Biomaterials 34:9803–9811. https://doi.org/10.1016/j.biomaterials.2013.08.089
125. Li X, Xiong Y (2022) Application of "click" chemistry in biomedical hydrogels. ACS Omega 7:36918–36928. https://doi.org/10.1021/acsomega.2c03931
126. Malafaia AP, Sobreiro-Almeida R, Rodrigues JMM, Mano JF (2025) Thiol-ene click chemistry: enabling 3D printing of natural-based inks for biomedical applications. Biomater Adv 167:214105. https://doi.org/10.1016/j.bioadv.2024.214105
127. Rehm TH (2020) Flow photochemistry as a tool in organic synthesis. Chem Eur J 26:16952–16974. https://doi.org/10.1002/chem.202000381

Top Heterocycl Chem (2026) 63: 161–204
https://doi.org/10.1007/7081_2025_96

Published online: 16 December 2025

"MCR/Tandem Click Chemistry": A Vital Tool to Access Biologically Privileged Functionalized 1,2,3-Triazole Analogues

Sanchayita Rajkhowa, Mangal S. Yadav, Anindra Sharma, and Vinod K. Tiwari

Contents

1 Introduction 163
2 Regioselectivity in "1,2,3-Triazole-Forming Click" and Its Compatibility with Multicomponent Reactions (MCRs) 163
3 Application of MCR Click Chemistry to Access Biologically Privileged Functionalized 1,2,3-Triazoles 165
3.1 MCR Click in the Development of Functionalized 1,2,3-Triazole Hybrid Analogues 165
3.2 MCR-Click Chemistry in Macrocyclization 184
3.3 CuAAC/MCRs in Sequential Fashion 189
3.4 "MCR Click Chemistry" in Bioconjugation 195
4 Conclusion and Future Perspective 198
References 199

Abstract Multicomponent click reactions are operationally simple, short, robust, and efficient reactions that have driven the process of development of a diverse range of biologically privileged functionalized analogues. This tool has emerged as a powerful synthetic strategy enabling the rapid access of 1,2,3-triazole-appended simple-to-complex molecular architectures starting from readily available building blocks under mild reaction conditions in a one-pot fashion. This tool facilitates the simultaneous formation of multiple bonds and rationalizes the development of

S. Rajkhowa (✉)
Department of Chemistry, Haflong Government College, Haflong, Assam, India
e-mail: Sanchayita.Rajkhowa@gmail.com

M. S. Yadav and V. K. Tiwari (✉)
Department of Chemistry, Institute of Science, Banaras Hindu University, Varanasi, Uttar Pradesh, India
e-mail: Vinod.Tiwari@bhu.ac.in

A. Sharma
Department of Chemistry, Lalit Narayan Mithila University, Darbhanga, Bihar, India

diverse structures including fused-ring systems, conjugates, and well-defined macrocycles. This modular approach offers exceptional selectivity and functional group compatibility in multiple CuAAC- and/or RuAAC-mediated 1,2,3-triazole-forming reactions. Our chapter highlights the notable features of multicomponent click chemistry, particularly its transformative impact on synthetic efficiency, modularity, molecular diversity, and emerging applications in medicinal chemistry, drug discovery, development of advanced functional materials for material science, and complex biomolecular assemblies for biology and medicine.

Keywords 1,2,3-Triazole · Alkyne · Azide · Click chemistry · CuAAC · Cyclization · MCR · RuAAC

Abbreviations

AAC	Azide-alkyne cycloaddition
AcCN	Acetonitrile
BSA	Bovine serum albumin
CFC	Continuous flow chemistry
CuAAC	Copper-catalyzed azide alkyne cycloaddition
DBU	Diazabicyclo[5.4.0]undec-7-ene
DCM	Dichloromethane
DIPEA	*N,N*-Diisopropylethylamine
DMAP	4-Dimethylaminopyridine
DMF	*N,N*-Dimethylformamide
DNA	Deoxyribonucleic acid
DOS	Diversity-oriented synthesis
HFIP	1,1,1,3,3,3-Hexafluoroisopropanol
HTS	High-throughput screening
MCRs	Multicomponent reactions
MES	Maximum electroshock
MW	Microwave
Na-Asc	Sodium ascorbate
PFEx	Phosphorus fluoride exchange
PIDA	Phenyliodonium diacetate
p-TSA	*Para*-toluene sulfonic acid
PTZ	Pentylenetetrazole
rt	Room temperature
RuAAC	Ruthenium-catalyzed azide alkyne cycloaddition
SPAAC	Strain-promoted azide alkyne cycloaddition
SuFEx	Sulfur fluoride exchange
TBAA	Tetrabutylammonium azide
TEMPO	Tetramethyl piperidinyloxy

1 Introduction

To address a set of robust, modular, and convenient organic reactions, Prof. Karl Barry Sharpless in 2001 coined a term "click chemistry" [1]. The following year, the concept of regioselectivity in 1,2,3-triazole-forming [3 + 2] azide-alkyne cycloaddition (AAC) reaction was introduced by utilizing CuI/DIPEA in DCM (Meldal's CuAAC tool) [2] or $CuSO_4$/NaAsc in BuOH/water (Sharpless's CuAAC tool) [3]. Since then, the excellent regioselectivity in CuAAC has been widely exploited in many emerging areas of science and technology [4–7]. In 2004, Prof. Carolyn Bertozzi brilliantly explored the Wittig's strain-promoted AAC (SPAAC) [8] method in living systems (known as Bertozzi's bio-orthogonal concept) [9]. In 2022, this powerful 1,2,3-triazole-forming tool was recognized with the Nobel Prize in Chemistry being awarded to Prof. Carolyn Bertozzi, Prof. Morten Meldal, and Prof. Karl B. Sharpless.

The ideal chemical process is that which an "one-armed operator can perform by pouring the reactants into a bathtub and collecting the pure product from the drain hole" – as quoted by Sir John Cornforth, 1975 Nobel laureate in chemistry [10]. Multicomponent reactions (MCRs) are examples of ideal chemical processes; they are robust and versatile and critical to synthetic and medicinal chemistry [11, 12]. The MCR tool combined with "click chemistry" in a sequential tandem/cascade fashion enables the easy access of 1,2,3-triazole-linked functionalized molecular frameworks for notable applications in drug development [13].

This chapter highlights the transformative impact of MCR/click chemistry in terms of its efficiency, modularity, and molecular diversity that has led to utility in various emerging fields such as development of simple-to-complex biomolecular assemblies of advanced functional molecules for drug discovery and development or material science.

2 Regioselectivity in "1,2,3-Triazole-Forming Click" and Its Compatibility with Multicomponent Reactions (MCRs)

Michael's first report (in 1893) [14] on AAC was extensively explored during the 1960s by R. Huisgen [15]. Although Huisgen's thermally induced AAC of terminal alkynes **1** and organic azides **2** was well utilized to access various 1,2,3-triazole-linked hybrids, the poor regioselectivity (as both 1,4-disubstituted 1,2,3-triazoles **3** and 1,5-disubstituted 1,2,3-triazoles **4** were obtained in almost equimolar amount which were difficult to isolate by column chromatography) posed a serious limitation [15]. Meldal [2] and Sharpless's [3] CuAAC tool afforded good-to-excellent yield of 1,4-disubstituted 1,2,3-triazoles **3** as the sole product with excellent regioselectivity. Similar AAC in the presence of a suitable ruthenium catalyst (RuAAC) gave exclusively the opposite regioisomer, i.e., 1,5-disubstituted 1,2,3-triazole **4** [16, 17]. The scope of classical 1,2,3-triazole-forming "click chemistry" and

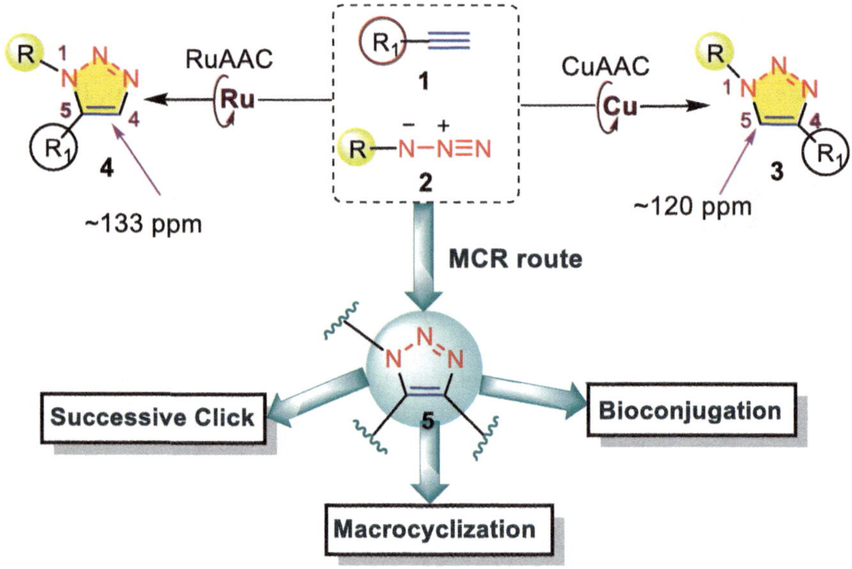

Scheme 1 The regioselective variant of classical 1,2,3-triazole-forming click reactions and their compatibility with multicomponent reactions (MCRs) to access diverse functionalized 1,2,3-triazole analogues

their compatibility with multicomponent reactions (MCRs) are depicted in Scheme 1.

The CuAAC reaction proceeds with enhanced rate of acceleration of kinetics than the uncatalyzed condition. The classical 1,2,3-triazole-forming reactions include the regioselective metal-catalyzed AAC (CuAAC [2, 3, 6] or RuAAC [16, 17]). Such AAC facilitates the development of a diverse range of functionalized 1,2,3-triazoles with notable regioselectivity [18–22], and the scope is further expanding exponentially [3]. Other seminal development on 1,2,3-triazole-forming click reaction includes SPAAC [9], photoinduced click [23], organo-catalyzed click [24–28], topochemical AAC [29], and thermally induced AAC under inter- and/or intramolecular fashion [30, 31]. The classical method for development of functionalized 1,2,3-triazoles follows a stepwise path that involves the initial formation of the 1,2,3-triazole core, which then underwent subsequent functionalization [7–31]. Conversely, there is increased interest to devise novel and proficient methods to streamline the associated issues including the mild reaction condition with requirement of lesser reagents avoiding isolation of intermediate involved herein.

The feature of multicomponent reactions (MCRs) in a parallel fashion is critical for robust synthesis as it results in the elimination of some of the synthetic steps. Considering the abovementioned features, MCR is well explored during the last three decades, and scope is further expanding as a promising tool for easy access of selective and functionalized drug candidate [11]. The notable features including one-step procedure without the isolation of involved intermediates, high yield, and mild reaction conditions make the protocol quite imperative in combinatorial chemistry and high-throughput screening (HTS) for its promising applications in pharmaceutics and clinics [32, 33]. In addition to CuAAC, a consecutive one-pot MCR is well explored in medicinal chemistry that enables easy access of 1,2,3-triazole-linked functionalized molecular architectures [34–36]. The robust, convenient, modular, and versatile nature of MCR tool is widely explored in organic synthesis [12]. For example, a number of MCRs, notably Ugi four-component reaction [37], Passerini three-component reactions [38], Biginelli three-component reactions [39], and other related synthetic tools including Strecker synthesis [40], Hantzsch reaction [41], Mannich reaction [42], and all together the "MCR click," have ample opportunity in organic and medicinal chemistry. The powerful "MCR tool" coupled with modular "click chemistry" has modernized the synthetic processes for drug discovery, which we cover in the next section of this chapter.

3 Application of MCR Click Chemistry to Access Biologically Privileged Functionalized 1,2,3-Triazoles

Click chemistry has emerged as an essential, innovative, and versatile method for synthesizing a wide range of 1,2,3-triazole-linked molecular structures, including heterocycles, carbocycles, sugars, and their fused systems, with high efficiency and precision [6, 7]. Its applications extend across various fields demonstrating its usefulness in glycoscience, pharmaceuticals, materials science, and other areas. This powerful approach has greatly enhanced the options available in synthetic chemistry for designing complex molecular architectures to meet the needs of applications in modern science.

3.1 MCR Click in the Development of Functionalized 1,2,3-Triazole Hybrid Analogues

The chemistry of heterocyclic molecules, particularly the 1,2,3-triazole moiety, has attracted considerable attention in medicinal chemistry in recent years [21]. These compounds have shown potential as antibacterial, anticancer, and antituberculosis agents [43]. In particular, 1,2,3-triazoles, appended with biocompatible carbocycles and heterocycles, have demonstrated therapeutic potential in treating diseases such as cancer, HIV, and Alzheimer's disease [43–45]. In addition to the numerous 1,2,3-triazole-based derivatives, glycohybrid-1,2,3-triazole analogues have gained importance across fields such as drug discovery and development. Therefore, it is crucial to establish synthetic approaches that are practical, simple, high-yielding, and compatible with biological systems [6, 46–55]. This section introduces the representative

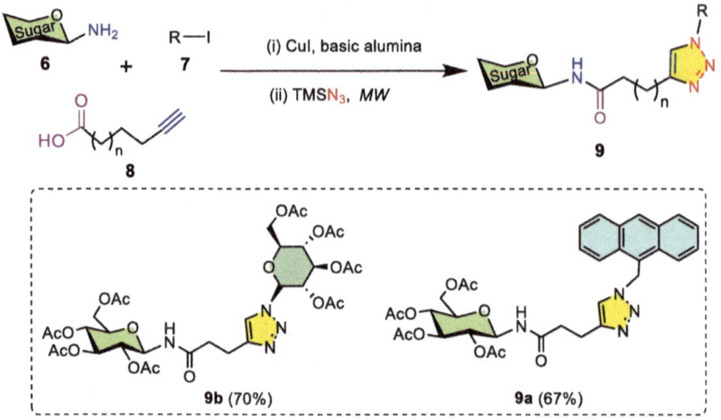

Scheme 2 MCR route for easy access of 1,2,3-triazolyl-*N*-carboxamides **9** using basic alumina as solid support

methods for easy access of a variety of 1,2,3-triazole-fused hybrid analogues using a classical variant of modular "MCR click chemistry."

Toward this end, Sen et al. [56] developed a noteworthy one-pot synthesis for the preparation of various 1,2,3-triazolyl *N*-carboxamides **9** using a straightforward approach, which involves reacting glycosyl amines **6** with different aryl or alkyl halides **7** and a propargyl source **8** in the presence of CuI as copper catalyst and basic alumina as a solid support. This process eliminates the need for solvent, base, and/or ligand required to facilitate the catalytic cycle (Scheme 2).

Arora et al. further extended this methodology for the synthesis of unsymmetrical *bis*-1,2,3-triazoles through a sequential "domino-click" strategy [57]. Thus, sugar azides on reaction with allenylmagnesium bromide in anhydrous THF at rt gave 1,5-disubstituted butynyl-*N*-glycosyl-1,2,3-triazoles. Subsequent CuAAC reaction with sugar azides under domino fashion gave a high yield of unsymmetrical *bis*-1,2,3-triazoles.

In most of the cases, the AAC protocol could be exploited by using a suitable copper catalyst; however, in parallel, a multicomponent condensation under one-pot mild reaction conditions has also been considered a high priority in recent years [13]. Mukherjee and group reported a three-component transformation reaction under one-pot fashion through sequential addition of glucose peracetate **10**, propargyl alcohol, D-glucal **11**, and $TMSN_3$ using catalytic amount $Cu(OTf)_2$ and Cu powder to deliver a high (>80%) yield of respective furan-based hydroxy 1,2,3-triazole **12** with high diastereomeric purity (Scheme 3) [36].

Furthermore, glycosyl azides **14** and glycosyl 1,2,3-triazole hybrids **15** can be synthesized rapidly from respective glycosyl bromides **13** using a two-step high-yielding procedure under one-pot condition. This transformation occurs in an aqueous medium using a modular click catalyst under one-pot mild condition (Scheme 4) [58].

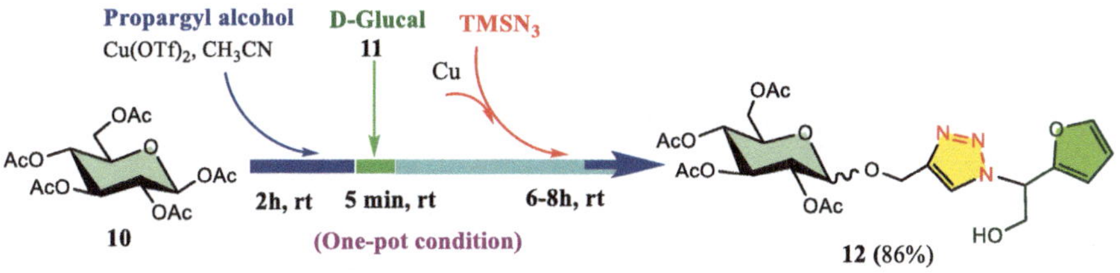

Scheme 3 Cu-catalyzed MCR cascade route to furan-linked glycohybrid-1,2,3-triazole **12** under sequential one-pot conditions, developed by Mukherjee et al. [36]

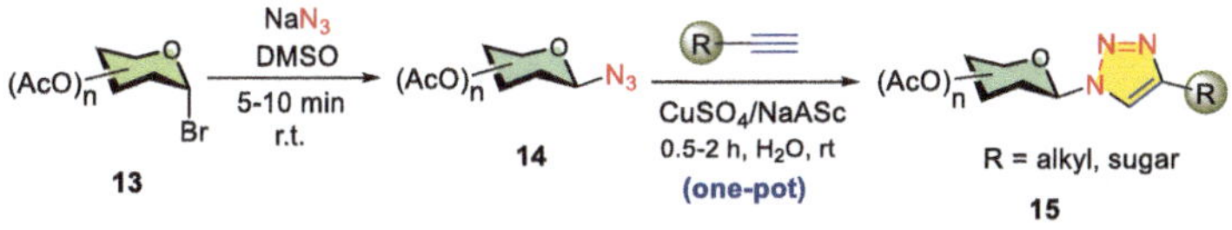

Scheme 4 CuAAC-inspired 4-substituted 1,2,3-triazole-glycoconjugates in one-pot fashion

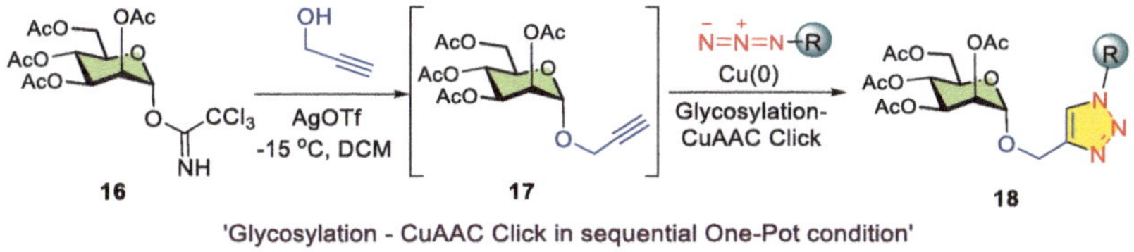

Scheme 5 Synthesis of 1,2,3-triazole-*O*-glycoconjugate **18** under one-pot fashion, reported by Sen et al. [59]

Sen et al. [59] have reported a straightforward method for the one-pot synthesis of 1,2,3-triazole-*O*-glycoconjugates **18** from the mannose imidate donor **16**. In this approach, AgOTf serves as both a promoter for glycosylation and a co-catalyst for the subsequent click reaction (one-pot condition via *O*-glycoside **17**). During the click reaction, AgOTf facilitates the in situ oxidation of Cu(0), enabling the reaction to proceed. This one-pot and high-yielding process simplifies the synthesis by eliminating time-consuming traditional multistep procedures (Scheme 5) [59].

Xu and Loh [60] cleverly explored a strain-release glycosylation method by utilizing a strained furanosyl **19** as donor with variety acceptors in the presence of a thiourea catalyst **20** to give high yield of respective *O*-glycosides **21** with notable anomeric selectivity. The resulting alkyne **21** then underwent CuAAC conjugation under standard condition separately with furanosylated glycosyl azides or pyranosyl sugar azides **22** to give the respective 1,4-disubstituted glycohybrid 1,2,3-triazoles **23**. Interestingly, both the glycosylation steps and click conjugation step were performed in one-pot condition, and the final 1,2,3-triazole was obtained in good-to-high yields. Alternatively, an opposite regioisomeric glycohybrid 1,2,3-triazole analogue **24** with 1,5-disubstituted fashion was obtained using a catalytic amount of $RuCp^{*}Cl(PPh_3)_2$ (RuAAC) (Scheme 6) [60].

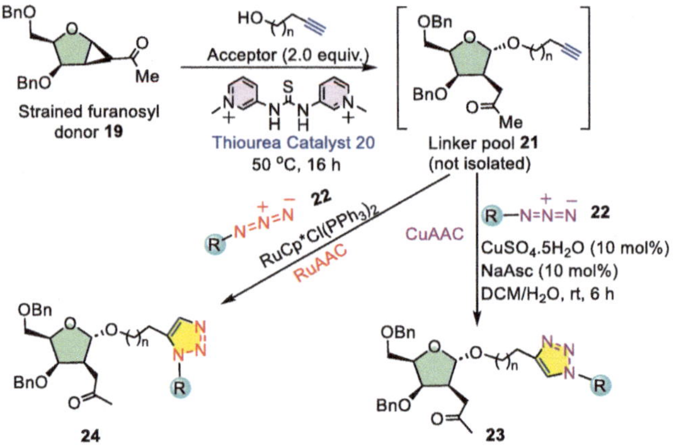

Scheme 6 The one-pot strain-release glycosylation to develop 1,2,3-triazolyl analogues developed by Xu and Loh [60]

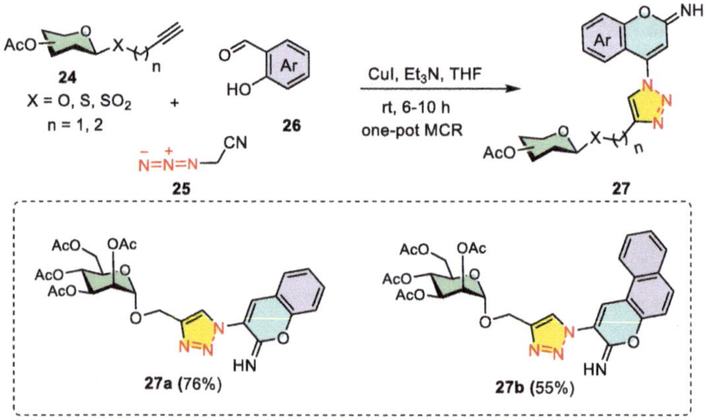

Scheme 7 CuI-induced MCR tool to access glycosylated-3-triazolyl-2-iminocoumarins **27**, developed by Mandal [61]

Mandal [61] prepared glycosylated 3-triazolyl-2-iminocoumarins **27** through a one-pot, three-component tandem reaction involving a "CuAAC-aldol-cyclization-dehydration" process. In this method, sugar alkynes **24** and 2-azidoacetonitrile **25** participated in a copper-catalyzed cycloaddition, generating a triazole derivative (in situ), which subsequently activated the adjacent methylene group. This activation initiated an aldol-cyclization-dehydration sequence in the presence of salicylaldehyde **26**, leading to the formation of the targeted glycosylated 3-triazolyl-2-iminocoumarin derivatives **27** in moderate-to-good yields (Scheme 7) [61]. The author described this MCR approach as a "click and activate nucleophile" strategy.

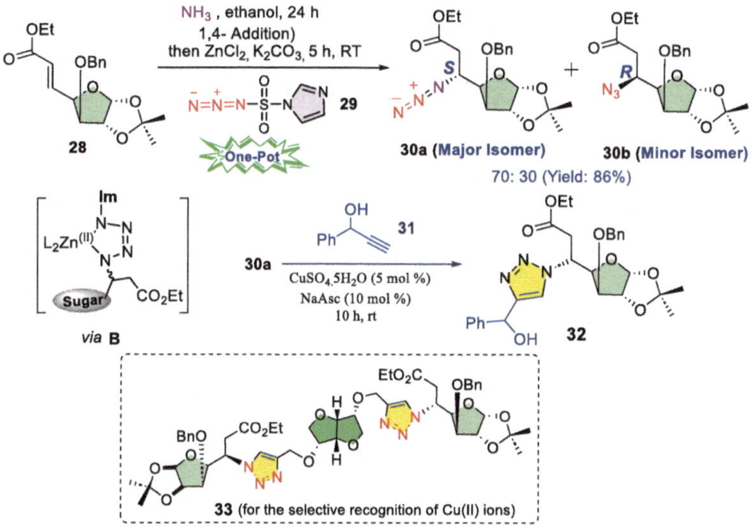

Scheme 8 Synthesis of glycosylated-*β*-azido ester **32** from α,β-unsaturated ester **28**, as reported by Tiwari et al. [64]

The Komnenos-Michael addition provides an alternative strategy for introducing azido groups at the β-position of α,β-unsaturated olefinic esters [62]. However, the protocol is unexpectedly challenging in some instances, particularly concerning the conjugate addition of the azido anion to glycosyl α,β-unsaturated olefinic esters. Due to the structural similarities between azides and diazoamines [63], amines can be efficiently transformed into their corresponding azido derivatives through a diazo-transfer approach. Tiwari et al. [64] developed a concise and proficient route for the synthesis of glycosyl-β-azido esters **30** through a one-pot reaction that combines Michael addition and diazo-transfer reaction (Scheme 8). In this synthetic pathway, α,β-olefinic ester **28** initially reacted with ethanolic ammonia for 24 h, producing a diastereomeric mixture of glycosyl *β*-amino esters. This mixture was then treated with imidazolyl-1-sulfonyl azide **29** in the presence of $ZnCl_2$ and K_2CO_3, leading to the formation of the desired glycosyl azides **30** [64]. The resulting azido esters **30a** and **30b** were purified via flash column chromatography (SiO_2), and their structures were confirmed through NMR as well as single-crystal X-ray diffraction.

The proposed mechanism indicates that the amine complex **A** undergoes nucleophilic addition to the electrophilic center of the azide followed by deprotonation to form a Zn-stabilized tetrazene intermediate **B**. This is then followed by a retro 1,3-dipolar cycloaddition that yields the target glycosyl β-azido esters **30a** and **30b**. Notably, the diastereomeric ratio remained constant during this stereoisomeric synthesis since the C-5 stereogenic center was not involved in the diazo-transfer process [64]. The CuAAC of azide **30a** with alkyne **31** provided the respective glycohybrid triazole **32** in 94% yield. 1,4-Disubstituted triazole **33** obtained by similar CuAAC is used for selective sensing of Cu(II) ions [65].

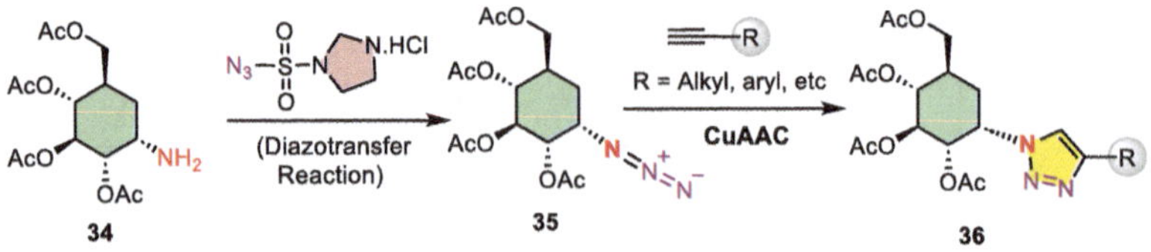

Scheme 9 Cu(I)-assisted tandem synthesis of aminocyclitol-linked 1,2,3-triazoles **36**

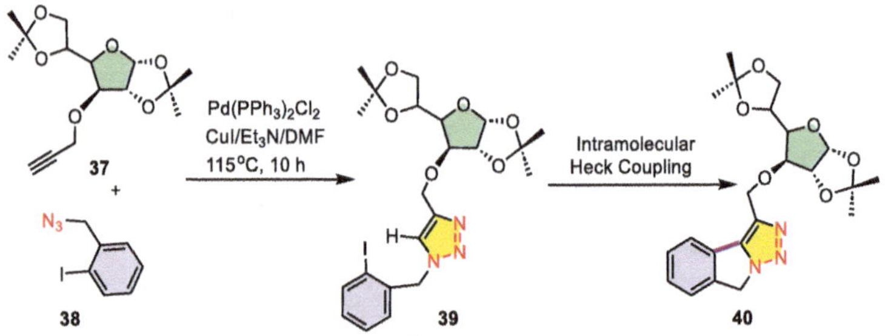

Scheme 10 The Pd/Cu-mediated synthesis of isoindoline-fused 1,2,3-triazole **40**, as reported by Choudhary et al. [68]

Currently, there are only a limited number of studies focusing on the diversification of aminocyclitols via CuAAC [66]. Qian et al. [67] devised a method for synthesizing triazolyl C_7-*N*-aminocyclitol scaffolds in two steps. In the initial step, diazo-transfer reactions were performed with C_7-*N*-aminocyclitol analogues **34** (such as validamine, valienamine, and valiolamine) using imidazole-1-sulfonyl azide hydrochloride **29** and a Cu(II) catalyst, which produced the azidocyclitol derivative **35**. In the subsequent step, the alkyne analogues underwent a [3 + 2] cycloaddition with organic azide **35** under modular click conditions to generate the corresponding C_7-*N*-aminocyclitol-derivatized 1,2,3-triazoles **36** (Scheme 9) [67].

Choudhary et al. [68] described a tandem strategy for synthesizing biocompatible isoindoline-fused 1,2,3-triazoles using a click-Heck coupling reaction catalyzed by a Pd-Cu catalytic system. In this process, an alkyne-functionalized glycosyl derivative **37** reacted with an azido analogue **38** in the presence of a catalytic combination of Pd $(PPh_3)_2Cl_2$ and CuI/Et_3N in DMF at 115°C, resulting in the formation of the 1,2,3-triazolyl glycohybrid product **39**. Subsequently, the triazolyl glycohybrid **39** underwent an intramolecular Heck coupling reaction to yield the target isoindoline-fused 1,2,3-triazoles **40** (Scheme 10).

Similarly, Sen et al. described an interesting method to synthesize triazole-fused tetracyclic glycosides **43** in an aqueous medium using sonication method. The desired glycosides **43** were obtained by reacting glycosylated azido-alkyne analogue **41** with aryl iodide **42** in the presence of Cu_2O nanoparticles and Cs_2CO_3 in aqueous medium. This process utilized both intramolecular CuAAC and Sonogashira coupling reactions to synthesize the final products **43**. Additionally, sonication helps to disperse the catalyst and increase its surface area. Cu_2O, a heterogeneous catalyst,

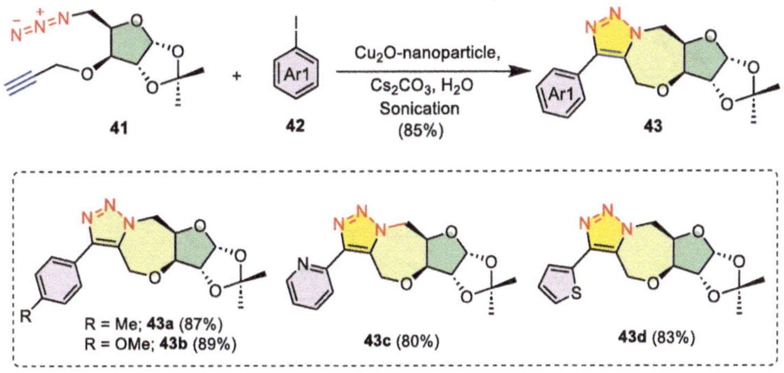

Scheme 11 Synthesis of polycyclic furanosides **43** in tandem one-pot fashion [69]

exhibited sustained catalytic activity over three consecutive reactions with negligible loss in efficiency (Scheme 11) [69].

Lieki's lab [70] first synthesized isocyanides in 1859, and since then, the isocyanide group has become central to modern organic chemistry. The study of isocyanides has significantly progressed, leading to the development of the Ugi-multicomponent reaction, a powerful method for creating complex, highly functionalized molecules in a single step. The Ugi reaction is widely recognized for its efficiency and versatility, especially in drug discovery and pharmaceutical research. By enabling the rapid assembly of intricate molecular architectures, this reaction has opened new avenues in medicinal chemistry, particularly for designing important therapeutic agents [70, 71]. To carry out the reaction, a one-pot, four-component Ugi reaction has been thoroughly described, involving glycosyl azido-aldehyde, alkyl isocyanide, benzylamine, and phenyl propionic acid. This approach generates a new class of sugar-based triazolodiazepines **49** with good yields and high selectivity (Scheme 12). The reaction begins with the formation of an iminium ion intermediate **I** from sugar azido-aldehyde **44** and benzylamine **45**. Isocyanides **48** and arylpropionic acid **47** then undergo nucleophilic addition on the iminium intermediate **I**, yielding the intermediate **II**. This intermediate isomerizes to form the Ugi adduct **48**. In the final step, heating the Ugi adduct **48** in refluxing DMF triggers an intramolecular [3 + 2] click cycloaddition reaction leading to the synthesis of sugar-tethered 1,2,3-triazole-diazepine analogues **49** [71].

Barlow and co-workers [72] explored a four-component Ugi reaction to access biologically relevant amino-triazoloazepinone **54** bearing di- and tripeptide structures (Scheme 13). In this MCR, β-azido-α-amino acid **50**, isocyanide **46**, propargylamine **51**, and aldehyde **52** were reacted in acetonitrile without any catalyst, simplifying the process. This high-yielding catalyst-free approach highlights the efficiency of the Ugi reaction in constructing complex molecular frameworks. The approach underscores the potential of Ugi reactions for synthesizing bioactive molecules, with implications for medicinal chemistry and drug discovery. In this one-pot process, the Ugi adducts **53** were initially formed and then underwent an intramolecular [3 + 2] cycloaddition at 70°C, yielding amino-triazoloazepinone

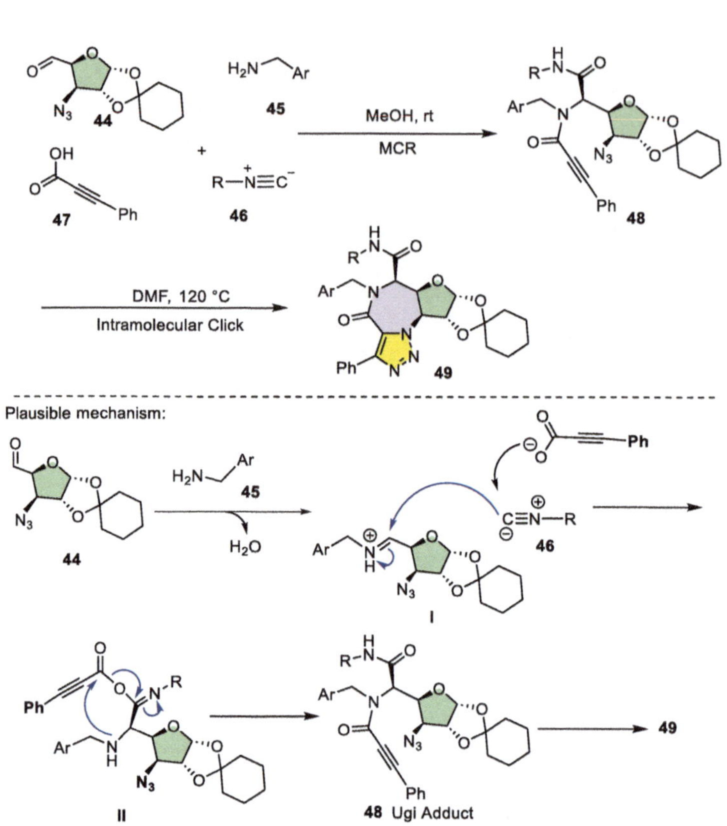

Scheme 12 MCR for synthesis of sugar-derived 1,2,3-triazolodiazepines **49** [71]

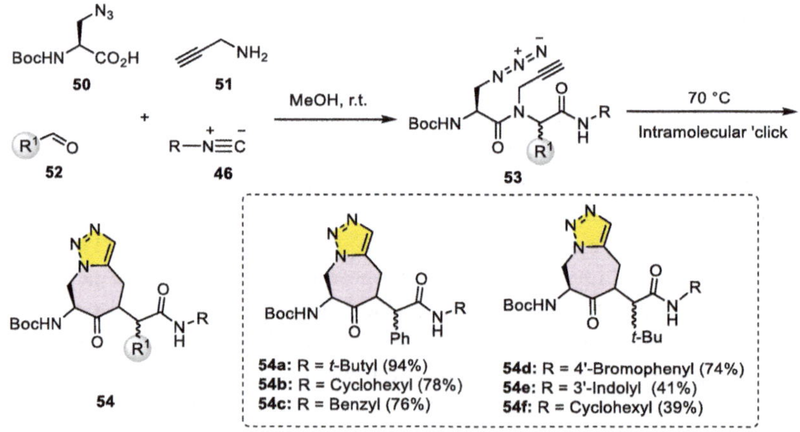

Scheme 13 Ugi four-component reaction to access bicyclic scaffolds **54** under one-pot condition, developed by Barlow et al. [72]

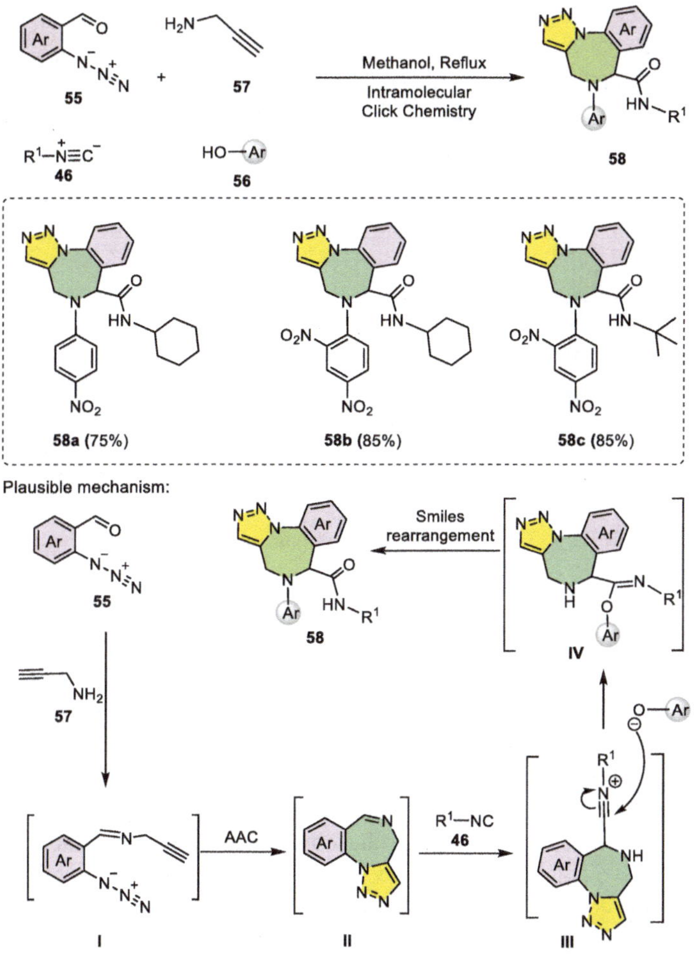

Scheme 14 Ugi-Smiles-type reaction to access 4,5-dihydro-1,2,3-triazolo[1,5-*a*][1, 4] benzodiazepines **58** [73]

54 as diastereoisomeric mixture (1:1) (Scheme 13). The study also revealed that substituents on the phenyl ring of the aromatic aldehyde did not significantly affect reaction yields or diastereoselectivity. Interestingly, the reactions with aromatic aldehydes demonstrated overall higher yields.

Saeedi et al. [73] reported a similar high-yielding four-component reaction involving isocyanides **46**, *o*-azidobenzaldehyde **55**, nitrophenols **56**, and propargyl amines **57** to synthesize fused-4,5-dihydro-1,2,3-triazolo[1,5-*a*][1,4]benzodiazepine **58**. This reaction follows a click-Ugi-Smiles pathway in refluxing methanol. Initially, propargyl amine **57** condenses with o-azidobenzaldehyde **55**, undergoing an intramolecular click reaction to form a tricyclic intermediate (**II**) (Scheme 14). This is followed by the nucleophilic addition of isocyanide and phenoxide ions (obtained

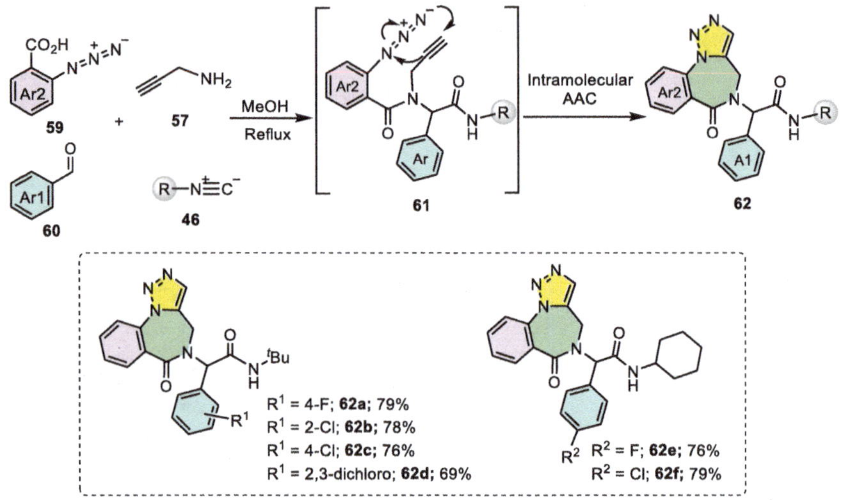

Scheme 15 Synthesis of 1,2,3-triazolo-benzodiazepines **62** via a Ugi-click strategy, as demonstrated by Shafie et al. [74]

from nitrophenol **56**), leading to intermediate **IV**. The subsequent Smiles rearrangement produces functionalized 1,2,3-triazole heterocycles **58** in good-to-excellent yields. Notably, the presence of electron-withdrawing groups, such as *o*-nitrophenol on the phenol ring, significantly enhances the reactivity of this multicomponent reaction (Scheme 14).

Shafie et al. [74] developed a practical and robust method for synthesizing a unique class of 1,2,3-triazolo-benzodiazepine analogues **62**, which have shown potent in vivo anticonvulsant activity. The synthesis involves reacting 2-azidobenzoic acids **59** with benzaldehydes **60** and propargylamine **57** in refluxing methanol, forming an adduct **61** that undergoes an intramolecular AAC click cycloaddition, producing the 1,2,3-triazole-benzodiazepine structure **62** (Scheme 15).

Additionally, this protocol demonstrates the versatility of click chemistry in creating biologically active compounds, in this case for treating convulsive disorders. The anticonvulsant properties of these compounds were tested in mice using pentylenetetrazole (PTZ) and maximum electroshock (MES)-induced seizure models. In the PTZ test, many of the compounds showed stronger anticonvulsant effects than diazepam (0.5 mg/kg). One of the compounds exhibited greater potency than diazepam in the PTZ screening, while **62e** and **62f** were highly effective in the MES screening. Further, compound **62f** showed less neurological side effects compared to diazepam, therefore offering improved anticonvulsant benefits with reduced neurological side effects. These findings provide valuable insights into neuropharmacology and could lead to the development of better treatments for seizure disorders.

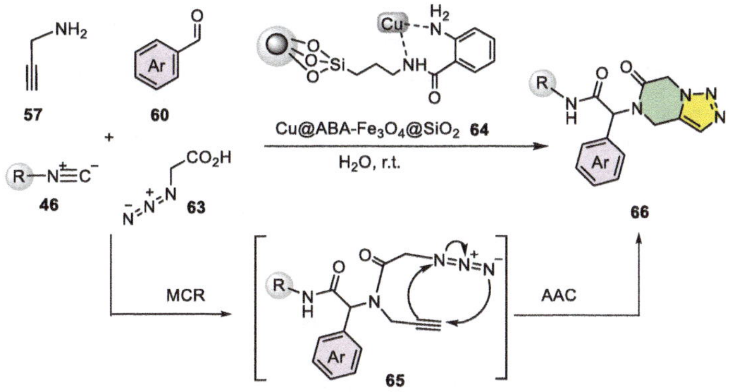

Scheme 16 Cu@ABA-Fe_3O_4@SiO_2-driven synthesis of triazolo-pyrazinyl acetamides **66** [75]

Mahdavi and co-workers [75] used Cu@ABA-Fe_3O_4@SiO_2 **64**, a magnetic silica nanoparticle-supported copper complex, for a4-component reaction of azides **63**, isocyanide **46**, aromatic aldehydes **60**, and propargylamine **57**, under one-pot conditions to provide 1,2,3-triazolo-fused pyrazinyl acetamides **66** in a high yield (Scheme 16).

This MCR proceeds through the formation of Ugi adduct **65** which, on subsequent intramolecular AAC reaction, results in triazolo-pyrazinyl-acetamides **66**. The method is versatile and high yielding, with wide substrate scope. Some of the resulted fused heterocyclic analogues displayed promising antibacterial potency against gram-positive and gram-negative bacterial strains, including *E. coli* and *Staphylococcus aureus*.

Some Cu-based halloysite clay catalysts were used to catalyze the CuAAC click chemistry. Toward this end, Kahangi and co-workers [76] reported the use of a halloysite nanoclay, for example, Hal-IMI-SO_3Cu **69**, as a suitable CuAAC catalyst for multicomponent reaction of alkyne **67**, α-halo ketone **68**, and sodium azide to give high-to-excellent yield of respective 1,2,3-triazole analogues **70** under one-pot condition (Scheme 17).

Other types of catalyst support have been investigated. For example, a multicomponent reaction of nitromethane, aromatic aldehydes, and NaN_3 with a catalytic amount of magnetically active $CuFe_2O_4$ yielded aryl-1*H*-1,2,3-triazoles in excellent yield [77]. An ionic liquid-based Cu(II) catalyst, prepared by mixing 1-(1-carboxymethyl)-3-methylimidazolium tetrafluoroborate with $Cu(OAc)_2$, was employed to access 1,2,3-triazoles under microwave (MW) irradiation in an aqueous solution for 20 min [78]. Furthermore, MW irradiation of 1-iodoalkynes and aryl azides using Cu(I) catalysts in the ionic liquid [BMIM]PF_6 under one-pot conditions can afford fused 1,2,3-triazole structures in high yield [79]. Likewise, "click" in MCR fashion by reacting alkynes, halides, and NaN_3 under a suitable catalytic system to access 1,2,3-triazoles as hybrid molecular architectures has been well explored in recent years. For example, Crowley et al. [80] demonstrated a one-pot MCR-CuAAC method, providing high yield of coveted polydentate 2-pyridyl-1,2,3-

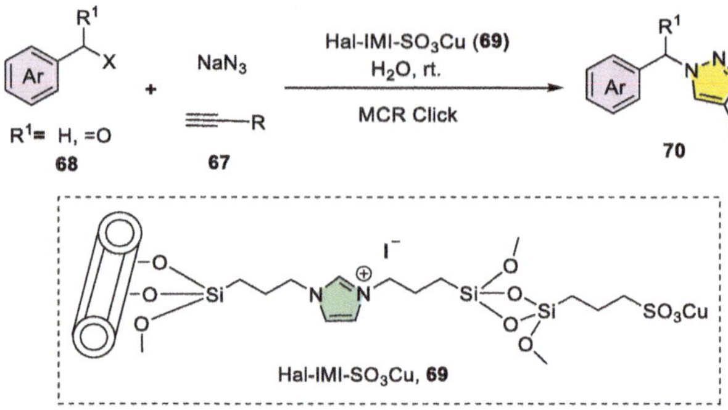

Scheme 17 Cu-based halloysite nanoclay-induced MCR to access 1,4-disubstituted 1,2,3-triazoles **70** under one-pot conditions [76]

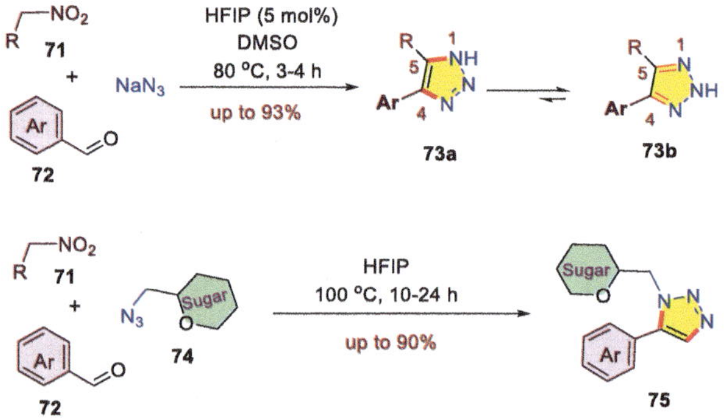

Scheme 18 HFIP-mediated MCR-click route for *N*-unsubstituted 1,2,3-triazoles **73** and *N*-substituted 1,2,3 triazoles **75**, developed by Tiwari et al. [81]

triazoles. These polydentate 2-pyridyl-1,2,3-triazoles after coordination with Ag (I) can be utilized as a suitable ligand to access distinct molecular and polymeric supramolecular metallo-architectures.

Recently, Tiwari and group [81] established an MCR route for synthesis of *N*-unsubstituted-1,2,3-triazoles **73** and *N*-substituted-1,2,3 triazoles **75** under 1,1,1,3,3,3-hexafluoroisopropanol (HFIP), a hydrogen bond-donating reaction medium. The authors performed the three-component reaction of aldehydes **72**, nitroalkanes **71**, and sodium azide or glycosyl azides **74** in 5 mol% of HFIP in DMSO (Scheme 18). This MCR tool has notable features including metal-free condition, high yields, functional group tolerance, wide substrate scope, operational simplicity, and ease of purification, scale-up, and one-pot mild conditions.

Tiwari's group [28] further extended the previous *p*-TSA-catalyzed [3 + 2] cycloaddition of nitroolefins **76** and sugar azides **74** to access respective 1,5-disubstituted 1,2,3-triazolyl glycoconjugates **75** [82]. The authors further used Schreiner thiourea **77** as an important organocatalyst for the successful cycloaddition of nitroolefins **76** and sugar azides **74** (Scheme 19). By applying organocatalysis followed by CuAAC,

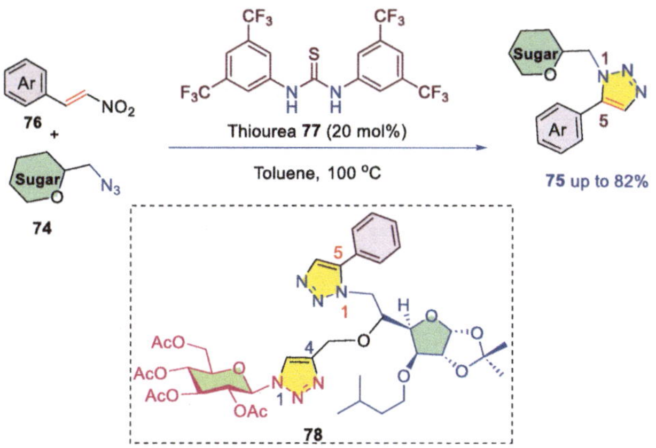

Scheme 19 Schreiner thiourea-catalyzed synthesis of 1,5-disubstituted 1,2,3-triazole glycoconjugates (**75, 78**)

the glycohybrid 1,2,3-triazole **78** was obtained in high yield. Catalyst **77** enables the required activation of nitroolefins **76** through double-hydrogen bonding. The method is straightforward, metal free, acid free, mild, operationally simple, and high yielding. This straightforward tool is orthogonal and regioselective and thus could be a complementary choice for the classical RuAAC method [82].

Furthermore, Tiwari and group recently used silicomolybdic acid as a versatile and biocompatible catalyst in one-pot protection of a diverse range of carbohydrates and also domino synthesis of bioactive heterocycles [83, 84]. Interestingly, the MCR of aldehydes, nitroalkanes, and sodium azide in the presence of 5 mol% of silicomolybdic acid can be used for the one-pot preparation of 4-aryl-*NH*-1,2,3-triazole analogues [85]. Although the characterization of regioisomeric 1,4-disubstituted 1,2,3-triazoles **3** and 1,5-disubstituted 1,2,3-triazoles **4** is difficult using ^{1}H NMR spectra, ^{13}C NMR can be used for assigning their chemical structure. For example, the appearance of C-5 signal in ^{13}C NMR at about δ 120 ppm for triazole **3** and C-4 signal at about δ 133 ppm for triazole **4** is quite helpful to distinguish between them [86]. The up-field shift of C-5 (δ 120 ppm) of 1,4-disubstituted 1,2,3-triazole compared to C-4 (δ 133 ppm) of 1,5-disubstituted 1,2,3-triazole was further validated by computational studies [87].

The Kurth group devised an $InCl_3$-catalyzed MCR of substituted propargylic amines **79**, α-diketones **80**, *o*-azidobenzaldehydes **55**, and NH_4OAc in refluxing methanol to give high yield of respective imidazo-1,2,3-triazolo-benzodiazepines **81** under one-pot condition (Scheme 20) [88]. This four-component MCR proceeds to the required azido-alkyne intermediate **V**. The reaction starts with the activation of *o*-azidobenzaldehydes **55** with $InCl_3$ and intermediate **I** and then reacts with NH_4OAC as ammonia source to give respective imine **II**. Consequent nucleophilic addition of propargylamine to imine **II** afforded intermediate **III** which on reaction with intermediate **IV** resulted in the desired azido-alkyne imidazole scaffold **V**. At the end, the

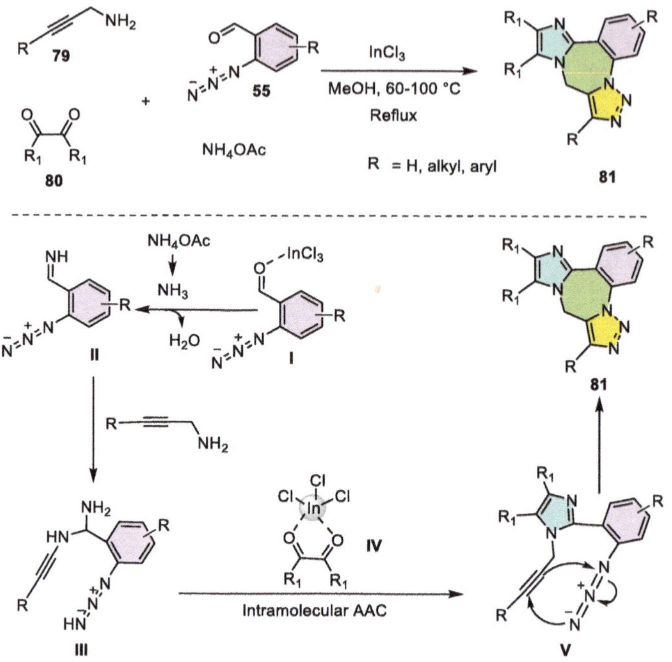

Scheme 20 $InCl_3$-induced MCR synthesis of imidazo-fused triazolobenzodiazepines **81**, demonstrated by Kurth et al. [88]

resulting azido-alkyne intermediate **V** underwent thermally induced intramolecular AAC, providing triazole-fused system **81** in a high yield. The method has some notable features including versatility, high reaction yield, broad substrate scope, one-pot mild conditions, and high atom economy.

Lee and co-workers developed a high-yielding two-step multicomponent coupling tool followed by intramolecular click chemistry to provide biocompatible tricyclic-fused pyrazine analogues in a diastereoselective manner [89]. The synthesis begins with an $AuBr_3$-promoted solvent-free reaction of cyclic amines **82**, aldehydes **60**, and terminal alkynes **83**, to obtain propargyl amine analogue **84**, which undergoes an intramolecular AAC click cycloaddition process in refluxing DMF to deliver the desired tricyclic-fused pyrazines **85** (Scheme 21). This MCR reaction has some notable features including an exceptional diastereoselectivity, versatility, and high reaction yields.

Likewise, Basso and co-workers extend classical alkylative Passerini three-component reaction of an isocyanide, a carbonyl compound, and a carboxylic acid under one-pot condition to produce an α-acyloxy amide [38] via a four-step MCR under one-pot conditions for the synthesis of tricyclic benzoxazepine **89** [90], as follows. The MCR is promoted by a TEMPO/PIDA combination, with the substrates consisting of propargyl bromides **87**, isocyanide **46**, and 2-azidobenzyl alcohols **86** at room temperature, to afford adduct **88** which can undergo a thermally induced

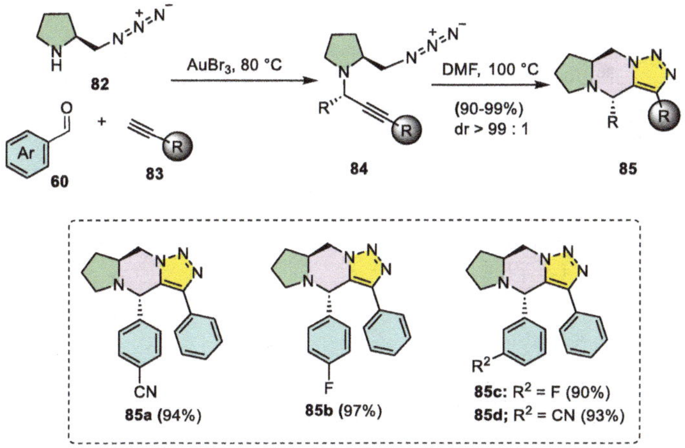

Scheme 21 Synthesis of tricyclic-fused pyrazines **85** via $AuBr_3$-promoted MCR approach

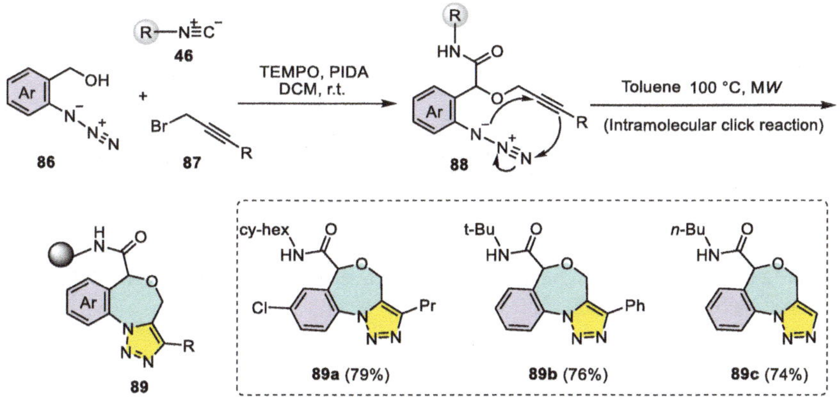

Scheme 22 MCR tool for the synthesis of benzoxazepines **89** under one-pot conditions, developed by Basso et al. [90]

intramolecular AAC click in refluxing toluene under MW exposure and can give the desired tricyclic benzoxazepine analogues **89** (Scheme 22). The method is practical and robust and nicely overcomes the challenges of the alkylative Passerini reaction.

A similar tandem reaction was developed by Arigela et al. [91] to provide a diverse class of 1,2,3-triazole-linked hybrid molecules in good yields without the need to isolate the intermediate involved during the reaction sequence. In this work, a sequential $Yb(OTf)_3$-catalyzed intermolecular Michael addition of 2-alkynyl indoles **91** to (*E*)-1-azido-2-(2-nitrovinyl)benzenes **90** in refluxing toluene gave respective azido-alkynes **92**, which – upon thermally induced intramolecular AAC under one-pot conditions – provided a robust approach to access indole-based tetracyclic-fused heterocycles **93** in high yield (Scheme 23).

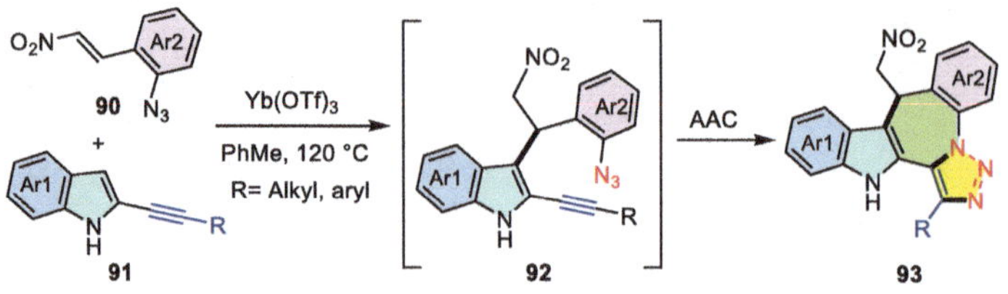

Scheme 23 Tandem synthesis of indole-based tetracyclic heterocycle **93** under one-pot condition, developed by Arigela et al. [91]

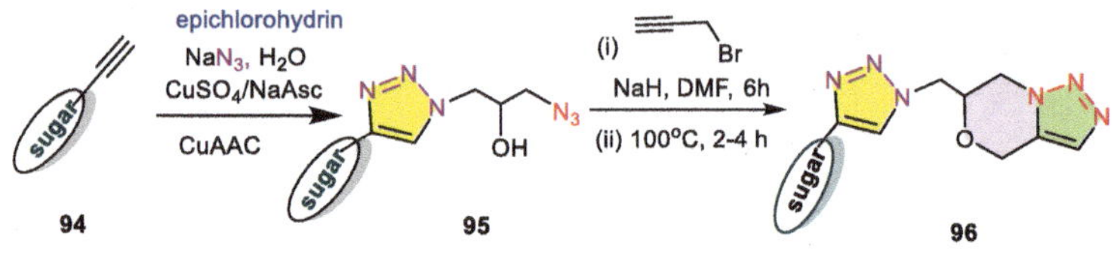

Scheme 24 One-pot synthesis of morpholine-fused triazole **96**, reported by Tiwari's group [92]

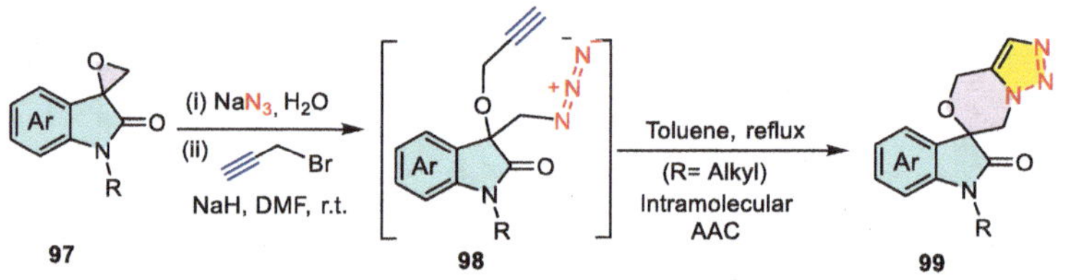

Scheme 25 One-pot synthesis of spiro-oxindole tetracyclic morpholine-fused 1,2,3-triazole **99** developed by Senwar et al. [93]

Furthermore, Mishra and Tiwari [92] developed a two-step click process enabling the facile synthesis of a series of morpholine-fused glycohybrid 1,2,3-triazoles **96**. In this process, glycosyl alkynes **94** were first treated with epichlorohydrin and NaN_3 to give the respective azido alcohol **95** [92]. The routine propargylation followed by successive thermally induced AAC as part of intramolecular click cyclization gave morpholine-fused glycohybrids **96** (Scheme 24).

Likewise, Senwar et al. [93] developed a thermally induced AAC of azido-alkyne **98**, obtained from respective spiro-oxirane **97**, to provide spiro-oxindole tetracyclic morpholine-fused 1,2,3-triazoles **99** in high yields (Scheme 25). The method used routine azidation via regiospecific ring opening, *O*-propargylation, and then intramolecular AAC in refluxing toluene under one-pot conditions. The spiro-triazole conjugate **99** displayed notable antiproliferative activity.

Interestingly, a Cu/Ag relay-induced step-efficient MCR tool has been established for easy access of optically pure triazole-pyrazine-fused system **104** [94]. Herein, a Cu/Ag-catalyzed asymmetric propargylic amination-hydroazidation of terminal alkyne **100** and amino-alkyne **102** using an *N,N,P*-ligand provided the

Scheme 26 MCR-click route to access 1,2,3-triazole pyrazines **104** under one-pot condition [94]

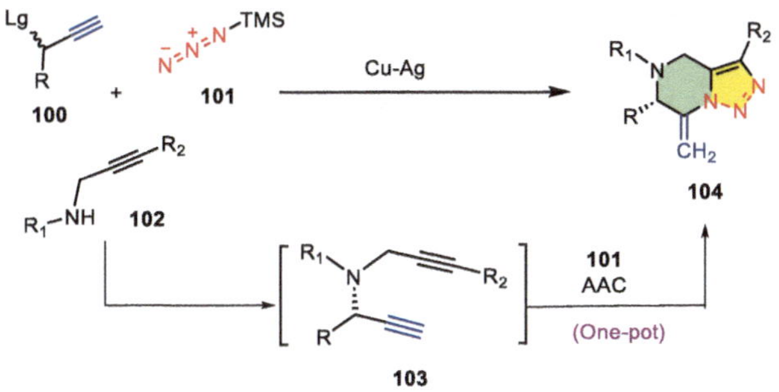

respective intermediate **103**, which on subsequent AAC with TMS-azide **101** gave the desired enantio-enriched 1,2,3-triazolo[1,5-*a*]pyrazine analogue **104** in a cascade fashion (Scheme 26) [94]. The method has notable features including one-pot conditions, regioselectivity, functional group tolerance, broad substrate scope, high yield, and moreover an excellent enantioselectivity as shown by enantiomeric excess (ee) values up to 97%.

Another interesting metal-free synthesis of functionalized 1,4,5-trisubstituted 1,2,3-triazoles **108** was devised by Gao et al. [95], in this case, to proceed in a regioselective manner. In this work, a three-component reaction of enolizable ketone **105** having α-CF_3 group, organic azides **107**, and amines, for example, pyrrolidine **106**, in THF/H_2O (3:1) at 50°C for overnight afforded polyfunctionalized 1,2,3-triazoles **108** (Scheme 27a). The mechanism proceeds through the base-catalyzed dehydro-fluorination of α-CF_3 ketone **105** that resulted in gem-difluoro vinyl ketone **I** which after successive nucleophilic addition-elimination reaction gave intermediate **II**. In case of R as an aryl or an alkyl, H_2O amalgamate with **II** gave the enolizable β-oxo amide intermediate **III**, which underwent AAC with aryl azide **107**, and the resulting intermediate **V** underwent removal of pyrrolidine **106** to give the desired 1,4,5-trisubstituted functionalized 1,2,3-triazoles **108** (Scheme 27a). Likewise, similar MCR using non-enolizable α-CF_3 esters **109** with azides **107** and pyrrolidine **106** under optimized conditions led to high yield of functionalized 5-amino-1,2,3-triazoles **110** exclusively as the post-functionalized complex bioactive candidates. Considering the mechanistic perspective, the intermediate **II** obtained from differently substituted α-CF_3 esters **109** underwent AAC with aryl azide **107**; the resulting triazoline intermediate **VI** gave high yield of the corresponding functionalized 1,2,3-triazole **110** after successive elimination of HF (Scheme 27b) [95].

Although C-H functionalization of 1,2,3-triazole is difficult, it is feasible [96]. Xu and co-workers nicely used MCR route for the modular synthesis of trisubstituted 1,2,3-triazoles **115** under one-pot mild reaction condition (Scheme 28) [97]. As a key step, vinyl copper intermediate **II** was resulted via an unprecedented Cu/Pd transmetalation relay process. The method has notable features as it allows an easy installation of required groups at three positions of 1,2,3-triazole heterocycle. The Cu(I) acetylide **I** on AAC with organic azide **112** under optimized condition gave cuprate-triazole **II**. Simultaneously, aryl halide **113** underwent oxidative addition to

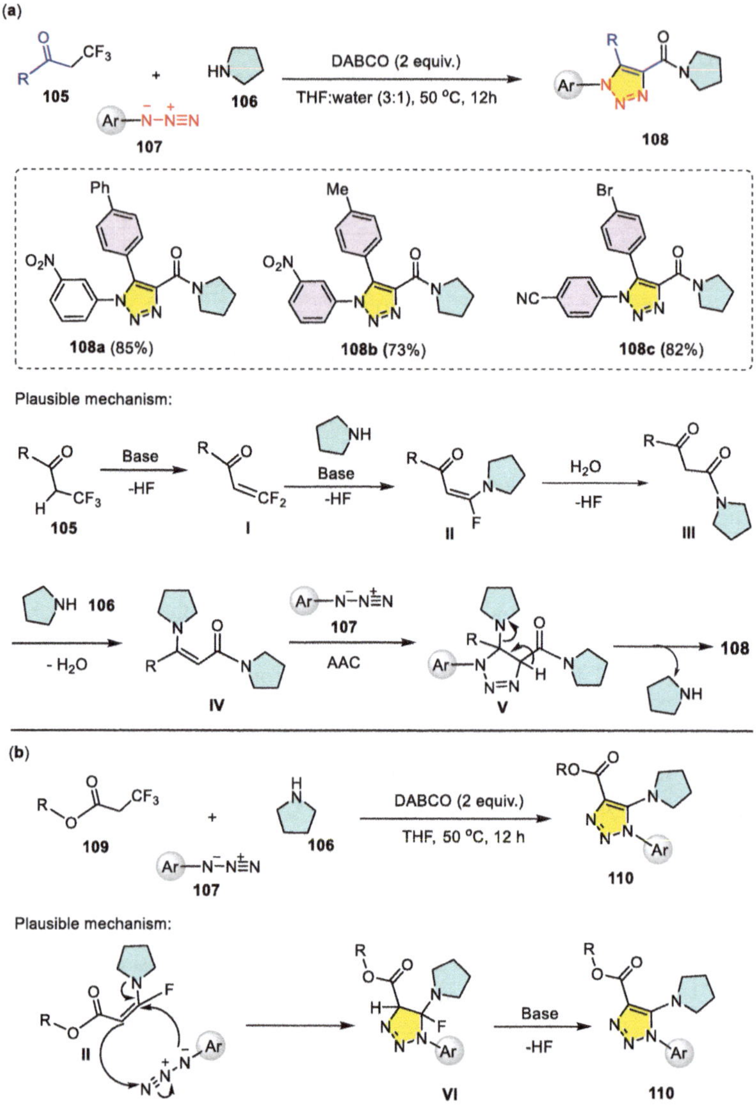

Scheme 27 (**a**) Organo-catalyzed MCR route to access functionalized 1,4,5-trisubstituted 1,2,3-triazoles **108** from α-trifluoromethyl ketones **105** and (**b**) 5-amino-1,2,3-triazoles **110** from α-CF_3 esters **109** [95]

a Pd(0) catalyst, generating the palladium intermediate **III** which was then transmetalated with intermediate **II**, resulting in the formation of the desired trisubstituted 1,2,3-triazole **115**. Finally, reductive elimination resulted in regeneration of the Pd(0) catalyst. On protonation, the cuprate-triazole intermediate **II** provides the undesired 1,4-disubstituted 1,2,3-triazole **116** (route a). Similar to a typical

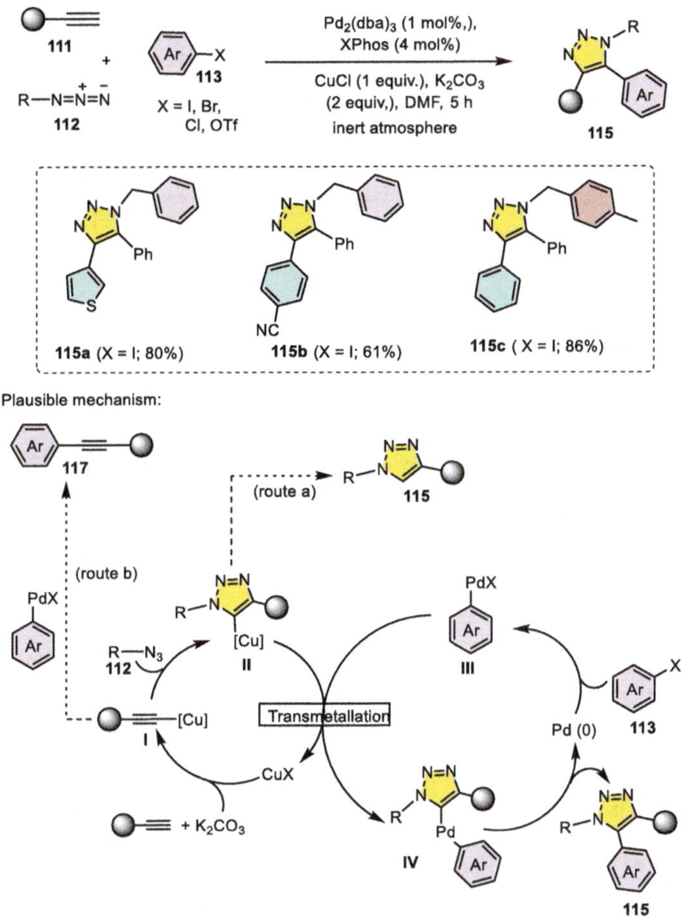

Scheme 28 MCR route to access trisubstituted 1,2,3-triazole **115** under Cu-Pd catalytic system [97]

Sonogashira coupling, the transmetalation of copper(I) acetylide **I** with Pd intermediate **III** may give the by-product **117**(route b) [97]. Since the two inevitable competing reactions may proceed, to promote the ideal catalytic cycle is challenging. To avoid route a, it is necessary to increase the rapidity of the transmetalation step and at the same time to eliminate all the possible sources of protons.

Xu et al. [98] reported a facile decarboxylative CuAAC method in tandem fashion to access high yield of functionalized 5-arylselanyl-1,2,3-triazoles **121**. Thus, an MCR of propiolic acid **119**, diselenide **120**, and organic azide **118** in the presence of $Cu(OAc)_2{\cdot}H_2O$ (10 mol%), K_2CO_3 (1.2 equiv.), and 1,10-Phen (20 mol%) in refluxing toluene for 6 h provided the desired functionalized 1,2,3-triazoles **121** with excellent regioselectivity and in high yields (Scheme 29) [98].

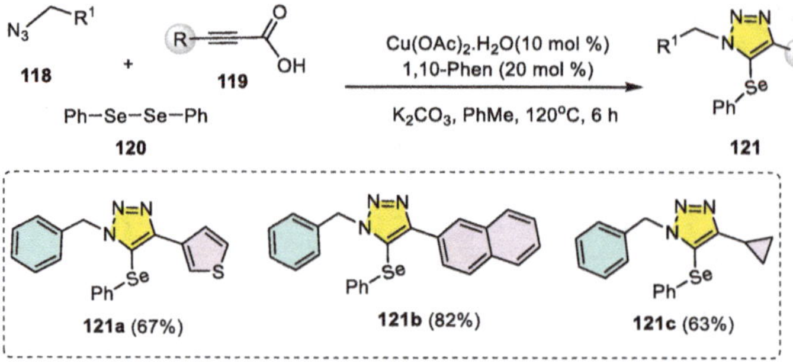

Scheme 29 Decarboxylative/click cascade tool for easy access of 5-arylselanyl-1,2,3-triazoles **121** [98]

Compounds having a 1,2,3-triazole ring are highly significant due to their biological relevance and their role as synthetic intermediates in various reactions [99–101]. The CuAAC-tandem MCR methods are currently recognized for their ability to rapidly generate 1,2,3-triazole-fused or isolated systems in multicomponent synthesis [11, 102, 103].

3.2 MCR-Click Chemistry in Macrocyclization

Macrocycles having diverse ring systems are an integral component of bioactive natural products and possess notable applications in various fields such as perfumery, metal complexation, sensing, host-guest supramolecular chemistry, and drug discovery and development [104]. The associated multifaceted nature of such well-defined macrocyclic systems has been widely explored in medicinal chemistry. For example, clarithromycin and their synthesis analogues displayed promising anticancer activity, whereas the erythromycin macrocycle is well known for their antibacterial potency [105–107]. Because of the associated diverse range of pharmacological activities with such macrocycles, their short and high-yielding synthesis is of paramount importance.

The combination of "click chemistry" and MCRs is an emerging approach in the field of macrocyclization. MCRs are reactions where three or more reactants combine in a single reaction step to form complex molecules. Several strategies such as CuAAC, isothiocyanate-based MCRs, aza-Michael addition, and cyclization via amidation and urethane formation have been developed to combine MCRs with click chemistry for the macrocyclization purpose. However, the most prominent one in recent years is the 1,2,3-triazole-forming click chemistry. The advantage of MCR-based click chemistry is that it allows for the rapid access of diverse macrocyclic structures from simple starting materials. Several groups have recently reported the synthesis of 1,2,3-triazole-containing macrocycles via an intramolecular variant of click chemistry [30]. In addition, the required azido-alkynes may be easily

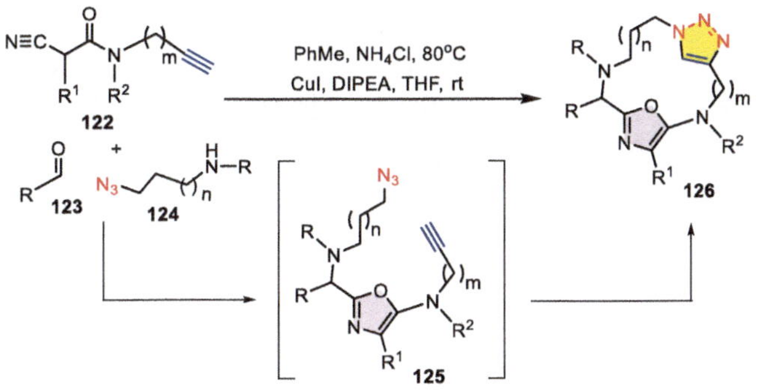

Scheme 30 The one-pot tandem synthesis of isoxazole-appended 1,2,3-triazole-containing macrocycle **126**, demonstrated by Pirali et al. [109]

achieved in high yield by exploiting the facile MCR concept. Thus, the established MCR method can offer advantages for the easy synthesis of a diverse range of sugar-containing macrocyclic architectures [108]. For example, in 2006, Pirali et al. [109] exemplified a click chemistry-mediated MCR as a convenient macrocyclization route and developed isoxazole-appended 1,2,3-triazole-containing macrocycles **126** under one-pot condition. MCR was explored for the sequential coupling of alkyne-bound acetamide-nitrile **122**, aldehydes **123**, and amino azides **124** using NH_4Cl in refluxing toluene that led to respective isoxazole-based azido-alkyne **125**, which then underwent intramolecular AAC using CuI/DIPEA in THF at rt to give high yield of desired macrocycle **126** (Scheme 30) [109].

Likewise, Prasanna and group devised a spontaneous MCR route followed by CuAAC-mediated macrocyclization, which was the key step to easy access of sugar-appended macrocycles **130**. Thus, a three-component reaction of *N*-alkyl azidoisatin **129** with secondary amino acid **128** and glucose-derived *O*-propargylenone **127** using CuI/DIPEA in acetonitrile at rt for 8 h followed by refluxing in the same pot for an additional 12 h gave high yield of the desired of 1,4-triazole-containing macrocycles **130** (Scheme 31a) [110]. The method was further extended to give biocompatible macrocycle **133** by reacting glucose-derived *O*-propargylenone **127** with *N*-benzyl azidoisatin **132** and heterocyclic amine **131** under optimized one-pot intramolecular click conditions (Scheme 31b) [110].

Furthermore, Rao and co-workers used azomethine ylides and developed a series of biocompatible triazole-linked glycol macrocycles under one-pot conditions [111]. In this work, a three-component reaction of *O*-propargyl nitroolefin **134**, salicylaldehyde-derived *O*-alkyl azides **135** and secondary amino acid **136** using CuI/DIPEA in refluxing AcCN provided 1,4-triazole-linked glycomacrocycles **137** in high yields (Scheme 32) [111].

Further, Miranda and group devised an MCR-click tool to access tryptamine-triazole-containing macrocycle **102** [112]. A four-component reaction of indole-derived amine-azide **138** (n = 1–5), isonitriles **46**, *O*-propargylated benzaldehyde **139**, and carboxylic acid **140** in methanol at rt gave the required Ugi adduct **141**, which was then subjected to MW-assisted macrocyclization using CuBr/DBU as the

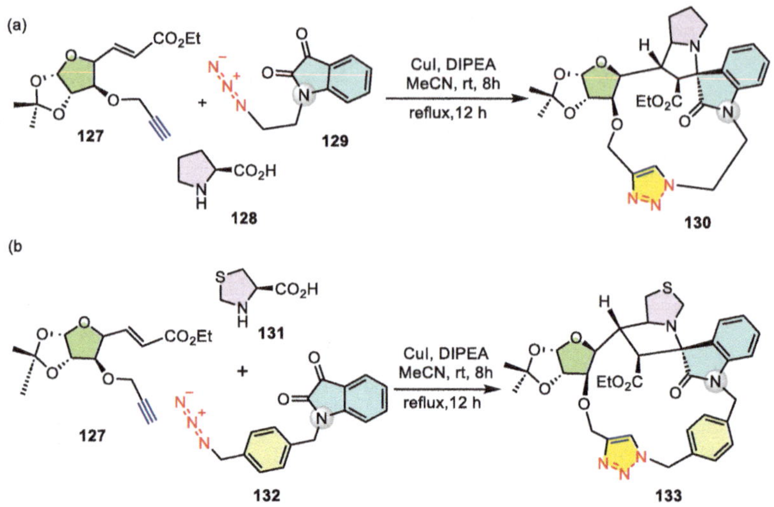

Scheme 31 The extended one-pot tandem synthesis of 1,2,3-triazole-containing macrocycles **130** and **133**

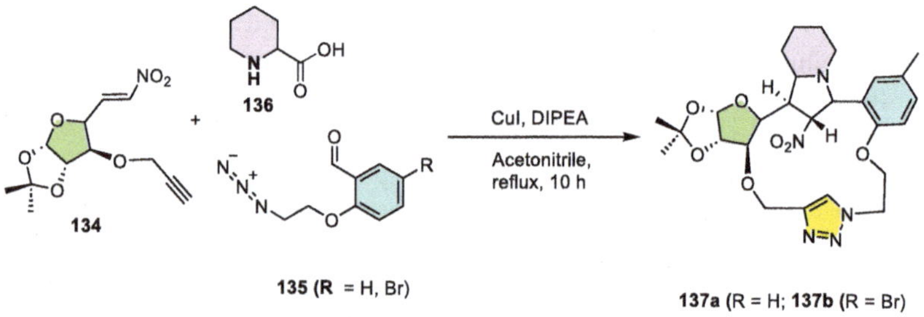

Scheme 32 A tandem AAC route for access of functionalized macrocycles **137**, demonstrated by Rao et al. [111]

catalytic system in toluene at 110°C, forming the desired tryptamine-based macrocyclic compounds **142** in high yields (Scheme 33) [112].

A Ugi-click tool similar to that depicted in Scheme 32 was developed by Zakharova et al. [113] to access diverse macrocyclic peptidomimetics through the incorporation of 1,2,3-triazole residues. The researchers demonstrated that azido-functionalized isocyanides and acetylenic functionality in acid components can be used to give respective bifunctional peptidomimetics via click-induced macrocyclization. For example, MCR of ethynylbenzoic acids **143**, azido-functionalized isocyanide **144**, DMB-amine **145**, and carbonyl compounds **166** in methanol at rt gave a bifunctional peptidomimetic **147**, which was then subjected to macrocyclization using CuI/DIPEA/lutidine as CuAAC catalytic system to provide a diverse range of 12-membered cyclomonomers **148** and 24-cyclodimers **149** in high yields (Scheme 34) [113]. The distance between the azide and the alkyne

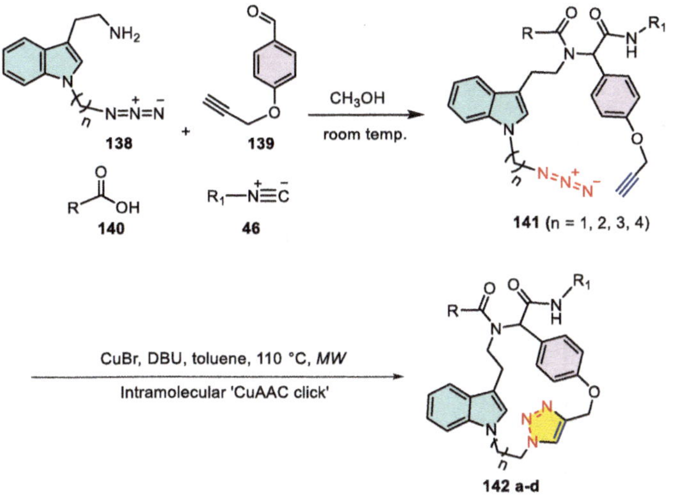

Scheme 33 Ugi-AAC reaction for easy access of tryptamine-based macrocycles **142**, developed by Miranda et al. [112]

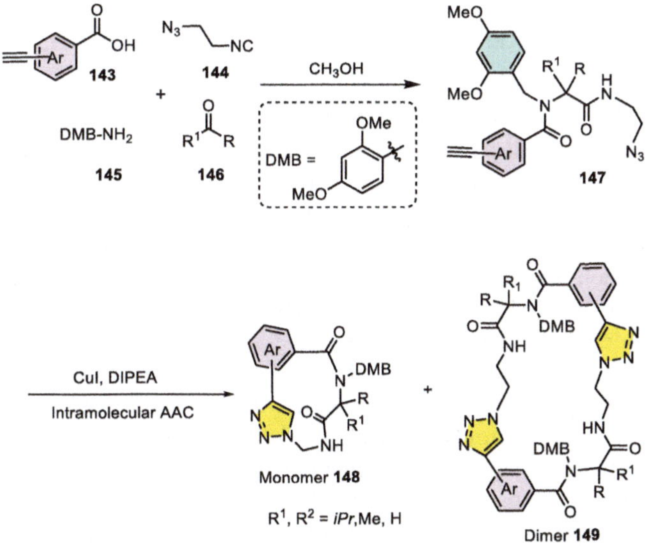

Scheme 34 Tandem Ugi-click for easy access of peptidomimetics (**148** and **149**), as demonstrated by Zakharova et al. [113]

functionalities in the Ugi adducts **147** controls the selectivity in macrocyclization step. The resulting macrocyclic architectures have notable significance in various emerging areas such as sensing and medicinal chemistry.

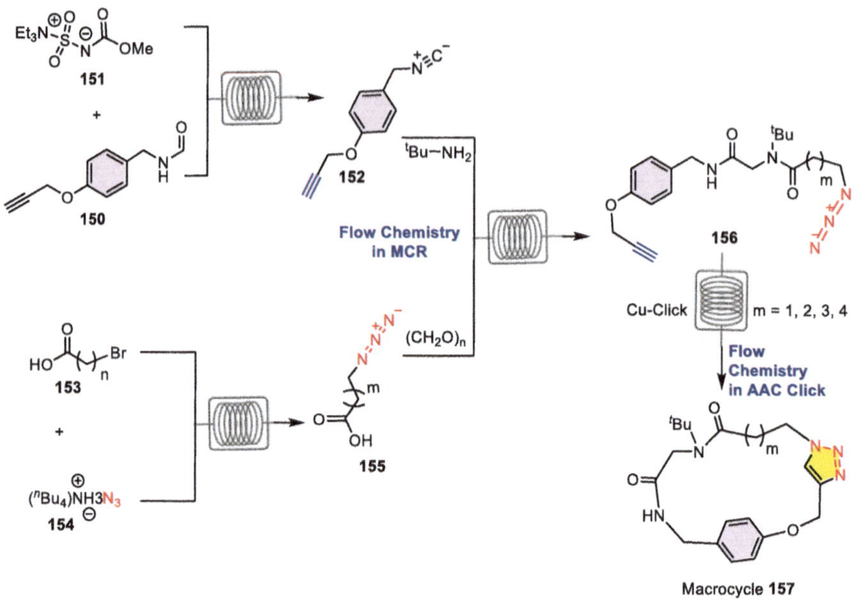

Scheme 35 MCR-intramolecular AAC in continuous flow to give cyclic peptides **157**, as demonstrated by Kappe and researchers [114]

Interestingly, macrocyclization through a similar MCR tool was performed using continuous flow chemistry (CFC) by Kappe and co-workers [114]. The researchers demonstrated this with linear peptoids **118** for AAC-mediated macrocyclization, resulting in the desired macrocyclic systems **119** in high yields. The method followed the dehydration of *N*-formamides **150** via nucleophilic substitution using *N*-(triethylammonium sulfonyl)carbamate (Burgess reagent) **151** that resulted in the continuous formation of desired isocyanide **152**. The azidation of bromides **153** with tetrabutylammonium azide (TBAA) **154** gave azide-functionalized carboxylic acid **155**. A combination of in situ-developed isocyanides **152** and azides **155**, respectively, with *t*-butyl amine and paraformaldehyde through Ugi-4CR at 100°C gave high yields of linear peptoids **156** (80–95%). At the end, intramolecular AAC in continuous flow reactor at 140°C resulted in the formation of desired monomeric or dimeric cyclic peptide macrocycle **157** in 80% yield (Scheme 35). This CFC route allows the development of desired cyclic peptide macrocycles in an efficient manner and offers a promising tool for notable application in academia and industry.

Undoubtedly, the MCR-click tool represents a promising intersection of two powerful synthetic strategies that can simplify the process of macrocyclic compound formation. By allowing the efficient and selective creation of diverse cyclic structures, this approach has a vast potential in drug discovery, materials science, and chemical synthesis. As new reactions and methodologies continue to be developed, the future of MCR-click chemistry in macrocyclization looks bright, with increasing applicability across various scientific disciplines. The synergy between MCR and click approaches offers a versatile toolset for creating new macrocyclic compounds with potential applications in medicine, biomaterials, and catalysis.

3.3 CuAAC/MCRs in Sequential Fashion

A sequential MCR-click reaction is an advanced synthetic method where multiple reactants are combined in a step-by-step manner to form complex, functional molecules. This approach typically involves a series of one-pot reactions that are executed sequentially, to furnish targeted compound. In a sequential MCR click reaction, the first stage typically involves a traditional MCR where three or more components react together to form a stable intermediate. Then, subsequent "click" reactions – often based on cycloaddition, click chemistry principles like the AAC reaction, or other robust bond-forming reactions – are carried out in a controlled sequence. This method ensures high yields, minimizes purification steps, and is ideal for creating libraries of compounds with diverse architectures, such as in the development of functional materials, pharmaceuticals, or biologically active molecules.

In addition to high demand of multicomponent reactions (MCRs) in click chemistry, there are several reports dealing with AAC in a sequential fashion to give the desired assembly of *bis*-, tris-, and poly-1,2,3-triazoles. *Bis*-1,2,3-triazoles can be used as ligands due to their conducive coordination ability with a variety of transition metals [115–118]. Toward this end, Simpson et al. used a sequential CuAAC tool and reported a facile method for one-pot access of 1,1′-disubstituted *bis*(1,2,3-triazole)s (Scheme 36) [119]. The method involves the AAC of benzyl azide and

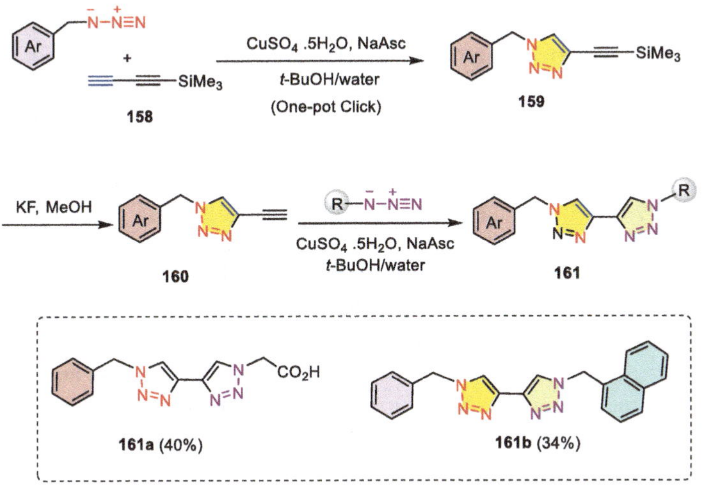

Scheme 36 Sequential CuAAC tool to access unsymmetrical 1,1′-disubstituted *bis*-1,2,3-triazoles **161**

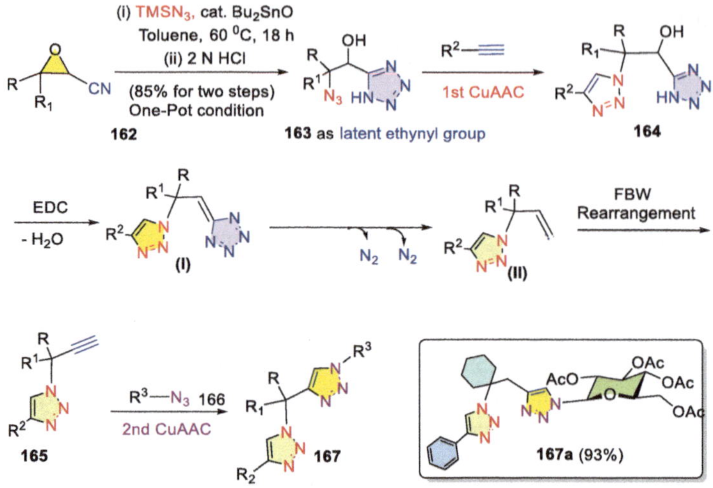

Scheme 37 Synthesis of α-hydroxy-β-azido tetrazoles and its use as double click for interesting triazolamers **167**

mono-silylbutadiene **158** to give TMS-protected ethynyl triazole **159** along with symmetrical *bis*-1,2,3-triazole as side product. TMS protection of ethynyl triazole **159** was removed by treating it with KF in MeOH, and the resulting alkyne **160** further underwent sequential AAC with various azides to give a high yield of unsymmetrical 1,4-disubstituted *bis*-1,2,3-triazole analogues **161** [119]. The method has notable features including reaction yield, ease of reaction to be feasible, and moreover sufficient stability of mono-silylbutadiene in comparison to TMS-protected butadiene.

Wright et al. [120] elegantly explored a double-CuAAC regioselective as one carbon staple for orthogonal tool and developed high yields of important triazolamer structures. For example, α-hydroxy-β-azido tetrazoles **163**, obtained from α,-β-epoxynitriles **162** as latent ethynyl residue in a one-pot fashion, underwent CuAAC with alkyne to give respective triazole **164**. This compound on reaction with EDC resulted in dehydration followed by extrusion of N_2, and then Fritsch–Buttenberg–Wiechell rearrangement (via vinylidene carbene intermediate **II**) gave the required alkyne **165**, which was on second consecutive CuAAC click with azide **166** that gave high-to-excellent yield of *bis*-triazole **167** (Scheme 37) [120].

Aromatic and aliphatic alkynes have difference in reactivity in CuAAC coupling with azides, and this reactivity difference was exploited by Zhu and co-workers to prepare a *bis*-triazole **171** through consecutive click reactions. The authors devised a three-component chemoselective CuAAC tool under one-pot condition for the successful coupling of two distinct alkynes **169** and **170** with unsymmetrical *bis*-azides **168** having one azide as the chelating one while another one to be a non-chelating residue. This chemistry also works in an aqueous solution at a pH = 7 and is useful therefore for bioconjugations (Scheme 38) [121].

Scheme 38 One-pot three-component chemoselective ligation for *bis*-triazole **171**

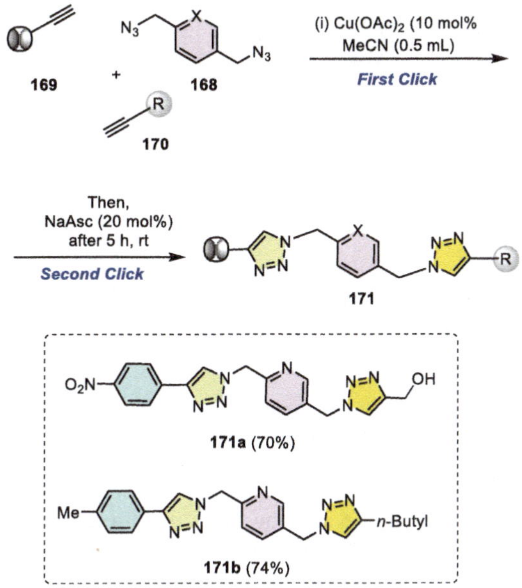

Scheme 39 Successive one-pot route for chemoselective access of *bis*-triazole-linked peptide **176**, demonstrated by Aucagne and co-workers [122]

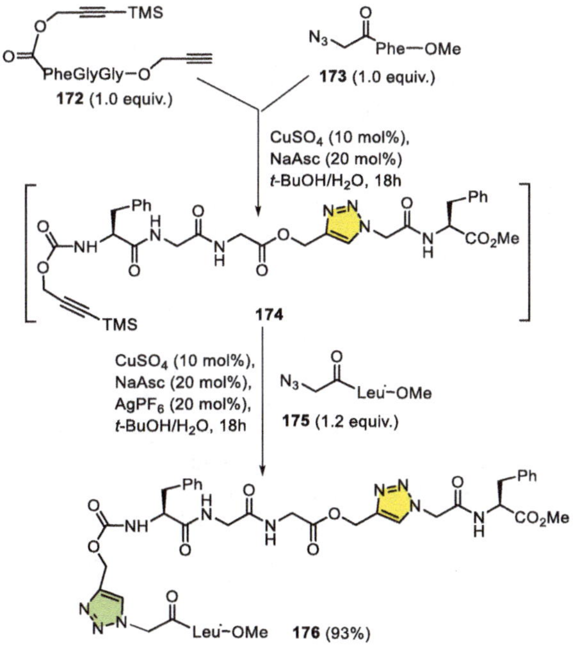

Aucagne and co-workers exemplified a regiospecific three-component sequential click tool to access 1,2,3-triazole-peptide ligation products in one-pot conditions (Scheme 39) [122]. The reaction of a functionalized tripeptide **172** having one

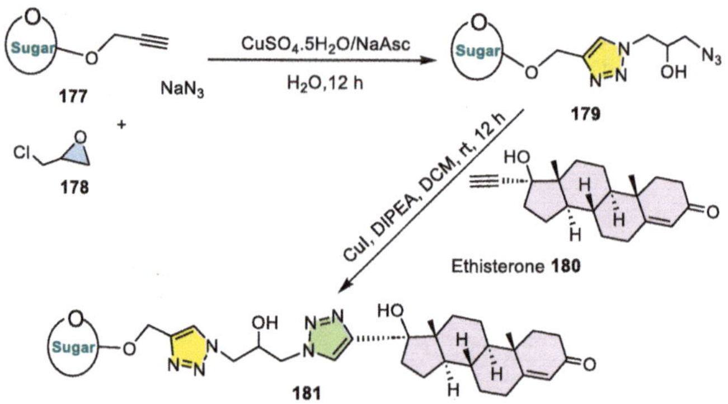

Scheme 40 CuAAC-induced synthesis of *bis*-triazolyl-ethisterone glycoconjugates **181** under one-pot MCR conditions

terminal alkyne and the other TMS-protected alkyne residue on CuAAC coupling with azide **173** under mild heating conditions proceeds regiospecifically, and the respective 1,2,3-triazole-linked ligation product **174** was obtained as the sole product. The azide **173** underwent chemoselective AAC with the terminal alkyne **172** using $CuSO_4$/NaAsc in *t*-BuOH/H_2O. The catalytic loading is important for the chemoselectivity as AAC performed in more amount of CuAAC catalyst brings the partial TMS deprotection, which may proceed for the subsequent $AgPF_6$-mediated CuAAC with another azide **175** to give high yield of desired *bis*-triazole **176** (Scheme 39) [122].

Tiwari group at BHU reported a one-pot three-component reaction of sugar alkyne **177**, epichlorohydrin **178**, and NaN_3 under CuAAC conditions to give 1,2,3-triazolyl azido-alcohols **179** via in situ azidation of oxirane analogue followed by CuAAC "click." The resulting azido alcohol **179** was then subjected to a second click conjugation with ethisterone **180** to give *bis*-triazolyl ethisterone glycoconjugates **181** (Scheme 40) [123]. The developed ethisterone conjugates could be an important candidate to lower levels of androgen receptor AR transcriptional activation than dihydrotestosterone and can also compete with AR binding; [123] however, a bioassay has not yet been reported for these compounds.

Jenson and co-workers coined a term "combining click reactions" [124] to address MCRs coupled with click chemistry. The authors first used photoinduced thiol-ene click of thiol **182** and alkyne **183** to incorporate desired trisaccharide mimic intermediate **184**, and then the resulting azide underwent regiospecific CuAAC with alkyne **185** under one-pot conditions to furnish the linear trisaccharide mimic **186** (Scheme 41) [125].

Sulfur(VI) fluoride exchange reaction (SuFEx Click) is another important variant of click chemistry [126], modular in nature and widely explored to obtain a diverse range of functional molecules of notable applications in chemistry, medicine, and materials science. Interestingly, the SuFEx click followed by accelerated PFEx click and then regioselective metal-catalyzed AAC (both CuAAC and RuAAC) and

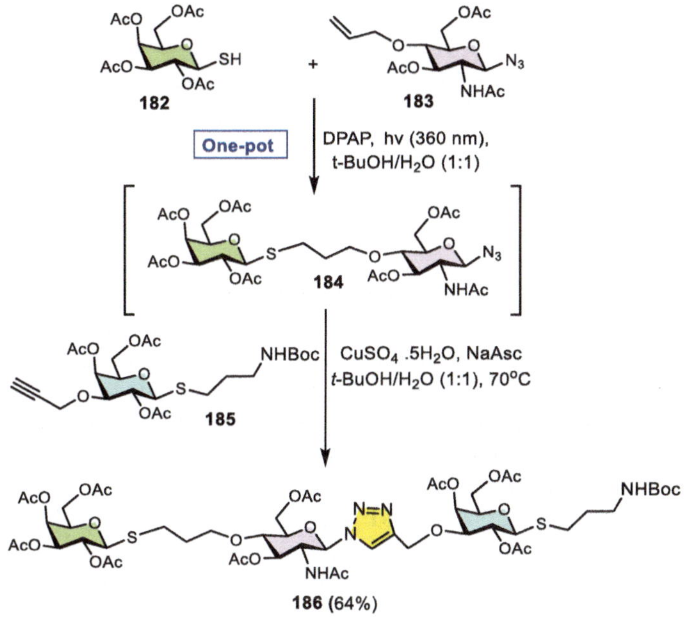

Scheme 41 CuAAC-mediated syntheses of the Gal-GlcNAc-Gal trisaccharide mimic **186**

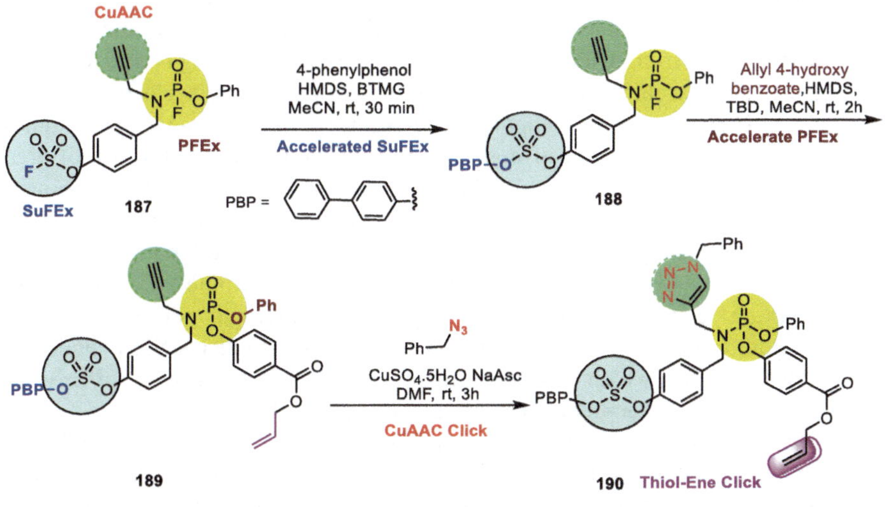

Scheme 42 Sequential SuFEx, PFEx, and CuAAC tool as a new variant of "click chemistry"

finally a thiol-ene click can be performed in consecutive fashion to give respective molecular architecture **190** of high demand in medicinal chemistry and functional material (Scheme 42) [127, 128].

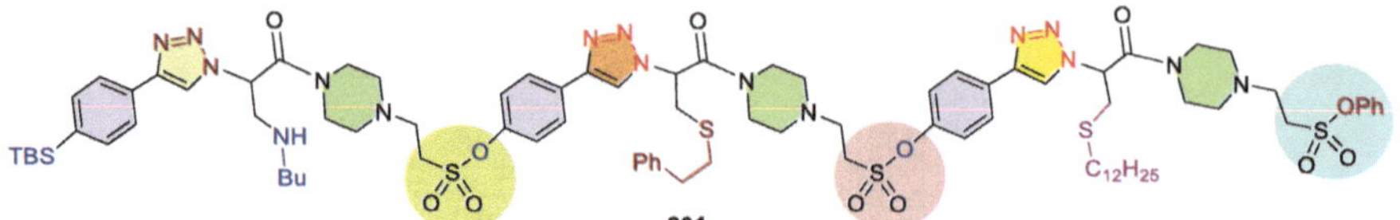

Fig. 1 Chemical structure of multi-1,2,3-triazole-functionalized molecular architecture **201** via iterative click conjugations

In recent years, there has been a lot of interest in multivalent platforms for the development of exceedingly functionalized molecular scaffolds through sequential click variant protocols such as acid-amine coupling, 1,2,3-triazole-forming click (e.g., CuAAC, RuAAC), SuFEx, PFEx, Michael addition, and thiol-ene click [129–141]. To this end, Yoshida and group have reported a highly expeditious synthesis of multi-triazole-functionalized molecular architecture **201** (Fig. 1) by utilizing a series of iterative click conjugation reactions [129].

Thus, a selective tris(triazole) **201** was obtained in good yield through the iterative click reactions starting from azido-functionalized trivalent platform **193**. This scaffold was obtained from respective acid and amine under standard acid-amine coupling conditions and served in the sequential click assembly by the SuFEx click with PhOH followed by CuAAC click conjugation with alkyne **195**, and then accelerated thiol-ene click reaction with dodecylthiol led to desired scaffold **197**. The resulting molecule was again reacted with trivalent platform **193** via accelerated SuFEx followed by modular CuAAC conjugation with alkyne **195** and thiol-ene click with 2-phenyl ethyl mercaptan to give *bis*-triazole **199**. This molecule was subjected to SuFEx click with compound **193** followed by regioselective CuAAC with alkyne **195**. At the end, compound **200** was subjected to aza-Michael addition with butyl to give high yield of desired functionalized tris(triazole) **201** of great significance as peptide bioisostere (Scheme 43) [129]. Interestingly, all the sequential synthetic steps such as SuFEx, CuAAC, and thia-Michael reactions were chemoselective and high yielding with good functional group tolerance.

Furthermore, Quin and group [130] developed a synthetic route to 2-azidoethane-1-sulfonylfluoride in 50 g scale in 87% yield from 2-chloroethane-1-sulfonyl fluoride. The author used this reagent to access 1,2,3-triazole-linked S(VI)-F analogues via modular CuAAC, which was then further explored for consecutive SuFEx using orthogonal click to generate dendrimeric architecture [130].

The abovementioned examples discussed in this section demonstrate the importance of sequential click reactions. A number of interesting trivalent and also tetravalent platforms have been well explored for the sequential click reactions [129–141]. Such sequential click reactions have widely explored in recent years to access a high yield of diverse range of fused heterocycles, conjugates, and macromolecular architectures of notable applications in drug discovery and materials science [112, 142–145].The novelty of this technique lies in its ability to combine the reliability and ease of click chemistry with the flexibility of multicomponent reactions, resulting in an effective tool for creating complex molecules with

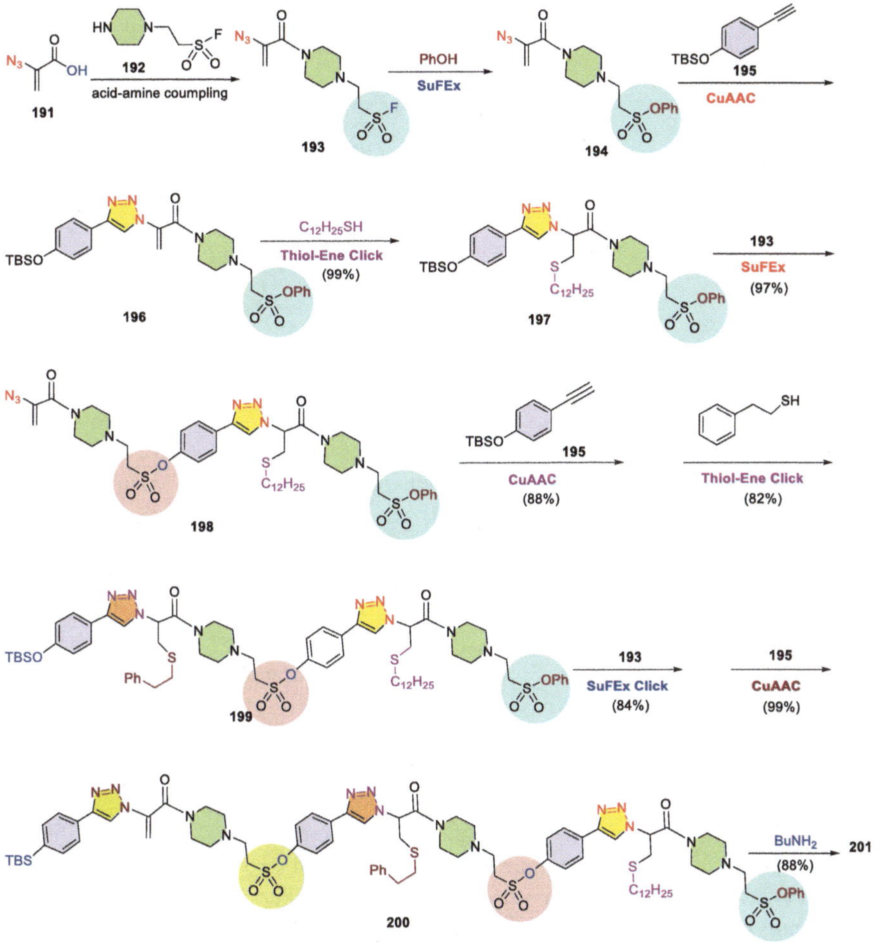

Scheme 43 Sequential iterative click reactions with azido-functionalized trivalent platform

controlled precision. By integrating multiple reaction steps, sequential MCR click reactions can offer increased synthetic efficiency, enhanced selectivity, and ability to create highly complex molecular structures with fewer synthetic routes. This concept has broad applications in drug discovery, materials science, and creation of bioconjugates.

3.4 *"MCR Click Chemistry" in Bioconjugation*

Bioconjugation refers to the process of chemically linking biomolecules (such as proteins, peptides, nucleic acids, and some small molecules) to other molecules such

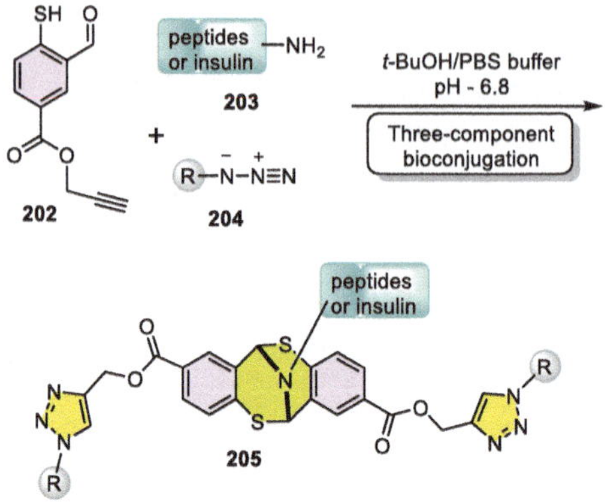

Scheme 44 Three-component one-pot bioconjugation via conventional click chemistry

as drugs, dyes, or diagnostic agents to enhance their therapeutic or diagnostic properties. This bioconjugation technology is pivotal in the development of targeted therapies, diagnostic tools, and personalized medicine. MCR-click chemistry is emerging as a versatile and powerful tool in bioconjugation due to its high efficiency, specificity, and mild reaction conditions. The integration of MCRs with "click chemistry" offers a simplified approach to functionalize biomolecules with various groups, facilitating more effective conjugation. While MCR-click chemistry is already being applied in several bioconjugation strategies, its future potential holds exciting possibilities.

The development of novel drugs can be facilitated by the chemical alteration of peptides and proteins in a sequential multicomponent fashion that may improve the investigation for the development of an economical and effective drug candidate [146, 147]. A notable example includes the chemoselective amine modification protocol by exploiting the unique reactivity of thiol group toward imines. The method is used to generate a required bifunctional linker **205** by utilizing a three-component bioconjugation of mercaptobenzaldehyde **202**, amino peptides **203**, and organic azide **204** under one-pot condition (Scheme 44). The mercaptobenzaldehyde-based bifunctional linker **205** lasts exceedingly and may be used for the selective amine labeling under biocompatible conditions. In addition to the ease of generation, the linker has additional feature as functional group tolerance. Notably, the presence of this biocompatible linker is not hampered by the action of modified proteins such as insulin [148].

A number of azide transfer reactions are well known for the successful in situ generation of desired azide-functional candidate of possible use in subsequent stage of reaction [149]. For example, Mancheno and co-workers explored a Cu-induced one-pot robust tool for required sp^3 C-H activation via azidation of unprotected reducing sugar example, tetrasaccharides **206**, and the resulting azidoglycoside (in situ) was then proceeded for CuAAC conjugation with propargylated glycine at the side chain of peptide **207** to give a high yield of respective 1,2,3-triazole glycopeptide conjugate **208** under one-pot condition (Scheme 45) [150].

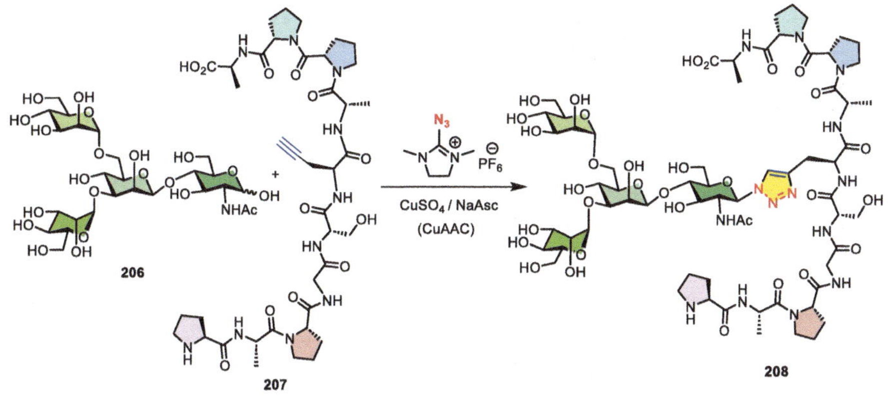

Scheme 45 One-pot synthesis of glycopeptide **208** via consecutive diazotransfer CuAAC

CuAAC click coupling of azide **209** and alkyne **210** has been explored to deliver good yields of nonionic xylo-nucleoside dimmers **211** [151]. Interestingly, a similar AAC of two complementary oligonucleotides resulted in complex molecular architecture of dsDNA catenanes. This structure comprises six helical turns with T-T mismatch on any part of ten base pairs in dsDNA catenane-based molecular architectures. Two strands having a clickable 5′-alkyne and also a clickable 3′-azide partner can easily proceed to stepwise click conjugation and deliver the title dsDNA catenane. Later, the Cu(I)-induced regioselective non-templated intramolecular AAC of the first oligonucleotide can easily undergo macrocyclization. The resulting macrocyclic system **212** can be utilized as a template for the consecutive click macrocyclization with linear ODN **213** leading to the formation of dsDNA catenane **214** (Scheme 46) [152].

Further, Sharpless and group exemplified a biocompatible thionyl tetrafluoride-derived connective hub for the convenient SuFEx bioconjugation of iminosulfuroxydifluorides to amine-tagged single-stranded DNA and BSA proteins. The biocompatible tool can be used in consecutive click reaction, and the connective hubs have a notable application in the biological system [153]. Despite seminal developments in MCRs for click diversification, the method still requires further modification to be a possible exploration in bioorthogonal fashion in living systems [154].

MCR-click chemistry could shape the future of bioconjugation such as advanced targeted drug delivery systems, precision radiopharmaceuticals, peptide and protein engineering, biomolecular imaging and diagnostics, optimized antibody-drug conjugates (ADCs), bioconjugation in vaccine development, synthetic biology and biocatalysis, and most notably high-throughput bioconjugation. The future of MCR-click chemistry in bioconjugation is very promising, with the potential to revolutionize a variety of fields, including drug delivery, diagnostics, vaccine development, and synthetic biology. By leveraging the power of MCR-click chemistry, it will be possible to create highly specific and scalable bioconjugates, opening new

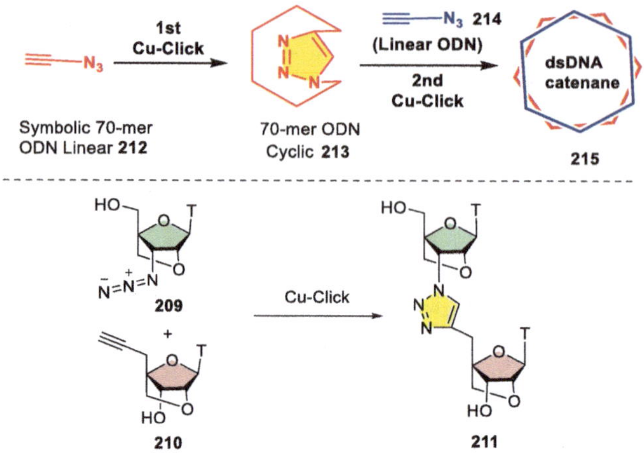

Scheme 46 Click chemistry in the synthesis of unnatural 1,2,3-triazole-nucleoside and dsDNA catenane

therapeutic, diagnostic, and industrial opportunities. As research continues to advance in this field, we can expect more precise, customizable, and multifunctional bioconjugates to emerge, bringing us closer to more effective treatments and diagnostic solutions.

4 Conclusion and Future Perspective

The copper(I)-catalyzed azide-alkyne cycloaddition (CuAAC) is identified as one of the most reliable tools for regioselective synthesis of 1,4-disubstituted 1,2,3-triazoles by AAC of an organic azide and an internal alkyne [7]. A number of magnetite nanocatalysts are explored for the regioselective triazole formation [155]. Similar AAC in the presence of a suitable Ru catalysis gave 1,5-disubstituted 1,2,3-triazoles [17]. Metal-catalyzed AAC was also explored by utilizing several nickel catalysts [156]. The resulting 1,2,3-triazole heterocycle is a privileged scaffold in medicinal chemistry. This method is regioselective, robust, clean, and versatile, as sometimes the chromatographic purification steps may be avoided. Once the "CuAAC click chemistry" is performed in a multicomponent fashion, it can make the protocol more efficient, rapid, robust, and diversity-oriented synthesis (DOS) of increased importance [157]. In addition to the wide impact of tandem click-mediated 1,2,3-triazoles in drug discovery, their applications in dendrimer synthesis [158], sensing [159], catalysis [160], and other fields have been further expanding in recent years. Copper (I)-catalyzed tandem one-pot route has been widely explored to access *bis*-1,2,3-triazoles, tetrakis-1,2,3-triazoles, and related molecular architectures [161]. Further, post-click approach through late-stage functionalization of 1,2,3-triazoles has ample opportunities in organic synthesis [162].

MCRs involve the one-step reaction of three or more reactants, deliver a single sole product without isolation of the intermediate involved therein, and have several advantages including DOS through incorporation of various readily available building blocks. MCRs in click diversification represent a foundation in the modern era of synthetic chemistry and offer exceptional opportunities for the rapid access of a diverse range of biologically privileged functionalized 1,2,3-triazole hybrid analogues. Azido-alkyne via MCR-click chemistry has been widely explored in macrocyclization through intramolecular AAC, and resulting 1,2,3-triazole-containing macrocycles have not only unlimited utilities in medicinal chemistry but also a great scope in host-guest supramolecular chemistry. Sequential "MCR-click" can be used for rapid access of dendrimers, drug conjugates, biocompatible glycosyl-triazole hybrids, and peptidomimetic 1,2,3-triazole analogues that provide a powerful platform for advancing the innovation in drug discovery research, chemical biology, and materials science. This tool can be equally explored in a parallel consecutive fashion with other regioselective variants of click chemistry, e.g., SuFEx click, PFEx click, thiol-yne click, SPAAC, thiol-ene click, and 4 + 2 cycloaddition reaction. In this direction, a great endeavor is required to devise convenient methodologies that can offer an enormous promise in the development of effective chemotherapeutics and diagnostic tools to address the future challenges in healthcare. The consecutive click reactions have also widespread applications in bioconjugation with carbohydrates, protein, and nucleic acids. The resulting imaging agents have notable features of cell surface labeling of potential applications in cellular biology. In addition, the standard tool can be well utilized to access functional materials with definite abilities including luminescence, conductivity, and drug delivery system. Beyond doubt, by virtue of reliability and regioselectivity, the modular click chemistry has again modernized the MCR tool. The combined robust method provides an ample opportunity for quick access of a diverse range of simple-to-complex molecular frameworks of stimulating applications in near future.

Acknowledgements The authors gratefully acknowledge Science and Engineering Research Board (SERB), New Delhi, for the financial support (Grant No.: CRG/2021/002776) and IoE-BHU for the incentive grant. MSY and SR sincerely thank IoE-BHU for RJP-PDF and Haflong Government College for the consistent support, respectively.

References

1. Kolb HC, Finn MG, Sharpless KB (2001) Angew Chem Int Ed 40:2004–2021
2. Tornøe CW, Christensen C, Meldal M (2002) J Org Chem 67:3057–3064
3. Rostovtsev VV, Green LG, Fokin VV, Sharpless KB (2002) Angew Chem Int Ed 41:2596–2599
4. Tiwari VK, Jaiswal MK, Rajkhowa S, Singh SK (2024) Click chemistry. Springer Nature Singapore, Singapore
5. Tiwari VK, Mishra BB, Mishra KB, Mishra N, Singh AS, Chen X (2016) Chem Rev 116: 3086–3240

6. Agrahari AK, Bose P, Jaiswal MK, Rajkhowa SS, Singh AS, Hotha, Mishra N, Tiwari VK (2021) Chem Rev 121:7638–7956
7. Meldal M, Tornoe CW (2008) Chem Rev 108:2952–3015
8. Wittig G, Krebs A (1961) Chem Ber 94:3260–3275
9. Agard NJ, Prescher JA, Bertozzi CR (2004) J Am Chem Soc 126:15046–15047
10. (a) Sir John Warcup "Kappa" Cornforth was an Australian-British Nobel Prize laureate in Chemistry (1975) for his brilliant research on the stereochemistry of enzyme-catalyzed reactions; (b) Hanson J, Cornforth J (2014) Nature 506:35
11. Younus HA, Al-Rashida M, Hameed A, Uroos M, Salar U, Rana S, Khan KM (2021) Expert Opin Ther Pat 31:267–289
12. Heravi MM, Zadsirjan V, Dehghani M, Ahmadi T (2018) Tetrahedron 74:3391–3457
13. Yadav MS, Rajkhowa S, Singh SK, Jaiswal MK, Tiwari VK (2024) ChemistrySelect 9: e202400776
14. Michael A (1893) J Prakt Chem 48:94–95
15. Huisgen R, Szeimies G, Möbius L (1967) Chem Ber 100:2494–2507
16. Zhang L, Chen X, Xue P, Sun HY, Williams ID, Sharpless KB, Fokin VV, Jia G (2005) J Am Chem Soc 127:15998–15999
17. Johansson JR, Beke-Somfai T, Stalsmeden AS, Kann N (2016) Chem Rev 116:14726–14768
18. Cao Y, Huang H, Wang L, Lin X, Yang J (2023) J Org Chem 88:4301–4308
19. Opsomer T, Dehaen W (2021) Chem Commun 57:1568–1590
20. Thomas J, Goyvaerts V, Liekens S, Dehaen W (2016) Chem Eur J 22:9966–9970
21. Sarma A, Agrahari AK, Rajkhowa S, Tiwari VK (2022) Eur J Med Chem 238:114454
22. Yadav S, Singh AK, Agrahari AK, Sharma K, Singh AS, Gupta MK, Tiwari VK, Prakash P (2020) Sci Rep 10:14204
23. Kumar GS, Lin Q (2021) Chem Rev 121:6991–7031
24. Ramachary DB, Barbas III CF (2004) Chem Eur J 10:5323–5331
25. Wang J (2011) Chem Eur J 17:3584–3587
26. Wang L, Peng S, Danence LJT, Gao Y, Wang J (2012) Chem Eur J 18:6088–6093
27. Ramachary DB (2014) Angew Chem Int Ed 53:10420–10424
28. Jaiswal MK, Gupta A, Yadav MS, Pandey VK, Tiwari VK (2023) Chem Eur J 29:e202301749
29. Hema K, Sureshan KM (2019) Acc Chem Res 52:3149–3163
30. Jaiswal MK, Tiwari VK (2023) Chem Rec 23:e2023090167
31. Jaiswal MK, Gupta A, Ansari FJ, Pandey VK, Tiwari VK (2024) Curr Org Syn 21:513–558
32. Tomaino E, Capecchi E, Ubertini V, Piccinino D, Bizzarri BM, Saladino R (2024) J Org Chem 89:2343–2350
33. Nandi R, Ajarul SK, Mandal PK, Manna AS, Kayet A, Maiti DK (2024) J Org Chem 89:2556–2570
34. Qian W, Amegadzie A, Winternheimer D, Allen J (2013) Org Lett 15:286–289
35. Alonso F, Moglie Y, Radivoy G, Yus M (2012) Heterocycles 84:1033–1044
36. Yousuf SK, Taneja SC, Mukherjee D (2010) J Org Chem 75:3097–3100
37. Ugi I (1962) Angew Chem Int Ed 1:8
38. Passernini M (1921) Gazz Chim Ital 51:181
39. Ranu BC, Hajra A, Jana U (2000) J Org Chem 65:6270–6272
40. Strecker A (1850) Ann Chem Pharm 75:27
41. Ducatti DRB, Massi A, Noseda MD, Duarte MER, Dondoni A (2009) Org Biomol Chem 7: 1980–1986
42. Mannich C, Krosche W (1912) Arch Pharm 250:647–667
43. Thirumurugan P, Matosiuk D, Jozwiak K (2013) Chem Rev 113:4905–4979
44. Agalave SG, Maujan SR, Pore VS (2011) Chem Asian J 6:2696–2718
45. Donikela S, Mainkar PS, Nayani K, Chandrasekhar S (2019) J Org Chem 84:10546–10553
46. Mishra N, Agrahari AK, Bose P, Singh SK, Singh AS, Tiwari VK (2020) Sci Rep 10:3586

47. Kushwaha D, Singh SK, Tiwari VK (2020) Carbo-click in drug discovery and development: opportunities and challenges. In: Tiwari VK (ed) Carbohydrates in drug discovery and development. Elsevier Inc., pp 403–450
48. Wojaczynska E, Steppeler F, Iwan D, Scherrmann M-C, Marra A (2021) Molecules 26:7291
49. Singh SK, Mishra N, Kumar S, Jaiswal MK, Tiwari VK (2022) ChemistrySelect 7: e202201314
50. Mishra N, Singh AS, Agrahari AK, Singh SK, Singh M, Tiwari VK (2019) ACS Comb Sci 21: 389–399
51. Mishra N, Singh SK, Singh AS, Agrahari AK, Tiwari VK (2021) J Org Chem 86:17884–17895
52. Kraft J, Schmollinger D, Maudrich J, Ziegler T (2015) Synthesis 47:199–208
53. Singh SK, Kumar S, Yadav MS, Tiwari VK (2022) J Org Chem 87(22):15389–15402
54. Singh SK, Kumar S, Yadav MS, Gupta A, Tiwari VK (2023) J Org Chem 88:13440–13453
55. Singh SK, Kumar S, Yadav MS, Bhattacharya S, Tiwari VK (2024) Synthesis 56:975–988
56. Pal R, Sarkar S, Chatterjee N, Sen AK (2013) Tetrahedron Lett 54:5642–5646
57. Arora BS, Shafi S, Singh S, Ismail T, Kumar HMS (2008) Carbohydr Res 343:139–144
58. Kumar R, Maulik PR, Misra AK (2008) Glycoconj J 25:595–602
59. Dutta S, Gupta SJ, Sen AK (2016) Tetrahedron Lett 57:3086–3090
60. Xu C, Loh CCJ (2018) Nat Commun 4057:1–12
61. Mandal PK (2014) RSC Adv 4:5803–5814
62. Guerin DJ, Horstmann TE, Miller SJ (1999) Org Lett 1:1107–1109
63. Witczak Z (2010) J Carbohydr Chem 36:176–193
64. Mishra A, Tiwari VK (2015) J Org Chem 80:4869–4881
65. Mishra A, Tiwari VK (2017) ChemistrySelect 2:9466–9471
66. Ji L, Zhang DF, Zhao Q, Hu SM, Qian C, Chen XZ (2013) Tetrahedron 69:7031–7037
67. Ji L, Zhou G-Q, Qian C, Chen X-Z (2014) Eur J Org Chem 2014:3622–3636
68. Chowdhury C, Mandal SB, Achari B (2005) Tetrahedron Lett 46:8531–8534
69. Chatterjee N, Pal R, Sarkar S, Sen AK (2015) Tetrahedron Lett 56:3886–3889
70. Lieke W (1859) Justus Liebigs Ann Chem 112:316–321
71. (a) Rocha RO, Rodrigues MO, Neto BAD (2020) ACS Omega 5:972–979; (b) Reddy BS, Majumder N, Rao TP (2014) Synthesis 46:3408–3414
72. Barlow T, Jida M, Tourwe D, Ballet S (2014) Org Biomol Chem 12:6986–6989
73. Saeedi M, Mahadavi M, Foroumadi A, Shafiee A (2013) Tetrahedron 69:3506–3510
74. Shafie A, Khanaposhtani MM, Asadi M, Rahimi N, Ranjbar PR, Ghasemi JB, Larijani B, Mahdavi M, Shafaroodi H, Dehpour AR (2020) Mol Divers 24:179–189
75. Asgari MS, Sepehri S, Bahadorikhalili S, Ranjbar PR, Rahimi R, Gholami A, Kazemi A, Khoshneviszadeh M, Larijani B, Mahdavi M (2020) Chem Heterocycl Comp 56:488–494
76. Sadjadi S, Heravi MM, Malmir M, Kahangi FG (2018) Appl Clay Sci 162:192–203
77. Bhuyan P, Bhorali P, Islam I, Bhuyan AJ, Saikia L (2018) Tetrahedron Lett 59:1587–1591
78. Saikia AA, Rao RN, Das S, Jena S, Rej S, Maiti BK, Chanda B (2020) Tetrahedron Lett 61: 152273
79. Narsimha S, Battula KS, Reddy YN, Nagavelli VR (2018) Chem Heterocycl Compd 54:1161–1167
80. Crowley JD, Bandeen PH (2010) Dalton Trans 39:612–623
81. Jaiswal MK, Yadav MS, Maurya S, Ansari D, Tiwari VK (2024) J Org Chem 89:17213–17227
82. Mishra KB, Tiwari VK (2016) ChemistrySelect 1:3693–3698
83. Jaiswal MK, Yadav MS, Singh M, Garai S, Tiwari VK (2024) J Org Chem 89:10698–10708
84. Yadav MS, Jaiswal MK, Kumar S, Singh SK, Garai S, Tiwari VK (2025) Chem Asian J 2025: e202401301
85. Ansari D, Jaiswal MK, Yadav MS, Pandey VK, Singh M, Tiwari VK (2025) Org Biomol Chem 23:4045–4057
86. Creary X, Anderson A, Brophy C, Crowell F, Funk Z (2012) J Org Chem 77:8756–8761
87. Creary X, Chromanski K, Peirats G, Renneburg C (2017) J Org Chem 82:5720–5730

88. Nguyen HH, Palazzo TA, Kurth MJ (2013) Org Lett 15:4492–4495
89. Dhondge AP, Afraj SN, Nuzlia C, Chen C, Lee GH (2013) Eur J Org Chem 2013:4119–4130
90. Moliner FD, Bigatti M, Banfi L, Riva R, Basso A (2014) Org Lett 16:2280–2283
91. Arigela RK, Mandadapu AK, Sharma SK, Kumar B, Kundu B (2012) Org Lett 14:1804–1807
92. Mishra KB, Tiwari VK (2014) J Org Chem 79:5752–5762
93. Senwar KR, Sharma P, Reddy TS, Jeengar MK, Nayak VL, Naidu V, Kamal A, Shankaraiah N (2015) Eur J Med Chem 102:413–424
94. Geng Z-Q, Zhao C, Qian H-D, Li S-J, Peng H, Xu H (2023) Org Lett 25:4504–4509
95. Gao G, Kuantao M, Lv L, Lia Z (2022) Adv Synth Catal 364:1402–1408
96. Ackermann L, Potukuchi HK, Landsberg D, Vicente R (2008) Org Lett 10:3081–3084
97. Wei F, Li H, Song C, Ma Y, Zhou L, Tung C-H, Xu Z (2015) Org Lett 17:2860–2863
98. Cui FH, Chen J, Mo Z-Y, Su S-X, Chen Y-Y, Ma X-L, Tang H-T, Wang H-S, Pan Y-M, Xu Y-L (2018) Org Lett 20:925–929
99. Yadav MS, Singh AS, Agrahari AK, Mishra N, Tiwari VK (2019) ACS Omega 4:6681–6689
100. Yadav MS, Singh SK, Agrahari AK, Singh AS, Tiwari VK (2021) Synthesis 53:2494–2502
101. Agrahari AK, Jaiswal MK, Yadav MS, Tiwari VK (2021) Carbohydr Res 508:108403
102. Tiwari VK (2021) Chem Rec 21:3029–3048
103. Yadav MS, Jaiswal MK, Kumar S, Singh SK, Ansari FJ, Tiwari VK (2022) Carbohydr Res 521:108674
104. Chen A, Wang G (2023) Synthesis of carbohydrate-containing macrocycles. In: Tiwari VK (ed) Synthetic strategies in carbohydrate chemistry, pp 679–673
105. Roxburgh CJ (1995) Tetrahedron 51:9767–9822
106. Parenty A, Moreau X, Campagne J-M (2006) Chem Rev 106:911–939
107. Martí-Centelles V, Pandey MD, Burguete MI, Luis SV (2015) Chem Rev 115:8736–8834
108. Yadav MS, Gupta A, Kumar S, Tiwari VK (2023) Emerging aspects of domino reactions in carbohydrate chemistry. In: Tiwari VK (ed) Synthetic strategies in carbohydrate chemistry, pp 679–673
109. Pirali T, Tron GC, Zhu J (2006) Org Lett 8:4145–4148
110. Prasanna R, Purushothaman S, Raghunathan R (2014) Org Biomol Chem 12:9375–9383
111. Rao JNS, Raghunathan R (2015) Tetrahedron Lett 56:2669–2673
112. Chavez-Acevedo L, Miranda LD (2015) Org Biomol Chem 13:4408–4412
113. Zakharova EA, Shmatova OI, Kutovaya IV, Khrustalev VN, Nenajdenko VG (2019) Org Biomol Chem 17:3433–3445
114. Salvador CEM, Pieber BU, Neu PM, Torvisco A, Kleber Z, Andrade C, Kappe CO (2015) J Org Chem 80:4590–4602
115. Struthers H, Mindt TL, Schibli R (2010) Dalton Trans 39:675–696
116. Yano M, Tong CC, Light ME, Schmidtchen FP, Gale PA (2010) Org Biomol Chem 8:4356–4363
117. Crowley JD, Bandeen PH, Hanton LR (2010) Polyhedron 29:70–83
118. Urankar D, Pinter B, Pevec A, De Proft F, Turel I, Kossmrl J (2010) Inorg Chem 49:4820
119. Doak BC, Scanlon MJ, Simpson JS (2011) Org Lett 13:537–539
120. Wright K, Quinodoz P, Drouillat B, Couty F (2017) Chem Comm 53:321–323
121. Yuan Z, Kuang G-C, Clark RJ, Zhu L (2012) Org Lett 14:2590–2593
122. Aucagne V, Leigh DA (2006) Org Lett 8:4505–4507
123. Mishra KB, Mishra BB, Tiwari VK (2014) Carbohydr Res 399:2–7
124. Brinko A, Risinger C, Lambert A, Blixt O, Grandjean C, Jensen HH (2019) Org Lett 21:7544–7548
125. Meng G, Guo T, Ma T, Zhang J, Shen Y, Sharpless KB, Dong J (2019) Nature 574:86–89
126. Dong J, Krasnova L, Finn MG, Sharpless KB (2014) Angew Chem Int Ed 53:9430–9448
127. Barrow AS, Smedley CJ, Zheng Q, Li S, Dong J, Moses JE (2019) Chem Soc Rev 48:4731–4758
128. Sun S, Homer JA, Smedley CJ, Cheng Q-Q, Sharpless KB, Moses JE (2023) Chem 9:2128
129. Orimoto G, Yoshida S (2024) Chem Commun 60:5824–5827

130. Zhang X, Moku B, Leng J, Rakesh KP, Qin HL (2019) Eur J Org Chem 8:1763–1769
131. Vaněk V, Pícha J, Fabre B, Buděšínský M, Lepšík M, Jiráček J (2015) Eur J Org Chem 2015: 3689
132. Fabre B, Picha J, Vanek V, Selicharova I, Chrudinova M, Collinsova M, Zakova L, Budesinsky M, Jirack J (2016) ACS Comb Sci 18:710
133. Knall A-C, Hollauf M, Saf R, Slugovc C (2016) Org Biomol Chem 14:10576
134. Yokoi T, Tanimoto H, Ueda T, Morimoto T, Kakiuchi K (2018) J Org Chem 83:12103
135. Yokoi T, Ueda T, Tanimoto H, Morimoto T, Kakiuchi K (2019) Chem Commun 55:1891
136. Terashima N, Sakata Y, Meguro T, Hosoya T, Yoshida S (2020) Chem Commun 56:14003
137. Meguro T, Sakata Y, Morita T, Hosoya T, Yoshida S (2020) Chem Commun 56:4720
138. Takemura S, Goto T, Hosoya S, Yoshida (2020) Chem Commun 56:15541
139. Maegawa K, Tanimoto H, Onishi S, Tomohiro T, Morimoto T, Kakiuchi K (2021) Org Chem Front 8:5793
140. Yoshida S, Sakata Y, Misawa Y, Morita T, Kuribara T, Ito H, Koike Y, Kii I, Hosoya T (2021) Chem Commun 57:899
141. Takemura H, Orimoto G, Kobayashi A, Hosoya T, Yoshida S (2022) Org Biomol Chem 20: 6007–6011
142. Knall A-C, Slugovc C (2013) Chem Soc Rev 42:5131
143. Zheng Z-J, Wang D, Xu Z, Xu L-W (2015) Beilstein J Org Chem 11:2557
144. Yoshida S (2020) Org Biomol Chem 18:1550
145. Sato D, Wu Z, Fujita H, Lindsey JS (2021) Organics 2:161
146. Gruyter JN, Malins LR, Baran PS (2017) Biochemistry 56:3863–3873
147. Tamura T, Hamachi I (2019) J Am Chem Soc 141:2782–2799
148. Dai Y, Weng J, George J, Chen H, Lin Q, Wang J, Royzen M, Zhang Q (2019) Org Lett 21: 3828–3833
149. Guo T, Meng G, Zhan X, Yang Q, Ma T, Xu L, Sharpless KB, Dong J (2018) Angew Chem Int Ed 57:2605–2610
150. Brandhofer T, Ozdemir A, Gini A, Mancheno OG (2019) Chem Eur J 25:4077–4086
151. Srivastava S, Singh SK, Sharma VK, Mangla P, Olsen CE, Prasad AK (2015) Nucleos Nucleot Nucl 34:388–399
152. Kumar R, El-Sagheer A, Tumpane J, Lincoln P, Wilhelmsson LM, Brown T (2007) J Am Chem Soc 129:6859–6864
153. Liu F, Wang H, Li S, Bare GAL, Chen X, Wang C, Moses JE, Wu P, Sharpless KB (2019) Angew Chem Int Edn 58:8029–8033
154. Agrahari AK, Rajkhowa S, Singh SK, Jaiswal MK, Tiwari VK (2025) Curr Org Chem 29: 495–517
155. Aslaheh HS, Payamifar S, Marjani AP (2025) J Organomet Chem 1036:123704
156. Aslaheh HS, Payamifar S, Marjani AP (2024) Appl Organomet Chem 38:e7692
157. Tiwari VK, Jaiswal MK, Rajkhowa S, Singh SK (2024) Tandem/MCR click chemistry in organic synthesis. In: Click chemistry. Springer, pp 603–635
158. Kumar S, Yadav MS, Singh SK, Rajkhowa S, Tiwari VK (2023) SynOpen 7:145–153
159. Rajkhowa S, Jaiswal MK, Singh SK, Chaubey S, Tiwari VK (2025) Chem Biodivers. 22(110): e00735
160. Singh SK, Sharma A, Mishra N, Tiwari VK (2025) Carbohydr Res 555:10957
161. Pandey VK, Jaiswal MK, Ansari D, Yadav MS, Maurya S, Tiwari VK (2025) Eur J Org Chem 2025:e202500491
162. Yadav MS, Pandey VK, Jaiswal MK, Singh SK, Sharma A, Singh M, Tiwari VK (2025) J Org Chem 90:5731–5762

Top Heterocycl Chem (2026) 63: 205–224
https://doi.org/10.1007/7081_2025_101

Published online: 01 February 2026

Bio-click Chemistry

Diego F. Rodríguez and Flavia C. Zacconi

Contents

1 Bio-click Chemistry 206
2 Recent Advances in Bio-click Chemistry: CuAAC 208
2.1 Bio-click Synthesis of Bioactive Oligosaccharides 208
2.2 Chimeric Bicyclic Peptide Antibiotics 210
2.3 Bio-click Chemistry in the Chemoenzymatic Synthesis of Bi- and Tetracyclic Peptides 211
3 Recent Advances in Bio-click Chemistry: Diels–Alder Reaction 214
3.1 Bio-click Approach to the Synthesis of Spirosorbicillinols 214
3.2 Late-Stage Functionalization of Fasamycin A 215
4 Recent Advances in Bio-click Chemistry: Epoxide Opening 216
5 Recent Advances in Bio-click Chemistry: Thio-Michael 218
6 Summary and Outlook 219
References 220

Abstract Selecting the most efficient synthetic processes remains a complex challenge, as it often requires balancing high reaction yields with sustainability considerations. In this context, bio-click chemistry has emerged as a powerful interdisciplinary approach that combines the efficiency of click chemistry with the inherent sustainability of enzymatic processes. This specialized category of chemoenzymatic reactions holds significant promise for the development of sustainable methods and the synthesis of bioactive compounds. This chapter reviews the

D. F. Rodríguez
Facultad de Química y de Farmacia, Pontificia Universidad Católica de Chile, Santiago, Chile

F. C. Zacconi (✉)
Facultad de Química y de Farmacia, Pontificia Universidad Católica de Chile, Santiago, Chile

Institute for Biological and Medical Engineering, Schools of Engineering, Medicine and Biological Sciences, Pontificia Universidad Católica de Chile, Santiago, Chile

Centro de Investigaciones en Nanotecnología y Materiales Avanzados, CIEN-UC, Pontificia Universidad Católica de Chile, Santiago, Chile
e-mail: fzacconi@uc.cl

most relevant bio-click chemistry reactions reported in recent years, with a particular focus on their applications in medicinal chemistry.

Keywords Biocatalysis · Chemoenzymatic synthesis · Click chemistry · Green chemistry

Abbreviations

AAC	Azide-alkyne cycloaddition
ADC	Antibody-drug conjugate
AMR	Antimicrobial resistance
BOC-ON	2-(Boc-oxyimino)-2-phenylacetonitrile
CuAAC	Cu(I)-catalyzed azide-alkyne cycloaddition
ee	Enantiomer excess
Equiv	Equivalent(s)
Kd	Dissociation constant
MIC	Minimum inhibitory concentration
MOM	Methoxymethyl ether
PMEN	Polymyxin E nonapeptide
pNP	4-Nitrophenyl
SPPS	Solid-phase peptide synthesis
TON	Turnover number

1 Bio-click Chemistry

Bio-click chemistry is a rapidly evolving field that integrates the reliability of click chemistry with the green, selective nature of biocatalytic transformations [1]. This specialized subset of chemoenzymatic reactions has found broad applications across diverse areas, ranging from drug discovery to materials science. Its success lies in the complementary – and at times synergistic – interplay between click chemistry and biocatalysis (Fig. 1). Notably, several studies have reported chemoenzymatic reactions in which click chemistry is combined with enzyme-catalyzed transformations to introduce chirality with high precision [1–4]. However, stereospecificity is not the only advantageous feature of this approach: bio-click chemistry also offers enhanced efficiency, selectivity, and functional group tolerance, significantly expanding its potential for innovative sustainable synthetic process.

The term *click chemistry* – which earned the Nobel Prize in Chemistry in 2022 – was introduced by Sharpless in 2001. It describes a group of highly efficient and reliable reactions that enable the rapid construction of diverse molecular structures and functions by joining small building blocks. These reactions are typically driven by strong thermodynamics (often greater than 20 kcal/mol); are modular and broad

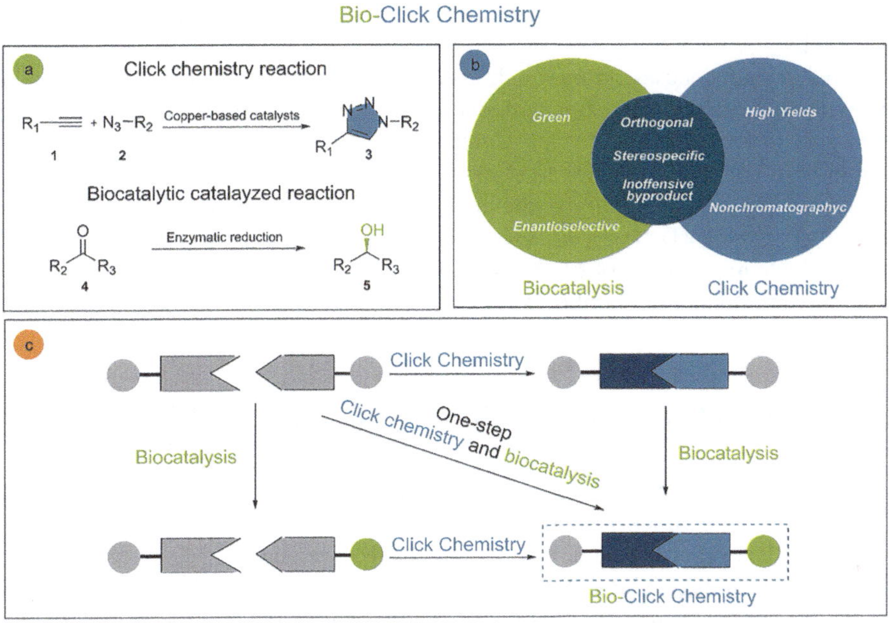

Fig. 1 (**a**) General click chemistry and biocatalysis reactions, (**b**) bio-click chemistry interface, and (**c**) schematic representation of bio-click chemistry cooperation for the synthesis of small molecules. (Reproduced from Ref. [1] with permission from the Royal Society of Chemistry)

in scope; produce high yields; and can be stereospecific (though not necessarily enantioselective). Ideal click reactions also feature simple conditions, readily available starting materials, minimal or benign solvents (e.g., water), and straightforward product isolation [5]. The concept is largely inspired by nature's "synthetic toolbox" [6], making its combination with biocatalysis particularly promising for synthesizing complex molecular structures [1].

Moreover, biocatalysis involves the use of purified enzymes, cell-free extracts, or whole microorganisms to catalyze a wide range of chemical reactions with remarkable efficiency [7]. Enzymatic processes enable the synthesis of complex molecules under highly sustainable and green conditions; in fact, biocatalysis is consistent with 10 of the 12 principles of green chemistry [8]. Although enzymes have been successfully implemented at the industrial level [9], their broader application in synthesis is often limited by factors such as low operational stability, substrate or product inhibition, narrow substrate scope with nonnatural compounds, poor compatibility with organic solvents, and short shelf life [10]. To address these challenges, innovative technologies such as protein engineering – particularly directed evolution, which received the Nobel Prize in Chemistry in 2018 – and enzyme immobilization have been developed. Many enzymes produced through protein engineering exhibit enhanced robustness, stability, and ability to utilize nonnatural substrates [11]. Additionally, immobilized enzymes typically offer superior operational stability, easier separation from reaction media, and possibility of reuse [12, 13].

This chapter focuses on bio-click chemistry reactions – synthetic methods that integrate biocatalytic processes with click reactions – and aims to expand upon the concepts and transformations presented in our previous review, with particular emphasis on developments reported in the past 4 years [1]. The chapter is organized around the four most employed click chemistry reactions in the synthesis of bioactive molecules: Cu(I)-catalyzed azide-alkyne cycloaddition (CuAAC), Diels–Alder, epoxide-opening, and thiol-Michael reactions. Special attention is given to recent advances in generating structurally diverse molecules with potent biological activity.

2 Recent Advances in Bio-click Chemistry: CuAAC

The copper-catalyzed azide-alkyne cycloaddition (CuAAC) is arguably the most widely recognized click reaction. It enables the regioselective synthesis of 1,2,3-triazoles from organic azides and alkynes in the presence of a copper(I) catalyst [14]. The resulting 1,2,3-triazole ring is a well-established pharmacophore and peptidomimetic, making it an invaluable structural motif in medicinal chemistry [15].

In the context of bio-click chemistry, the integration of CuAAC with oxidoreductases and hydrolases has been extensively explored [1]. More recently, in the past 2 years, there has been a growing interest in applying this combined approach to the synthesis of bioactive oligosaccharides and structurally complex peptides, highlighting its versatility and expanding its relevance in the development of functional biomolecules.

2.1 Bio-click Synthesis of Bioactive Oligosaccharides

Oligosaccharides are carbohydrates composed of up to ten monosaccharide units linked by glycosidic bonds [16]. These short sugar chains play key roles in a wide range of biological processes, including cell recognition [17], inflammation [18], and immune response [19], among others [20]. Due to their broad functional diversity – often associated with their structural complexity – the synthesis of oligosaccharides has been the focus of research for over a century [21]. As a result, chemical, enzymatic, and chemoenzymatic strategies have evolved significantly, enabling the production of structurally diverse glycan architectures [22].

Within this dynamic landscape, the convergence of click chemistry and biocatalysis has emerged as a powerful strategy for the synthesis of novel oligosaccharides with potent biological activity. The complementarity of these approaches – combined with their high yields, regioselectivity, and stereospecificity – frequently results in shorter synthetic routes by eliminating the need for protection and deprotection steps, thereby improving overall efficiency and sustainability.

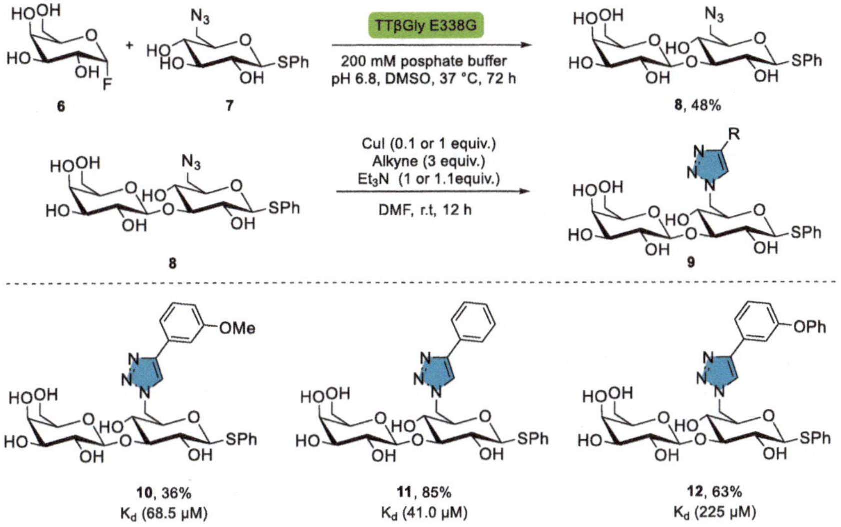

Scheme 1 Bio-click approach to the synthesis of galectin-3 inhibitors

The chemoenzymatic synthesis of galectin-3 inhibitors represents an innovative approach toward the development of carbohydrate-based therapeutics with potential applications in oncology, fibrosis, and immunomodulation [23–25]. By employing a chemoenzymatic approach (Scheme 1), Dussouy et al. [26] reported a stereospecific and regioselective synthesis of new galectin-3 inhibitors. A glycosynthase mutant from *Thermus thermophilus* glycoside hydrolase (TtβGly E338G) was successfully used to catalyze the glycosylation between α-galactopyranosyl fluoride **6** and phenyl 6-azido-6-deoxy-1-thio-β-D-glucopyranoside **7** to give phenyl α-D-galactopyranosyl-(1 → 3)-1-thio-α-D-glucopyranoside (**8**, 48%), subgram scale (141 mg). This intermediate served as a versatile building block for click chemistry transformations, most notably the CuAAC, enabling the introduction of triazole-linked bioactive motifs with an overall reaction efficiency of 61%. The resulting compounds exhibited promising inhibitory activity against galectin-3, with IC_{50} values in the micromolar range.

In 2021, Gorantla et al. reported the synthesis of α-glucosyl triazoles as a valuable approach for developing potent inhibitors of lysosomal α-glucosidase, a target relevant to diabetes management [27]. In this study, a glycosynthase strategy was employed using the *Thermoanaerobacterium xylanolyticum* glucosyl hydrolase 116 β-glucosidase (TxGH116) E441G nucleophile mutant, which catalyzed the regioselective formation of α-D-glucosylazide (αGA1) **14** from *p*-nitrophenyl β-*D*-glucoside (pNPGlc) **13** and sodium azide in aqueous medium. The reaction yielded **14** αGA1 in 68%, subgram scale (280 mg) (Scheme 2). These azide-functionalized glucosides were further derivatized via copper-catalyzed azide-alkyne cycloaddition (CuAAC), affording a series of 13 new 1,4-substituted 1,2,3-triazole-α-*D*-glucoside derivatives in yields ranging from 40% to 78% under mild, protecting-group-free

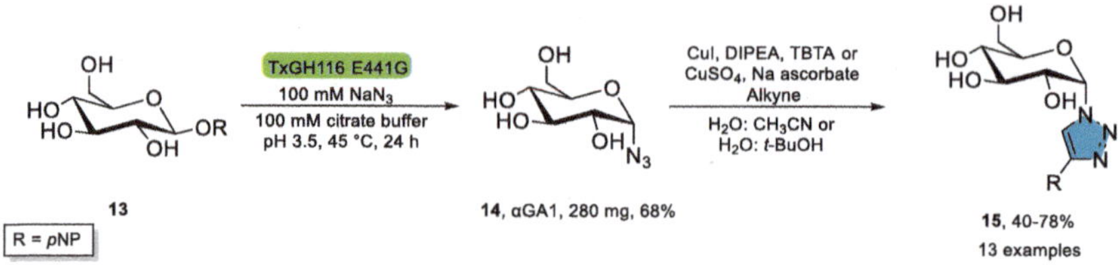

Scheme 2 Chemoenzymatic and protecting-group-free synthesis of α-glycosyl triazoles

conditions. Several compounds demonstrated potent inhibitory activity against human lysosomal α-glucosidase, exhibiting IC_{50} values between 18 and 17 μM, which are over 60-fold lower than those of the standard inhibitor acarbose (IC_{50} = 1.10 mM). Despite their promising activity, further studies are needed to evaluate their selectivity, metabolic stability, and potential to reduce glycemic effects in diabetes.

2.2 *Chimeric Bicyclic Peptide Antibiotics*

The pressing challenge of antimicrobial resistance (AMR) underscores the urgent need for new antibiotics that operate through previously unexploited mechanisms of action. Recent drug discovery efforts have focused on leveraging natural products from unculturable bacteria and on designing synthetic derivatives inspired by natural scaffolds [28]. Notably, among the most promising antibiotic candidates targeting novel bacterial pathways, macrocyclic peptides (MCPs) stand out as valuable sources for new drug discovery due to their constrained conformations and remarkable chemical stability [29]. In this context, the chemoenzymatic synthesis of chimeric macrocyclic peptide antibiotics has emerged as an efficient strategy for generating compounds with antibacterial activity comparable to, or even superior to, colistin, a last-line treatment option for multidrug-resistant Gram-negative bacteria.

In 2021, Wood et al. described a convergent strategy that combined biocatalysis and CuAAC ligation to generate chimeric bicyclic peptide antibiotics with potent activity against Gram-negative pathogens (Scheme 3) [30]. Initially, the polymyxin E nonapeptide (PMEN) was obtained through enzymatic degradation of colistin using ficin, followed by selective Boc protection of its Dab residues and subsequent coupling to azido-carboxylic acids of varying lengths. In parallel, the β-hairpin macrocyclic peptide was synthesized via solid-phase peptide synthesis (SPPS), incorporating orthogonally protected glutamic acid for later functionalization via CuAAC. The antibacterial activity of the resulting triazole conjugates was assessed against a range of Gram-negative pathogens including multidrug-resistant isolates. Notably, the conjugate **20a**, with the shortest linker, consistently demonstrated strong antibacterial activity, with minimum inhibitory concentration (MIC) values ranging from 0.25 to 2.0 μg/mL. These findings highlight the efficiency of the

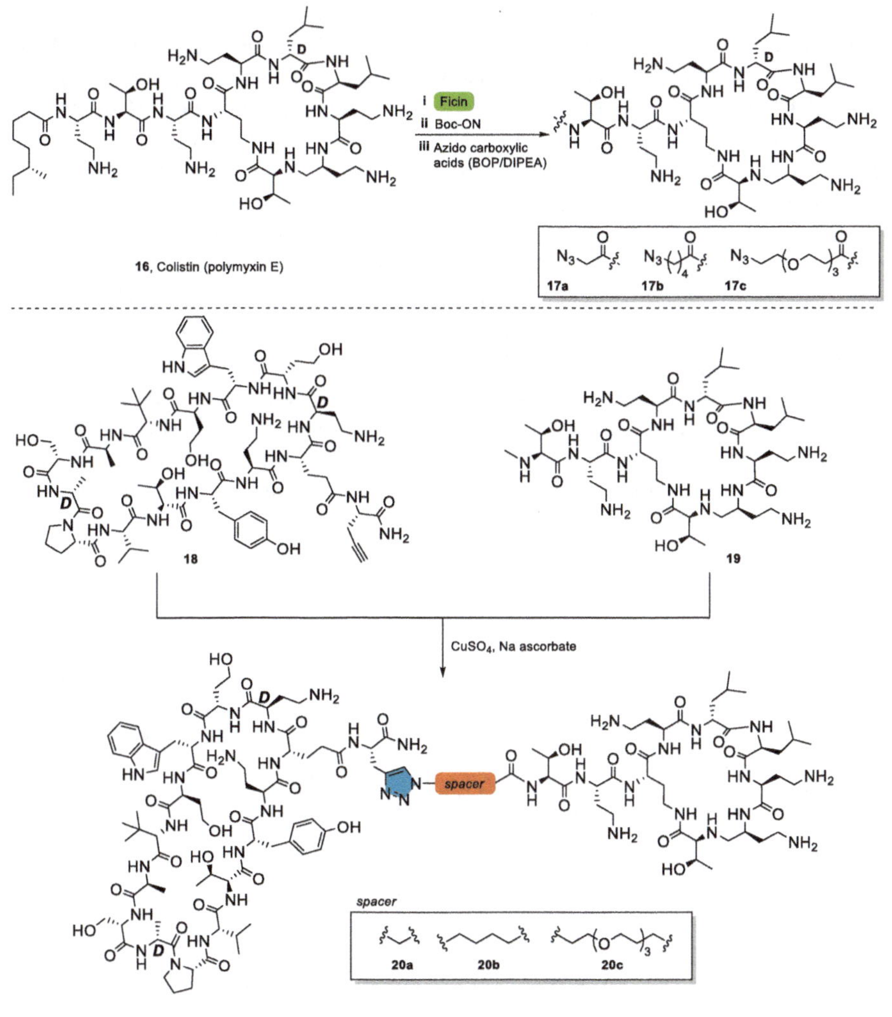

Scheme 3 Synthesis of triazole-linked bicyclic conjugates

bio-click approach in generating structurally diverse PMEN conjugates with promising antibacterial properties.

2.3 Bio-click Chemistry in the Chemoenzymatic Synthesis of Bi- and Tetracyclic Peptides

In the pursuit of improving the biopharmaceutical properties of linear peptides – such as enhancing metabolic stability and biological activity – Richelle et al. [31]

developed an innovative approach for synthesizing highly constrained tetracyclic peptides (Scheme 4). This one-pot "triple-C" methodology combines chemoenzymatic peptide synthesis (CEPS), chemical linkage of peptide onto scaffolds (CLIPS), and copper-catalyzed azide-alkyne cycloaddition (CuAAC) to achieve efficient macrocyclization. The process begins with CEPS, where Omniligase-1 catalyzes head-to-tail cyclization of linear peptides containing 13 or more amino acids under mild aqueous conditions. Upon completion of cyclization, a tetravalent T4 scaffold is introduced, followed by the addition of a CuAAC reaction mixture containing $CuSO_4$, sodium ascorbate, and tris-hydroxypropyltriazolylmethylamine (THPTA). This final step enables the selective incorporation of triazole-linked constraints, reinforcing the structural rigidity and stability of the peptide framework. To demonstrate the potential of this technology, the resulting isomerically pure tetracyclic peptides were evaluated as inhibitors of urokinase plasminogen activator (uPA) and coagulation factor XIIa (FXIIa), two key enzymes involved in cancer progression and thrombosis, respectively. The most potent tetracyclic peptides exhibited inhibition values in the micromolar range.

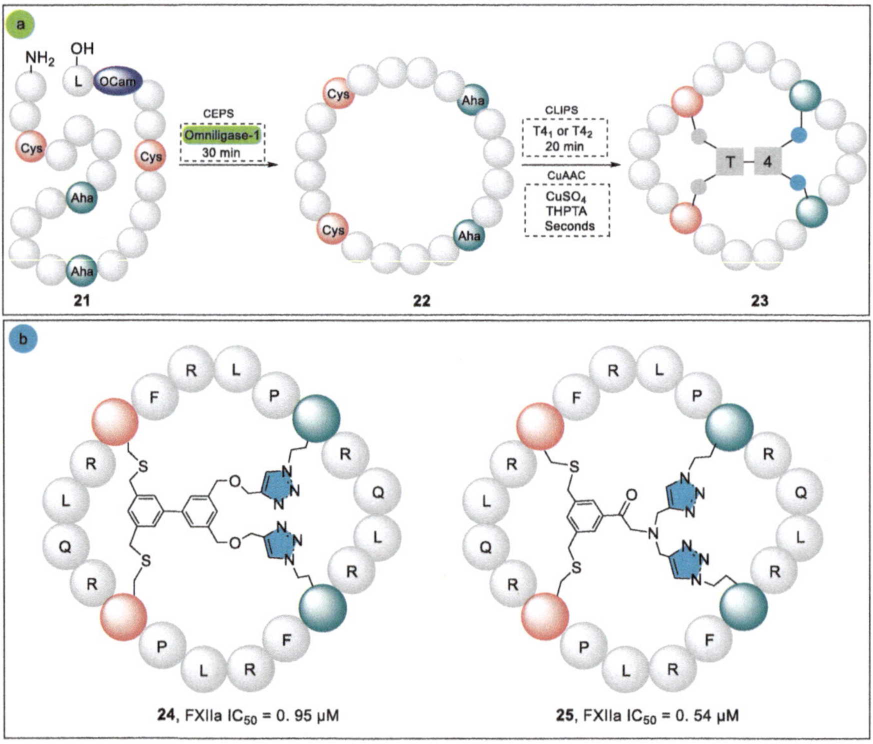

Scheme 4 (**a**) One-pot "triple-C" multicyclization methodology for the synthesis of highly constrained isomerically pure tetracyclic peptides, (**b**) schematic structures of the most active tetracyclic constructs evaluated against FXIIa

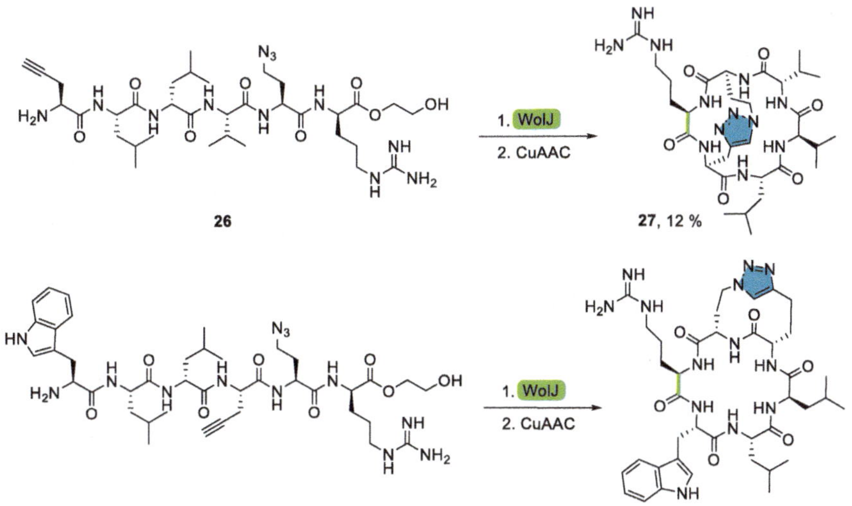

Scheme 5 Chemoenzymatic tandem cyclization for the facile synthesis of bicyclic peptides

Overall, this one-pot chemoenzymatic approach for tetracyclic peptide synthesis offers several advantages over conventional methods for constrained peptide synthesis. These benefits include high efficiency, shorter reaction times, precise regioselectivity, and ability to perform reactions under mild conditions without protecting-group strategies.

Despite the success of the chemoenzymatic synthesis of macrocyclic peptides, one limitation is that the ligation enzymes, such as Omniligase-1, recognize sequences longer than ~13 residues [32]. This restriction excludes small bicyclic peptides, which have significant therapeutic potential due to their improved bioavailability and lower immunogenicity. An alternative to overcome this limitation is the use of penicillin-binding protein-type thioesterases (PBP-type TEs), which would potentially allow the cyclization of small peptides, albeit with a narrow substrate scope. In 2024, Kobayashi et al. [33] reported a one-pot chemoenzymatic tandem cyclization strategy using a penicillin-binding protein-like thioesterase (WolJ) for selective head-to-tail cyclization of peptides, followed by CuAAC to generate bicyclic peptides. The WolJ-mediated cyclization was carried out at 25 °C in Tris-HCl buffer (pH 7.5) with 5 mM peptide substrate and 10 μM enzyme, allowing efficient macrocyclization. The subsequent CuAAC step employed $CuSO_4$, sodium ascorbate, and 1.5 equivalents of an azide-containing linker at room temperature for 12 h, yielding bicyclic peptides (Scheme 5). This approach enabled the synthesis of bicyclic scaffolds from peptides with different sizes (6–11 amino acid residues). The tandem process streamlined the synthesis by eliminating intermediate purification steps while maintaining high selectivity and efficiency.

3 Recent Advances in Bio-click Chemistry: Diels–Alder Reaction

The Diels–Alder (DA) reaction is widely regarded as one of the most powerful and versatile transformations in organic synthesis. This pericyclic reaction enables the atom-economical and stereoselective construction of complex molecular architectures through the [4 + 2] cycloaddition of a diene and a dienophile, leading to six-membered functionalized carbocyclic compounds [34, 35]. The potency of Diels–Alder reactions was validated by Woodward et al. [36] in 1952 when they reported the total synthesis of cortisone and cholesterol, followed by reserpine in 1958 [37], all of which involved the DA reaction.

Although the DA reaction can proceed without a catalyst, the use of Lewis acids enhances its efficiency by lowering the activation energy, thereby increasing the reaction rate and reducing the required temperature [38, 39]. Additionally, electronic effects play a crucial role in optimizing reaction efficiency, with the incorporation of electron-withdrawing groups (EWGs) on the dienophile and electron-donating groups (EDGs) on the diene [40]. These effects combined with the hydrophobic acceleration of Diels–Alder reactions in aqueous media, reported by Rideout et al. in 1980 [41], have facilitated their seamless integration with enzymatic catalysis, enabling the development of highly efficient one-pot synthetic protocols.

3.1 Bio-click Approach to the Synthesis of Spirosorbicillinols

Sorbicillinoids represent a structurally diverse class of secondary metabolites with significant biomedical relevance, exhibiting anti-inflammatory, antimicrobial, cytotoxic, phytotoxic, and enzyme-inhibitory properties. These characteristics make sorbicillinoids promising candidates for pharmaceutical and agrochemical applications. Sorbicillinoids can be classified according to the synthetic transformations employed in their synthesis, with hybrid derivatives falling into two major categories: Diels–Alder-type and Michael-type sorbicillinoids [42]. In this context, the Gulder group has utilized chemoenzymatic approaches for sorbicillinoid derivatization, and more recently, Milzarek et al. [43] reported the convergent chemoenzymatic total synthesis of spirosorbicillinols (Scheme 6). The key step in this process involves the oxidoreductase SorbC, which catalyzes the oxidative dearomatization of sorbicillin to generate the highly reactive intermediate sorbicillinol. This intermediate readily undergoes a Diels–Alder reaction with a shikimic acid-derived dienophile, affording a structurally complex family of six spirosorbicillinols in yields ranging from 4% to 35%. The main limitation of the study is the high reactivity of sorbicillin, which reduced overall reaction yields. Nevertheless, this chemoenzymatic strategy currently provides the shortest synthetic route, with superior overall efficiency and near-perfect stereocontrol.

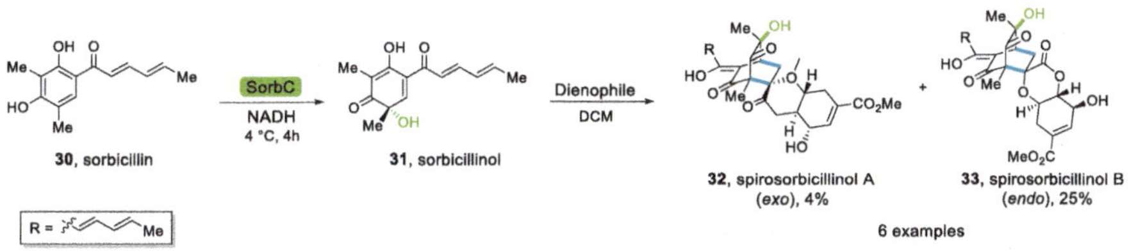

Scheme 6 Chemoenzymatic synthesis of the spirosorbicillinols

3.2 Late-Stage Functionalization of Fasamycin A

Over the past decade, late-stage functionalization (LSF) has emerged as a powerful tool in drug discovery, promoting the rapid diversification of complex molecules. In particular, C-H functionalization of drug-like compounds reduces synthetic efforts by eliminating the need for preinstalled functional groups. Nevertheless, the inherent challenge lies in achieving high chemoselectivity and site selectivity in structurally intricate molecules [44]. Consequently, LSF strategies must exhibit broad functional group tolerance while remaining compatible with polar substituents commonly found in bioactive compounds, such as basic amines, protic acids, and hydrogen-bond donors or acceptors [45]. Thus, late-stage enzymatic modifications have emerged as an effective alternative to chemical catalysis, supported by advances in protein engineering to develop highly selective biocatalysts. These engineered enzymes enable precise site-selective transformations, even in structurally complex molecules, by distinguishing between closely related reaction centers with remarkable specificity [46, 47].

In this context, Li and Renata developed a concise chemoenzymatic synthesis of fasamycin A, an antibacterial compound with potent activity against Methicillin-resistant *Staphylococcus aureus* (MRSA) and *Enterococcus faecalis* EF16 (VRA) [48]. The bio-click approach to fasamycin A synthesis began with a Diels–Alder reaction between diene **34** and alkyne **35**, yielding lactone **36** through sequential cycloaddition and acid-induced lactonization. Subsequent transformations of **36** enabled the synthesis of naphthacemycin B1 with a 32% overall yield over four steps (Scheme 7). Using naphthacemycin B1 as a substrate, various enzymatic systems (halogenase/reductase) were evaluated for late-stage enzymatic halogenation. The FasV/CtcQ system (a flavin-dependent dehydrogenase enzyme involved in the biosynthesis of fasamycin/chlortetracycline) successfully catalyzed the transformation of B1 into fasamycin A with a 5% yield. Although the enzymatic halogenation exhibited low efficiency, the discovery of FasV/CtcQ, capable of selectively catalyzing C–H chlorination, provides a basis for future optimization through enzyme engineering of enhanced halogenation strategies for naphthacemycin B1 [49].

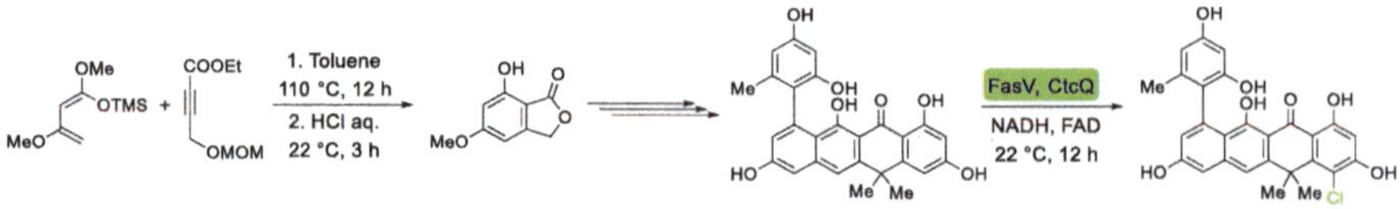

Scheme 7 Chemoenzymatic synthesis of fasamycin A

4 Recent Advances in Bio-click Chemistry: Epoxide Opening

Epoxides are valuable and versatile building blocks, widely used as synthetic intermediates in the construction of numerous natural products with significant biological activities, including alkaloids, steroids, terpenoids, biopolymers, carotenoids, polyketides, and flavonoids [50]. Their high reactivity primarily arises from the unique geometry and angle strain of the three-membered ring, which provides a favorable thermodynamic driving force for ring-opening reactions (typically 24–-28 kcal mol^{-1}) [51, 52]. Epoxide ring opening offers an efficient route for the incorporation of new functionalities – typically carbon-carbon or carbon-heteroatom bonds – while also delivering key advantages, such as reliability, stereospecificity, and regioselectivity [5].

The application of the bio-click strategies to epoxide ring opening is a growing area of interest and has been explored extensively – even prior to the formal introduction of the term "click chemistry" by Sharpless in 2001. Generally, bio-click approaches involving epoxides follow two main pathways: (a) the selective enzymatic oxidation of alkenes to generate epoxides under environmentally friendly conditions, followed by their ring opening, and (b) the oxidation of alkenes, followed by the enzymatic resolution of racemic epoxides and the subsequent ring opening of the desired chiral epoxide [1] (Scheme 8).

Over the past 2 years, engineered enzymes such as epoxide hydrolase, styrene oxide isomerase, and benzaldehyde lyase have been employed in the sustainable chemoenzymatic synthesis of nonsteroidal anti-inflammatory drugs (NSAIDs) and fungicides (Scheme 9). The integration of a bio-click approach in these syntheses enables shorter synthetic routes while maintaining high enantiopurity. Furthermore, asymmetric synthesis benefits from the reliable ring opening of chiral epoxides in conjunction with substrate-specific protein engineering, which collectively enhance process efficiency, as reflected in high yields, improved atom economy, and excellent enantiomeric excess (ee).

A notable green process was reported by See et al. [53] in 2024, describing an optimized chemoenzymatic synthesis of (*S*)-flurbiprofen (Scheme 9a). The approach involved an enantioselective chemical epoxidation of 3-fluoro-4-bromo-α-methylstyrene via Sharpless dihydroxylation, followed by ring closure to afford (*S*)-**44** with 99% ee. A subsequent cascade transformation utilizing an

Scheme 8 General bio-click approach in epoxide-opening reactions. (Reproduced from Ref. [1] with permission from the Royal Society of Chemistry)

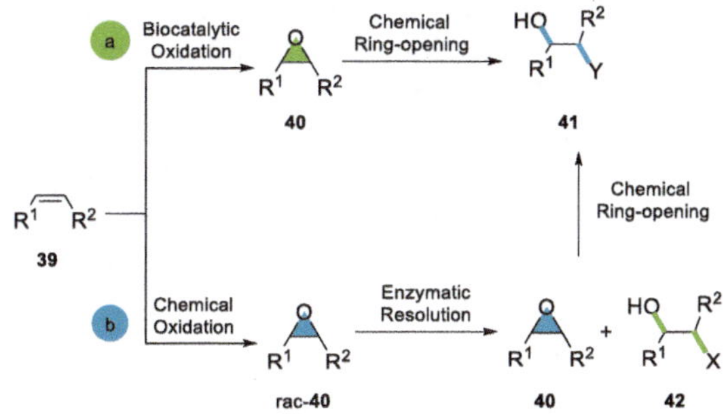

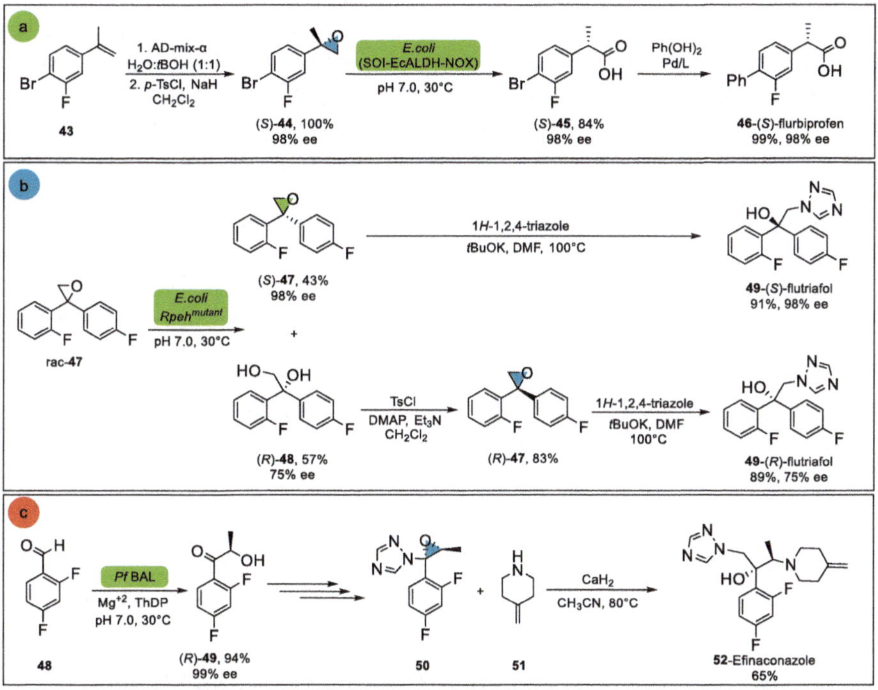

Scheme 9 (**a**) Chemoenzymatic synthesis of (*S*)-flurbiprofen, (**b**) chemoenzymatic asymmetric synthesis of (*S*)- and (*R*)-flutriafol, (**c**) chemoenzymatic approach for the synthesis of efinaconazole

engineered *E. coli* strain co-expressing styrene oxide isomerase double mutant (SOI_{DM}), styrene monooxygenase (SMO), aldehyde dehydrogenase (ALDH), and NADH oxidase (NOX) enabled the efficient conversion of *(S)*-**44** while maintaining 98% ee. The resulting intermediate *(S)*-**44** then underwent Pd-catalyzed arylation, furnishing (*S*)-flurbiprofen in 83% yield and 98% *ee* on a gram scale.

Flutriafol is a chiral pesticide commonly used as a racemic mixture. However, its individual enantiomers display distinct bioactivity and toxicity profiles against target species, as well as differing environmental behaviors and acute toxicity toward nontarget organisms [54]. Given the potential side effects associated with racemic mixtures, considerable research efforts have focused on the separation of chiral agrochemicals and pharmaceuticals. Nonetheless, the low efficiency and high costs of conventional chiral resolution techniques underscore the need for the development of enantioselective synthetic methodologies [55].

In this context, Hu et al. [56] reported an enantioselective synthesis that enables the production of both flutriafol enantiomers in high yields and with excellent enantiomeric excess. Their strategy involved the protein engineering of an epoxide hydrolase from *Rhodotorula paludigensis*, which was optimized to carry out the kinetic resolution of racemic-**47** intermediate. The resulting chiral precursors were subsequently subjected to regioselective ring opening using 1*H*-1,2,4-triazole, affording the (*S*)-flutriafol and (*R*)-flutriafol with high enantiomeric excess.

Efinaconazole (Jublia) is a third-generation triazole antifungal agent approved by the Food and Drug Administration (FDA) in 2014 for treating onychomycosis [57]. The synthesis of efinaconazole presents a significant challenge due to the presence of two contiguous chiral centers. A common synthetic strategy employed for this class of compounds involves the nucleophilic ring opening of a trisubstituted epoxide. However, the steric and electronic complexity introduced by the adjacency of two chiral centers adds an additional layer of difficulty to achieving an efficient enantioselective synthesis [58]. In 2023, Ju et al. [58] addressed this synthetic challenge using a highly efficient bio-click approach to generate the two contiguous chiral centers. The key step in this novel strategy involved a benzoin condensation catalyzed by a benzaldehyde lyase enzyme from *Pseudomonas fluorescens biovar* I, facilitating the enantioselective formation of the first chiral center with excellent yield (94%) and high enantiomeric excess (99% *ee*). This was followed by an epoxide ring opening with 4-methylene piperidine, ultimately yielding efinaconazole in 38% overall yield over six steps with a high optical purity. More importantly, given the structural similarity among various triazole antifungal agents, this synthetic approach offers a versatile platform for the efficient synthesis of related compounds, including ravuconazole and voriconazole.

5 Recent Advances in Bio-click Chemistry: Thio-Michael

The thiol-Michael click reaction is highly valued for its efficient formation of carbon-sulfur (C–S) bonds from thiols and alkenes. This reaction typically showcases high efficiency, operates under mild conditions, and exhibits both high regioselectivity and atom economy [59]. In fact, the thiol-maleimide click chemistry reaction is the most commonly employed method for synthesizing FDA-approved antibody-drug conjugates (ADCs) [60]. Despite the clear advantages of the thiol-Michael click reaction, there are still relatively few bio-click reaction approaches available to date [61].

In 2022, Lu et al. [62] reported an efficient and enantioselective one-pot biocatalytic approach for the synthesis of γ-hydroxy sulfides. This study highlighted the

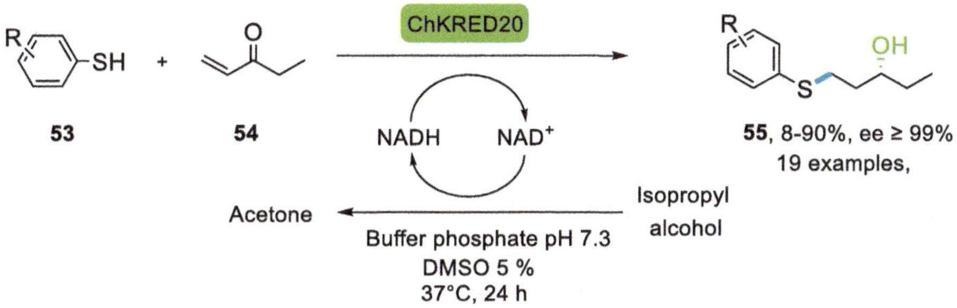

Scheme 10 One-pot biocatalytic synthesis of γ-hydroxy sulfides

enantioselective reduction of intermediate ketones, facilitated by ketoreductase 20 (ChKRED20) from *Chryseobacterium* sp. CA49, an NADH-dependent ketoreductase known for its high selectivity toward (*R*)-configured secondary alcohols. The researchers explored the role of thiophenol substituents in modulating enzyme activity and found that the enzyme does not tolerate large sterically hindered groups at the para positions, which is reflected in decreased yields. Nevertheless, the successful transformation of 18 structurally diverse substrates demonstrated the broad applicability of this method, yielding (*R*)-γ-hydroxy sulfides with excellent enantioselectivity [62] (Scheme 10).

6 Summary and Outlook

In recent years, bio-click chemistry reactions have expanded significantly owing to the development of more powerful and selective enzymes. These enzymes are often generated through enzyme engineering techniques, such as directed evolution, which enhance the ability to synthesize a wide variety of functionalized molecules, including antibiotics, antifungals, and anticoagulants.

A notable example is the chemoenzymatic synthesis of bicyclic peptides, which combines the bioorthogonality of copper-catalyzed azide-alkyne cycloaddition (CuAAC) with the high selectivity of engineered enzymes. This approach underscores the benefits of bio-click chemistry, enabling one-pot procedures that minimize purification steps and improve overall yields.

Despite the broad potential of bio-click chemistry, a comprehensive assessment of its true sustainability impact remains essential. A detailed life cycle analysis incorporating both click and enzymatic reactions is necessary to rigorously evaluate their environmental footprint. Nevertheless, the existing literature consistently highlights that newly developed bio-click methodologies provide significant advantages over conventional chemical approaches, offering a greener alternative for modern synthetic chemistry.

Acknowledgements FCZ is grateful to ANID/CONICYT/FONDECYT Regular N° 1250920.

References

1. Rodríguez DF, Moglie Y, Ramírez-Sarmiento CA, Singh SK, Dua K, Zacconi FC (2022) Bio-click chemistry: a bridge between biocatalysis and click chemistry. RSC Adv 12(4): 1932–1949. https://doi.org/10.1039/D1RA08053A
2. Ankati H, Yang Y, Zhu D, Biehl ER, Hua L (2008) Synthesis of optically pure 2-azido-1-arylethanols with isolated enzymes and conversion to triazole-containing β-blocker analogues employing click chemistry. J Org Chem 73(16):6433–6436. https://doi.org/10.1021/jo8009616
3. Slagman S, Fessner W-D (2021) Biocatalytic routes to anti-viral agents and their synthetic intermediates. Chem Soc Rev 50(3):1968–2009. https://doi.org/10.1039/D0CS00763C
4. Carrau G, Bellomo AI, Suescun L, Gonzalez D (2019) Chemoenzymatic synthesis of hygromycin aminocyclitol moiety and its C2 Epimer. Eur J Org Chem 2019(4):788–802. https://doi.org/10.1002/ejoc.201801424
5. Kolb HC, Finn MG, Sharpless KB (2001) Click chemistry: diverse chemical function from a few good reactions. Angew Chemie Int Ed 40(11):2004–2021. https://doi.org/10.1002/1521-3773(20010601)40:11<2004::AID-ANIE2004>3.0.CO;2-5
6. Kolb HC, Sharpless KB (2003) The growing impact of click chemistry on drug discovery. Drug Discov Today 8(24):1128–1137. https://doi.org/10.1016/S1359-6446(03)02933-7
7. Greg H, Lewis CJ (2018) Introduction: biocatalysis in industry. Chem Rev 118(1):1–3. https://doi.org/10.1021/acs.chemrev.7b00741
8. Sheldon RA (2024) Green chemistry and biocatalysis: engineering a sustainable future. Catal Today 431:114571. https://doi.org/10.1016/j.cattod.2024.114571
9. Alcántara AR, Domínguez de María P, Littlechild JA, Schürmann M, Sheldon RA, Wohlgemuth R (2022) Biocatalysis as key to sustainable industrial chemistry. ChemSusChem 15(9):e202102709. https://doi.org/10.1002/cssc.202102709
10. Reetz MT (2016) What are the limitations of enzymes in synthetic organic chemistry? Chem Rec 16(6):2449–2459. https://doi.org/10.1002/tcr.201600040
11. Kim S, Ga S, Bae H, Sluyter R, Konstantinov K, Shrestha LK, Kim YH, Kim JH, Ariga K (2024) Multidisciplinary approaches for enzyme biocatalysis in pharmaceuticals: protein engineering, computational biology, and nanoarchitectonics. EES Catal 2(1):14–48. https://doi.org/10.1039/D3EY00239J
12. Maghraby YR, El-Shabasy RM, Ibrahim AH, Azzazy HME-S (2023) Enzyme immobilization technologies and industrial applications. ACS Omega 8(6):5184–5196. https://doi.org/10.1021/acsomega.2c07560
13. Toro EC, Rodríguez DF, Morales N, García LM, Godoy CA (2019) Novel combi-lipase systems for fatty acid ethyl esters production. Catalysts. https://doi.org/10.3390/catal9060546
14. Tornøe CW, Christensen C, Meldal M (2002) Peptidotriazoles on solid phase: [1,2,3]-triazoles by regiospecific copper(I)-catalyzed 1,3-dipolar cycloadditions of terminal alkynes to azides. J Org Chem 67(9):3057–3064. https://doi.org/10.1021/jo011148j
15. Rodríguez DF, Durán-Osorio F, Duarte Y, Olivares P, Moglie Y, Dua K, Zacconi FC (2022) Green by design: convergent synthesis, computational analyses, and activity evaluation of new FXa inhibitors bearing peptide Triazole linking units. Pharmaceutics:33. https://doi.org/10.3390/pharmaceutics14010033
16. Hanau S, Almugadam SH, Sapienza E, Cacciari B, Manfrinato MC, Trentini A, Kennedy JF (2020) Schematic overview of oligosaccharides, with survey on their major physiological effects and a focus on Milk ones. Carbohydr Polym Technol Appl 1:100013. https://doi.org/10.1016/j.carpta.2020.100013
17. Varki A (1993) Biological roles of oligosaccharides: all of the theories are correct. Glycobiology 3(2):97–130. https://doi.org/10.1093/glycob/3.2.97
18. Nobre C, Simões LS, Gonçalves DA, Berni P, Teixeira JA (2021) Fructooligosaccharides Production and the Health Benefits of Prebiotics. In: Rai AK, Singh SP, Pandey A, Larroche C et al (eds) Current developments in biotechnology and bioengineering. Elsevier, pp 109–138. https://doi.org/10.1016/B978-0-12-823506-5.00002-3

19. Kulinich A, Liu L (2016) Human milk oligosaccharides: the role in the fine-tuning of innate immune responses. Carbohydr Res 432:62–70. https://doi.org/10.1016/j.carres.2016.07.009
20. Li W, Yu S, Zhang T, Jiang B, Stressler T, Fischer L, Mu W (2015) Efficient biosynthesis of Lactosucrose from sucrose and lactose by the purified recombinant Levansucrase from Leuconostoc Mesenteroides B-512 FMC. J Agric Food Chem 63(44):9755–9763. https://doi.org/10.1021/acs.jafc.5b03648
21. Seeberger PH, Werz DB (2007) Synthesis and medical applications of oligosaccharides. Nature 446(7139):1046–1051. https://doi.org/10.1038/nature05819
22. Tseng H-W, Tseng H-K, Ooi K-E, You C-E, Wang H-K, Kuo W-H, Ni C-K, Manabe Y, Lin C-C (2024) Controllable enzymatic synthesis of natural asymmetric human milk oligosaccharides. JACS Au 4(11):4496–4506. https://doi.org/10.1021/jacsau.4c00830
23. Dumic J, Dabelic S, Flögel M (2006) Galectin-3: an open-ended story. Biochim Biophys Acta - Gen Subj 1760(4):616–635. https://doi.org/10.1016/j.bbagen.2005.12.020
24. Johannes L, Jacob R, Leffler H (2018) Galectins at a glance. J Cell Sci 131(9):jcs208884. https://doi.org/10.1242/jcs.208884
25. Stegmayr J, Zetterberg F, Carlsson MC, Huang X, Sharma G, Kahl-Knutson B, Schambye H, Nilsson UJ, Oredsson S, Leffler H (2019) Extracellular and intracellular small-molecule galectin-3 inhibitors. Sci Rep 9(1):2186. https://doi.org/10.1038/s41598-019-38497-8
26. Dussouy C, Téletchéa S, Lambert A, Charlier C, Botez I, De Ceuninck F, Grandjean C (2020) Access to Galectin-3 inhibitors from chemoenzymatic synthons. J Org Chem 85(24): 16099–16114. https://doi.org/10.1021/acs.joc.0c01927
27. Gorantla JN, Maniganda S, Pengthaisong S, Ngiwsara L, Sawangareetrakul P, Chokchaisiri S, Kittakoop P, Svasti J, Ketudat Cairns JR (2021) Chemoenzymatic and protecting-group-free synthesis of 1,4-substituted 1,2,3-triazole-α-d-glucosides with potent inhibitory activity toward lysosomal α-glucosidase. ACS Omega 6(39):25710–25719. https://doi.org/10.1021/acsomega.1c03928
28. Javed A, Slingerland CJ, Wood TM, Martin NI, Broere F, Weingarth MH, Veldhuizen EJA (2023) Chimeric peptidomimetic antibiotic efficiently neutralizes lipopolysaccharides (LPS) and bacteria-induced activation of RAW macrophages. ACS Infect Dis 9(3):518–526. https://doi.org/10.1021/acsinfecdis.2c00518
29. Wang W-J, Dong X-M, Li G-B (2024) Macrocyclic peptides: up-and-coming weapons to combat antimicrobial resistance. Signal Transduct Target Ther 9(1):81. https://doi.org/10.1038/s41392-024-01813-4
30. Wood TM, Slingerland CJ, Martin NI (2021) A convenient chemoenzymatic preparation of chimeric macrocyclic peptide antibiotics with potent activity against gram-negative pathogens. J Med Chem 64(15):10890–10899. https://doi.org/10.1021/acs.jmedchem.1c00176
31. Richelle GJJ, Schmidt M, Ippel H, Hackeng TM, van Maarseveen JH, Nuijens T, Timmerman P (2018) A one-pot "triple-C" multicyclization methodology for the synthesis of highly constrained isomerically pure tetracyclic peptides. Chembiochem 19(18):1934–1938. https://doi.org/10.1002/cbic.201800346
32. Schmidt M, Toplak A, Quaedflieg PJLM, Ippel H, Richelle GJJ, Hackeng TM, van Maarseveen JH, Nuijens T (2017) Omniligase-1: a powerful tool for peptide head-to-tail cyclization. Adv Synth Catal 359(12):2050–2055. https://doi.org/10.1002/adsc.201700314
33. Kobayashi M, Onozawa N, Matsuda K, Wakimoto T (2024) Chemoenzymatic tandem cyclization for the facile synthesis of bicyclic peptides. Commun Chem 7(1):67. https://doi.org/10.1038/s42004-024-01147-w
34. Takao K, Munakata R, Tadano K (2005) Recent advances in natural product synthesis by using intramolecular Diels–Alder reactions. Chem Rev 105(12):4779–4807. https://doi.org/10.1021/cr040632u
35. Lichman BR, O'Connor SE, Kries H (2019) Biocatalytic strategies towards [4+2] cycloadditions. Chem - A Eur J 25(28):6864–6877. https://doi.org/10.1002/chem.201805412
36. Woodward RB, Sondheimer F, Taub D, Heusler K, McLamore WM (1952) The total synthesis of steroids 1. J Am Chem Soc 74(17):4223–4251. https://doi.org/10.1021/ja01137a001

37. Woodward RB, Bader FE, Bickel H, Frey AJ, Kierstead RW (1958) The total synthesis of reserpine. Tetrahedron 2(1):1–57. https://doi.org/10.1016/0040-4020(58)88022-9
38. Forman MA, Dailey WP (1991) The lithium perchlorate-diethyl ether rate acceleration of the Diels-Alder reaction: Lewis acid catalysis by lithium ion. J Am Chem Soc 113(7):2761–2762. https://doi.org/10.1021/ja00007a065
39. Portela S, Fernández I (2022) Origin of catalysis and selectivity in lewis acid-promoted diels–alder reactions involving vinylazaarenes as dienophiles. J Org Chem 87(14):9307–9315. https://doi.org/10.1021/acs.joc.2c01035
40. Dewar MJS, Olivella S, Stewart JJP (1986) Mechanism of the Diels-Alder reaction: reactions of butadiene with ethylene and cyanoethylenes. J Am Chem Soc 108(19):5771–5779. https://doi.org/10.1021/ja00279a018
41. Rideout DC, Breslow R (1980) Hydrophobic acceleration of Diels-Alder reactions. J Am Chem Soc 102(26):7816–7817. https://doi.org/10.1021/ja00546a048
42. Milzarek TM, Gulder TAM (2025) The fungal natural product class of the sorbicillinoids: structures, bioactivities, biosynthesis, and synthesis. Nat Prod Rep. https://doi.org/10.1039/D4NP00059E
43. Milzarek TM, Gulder TAM (2023) Chemo-enzymatic total synthesis of the spirosorbicillinols. Commun. Chem. 6(1):187. https://doi.org/10.1038/s42004-023-00996-1
44. Castellino NJ, Montgomery AP, Danon JJ, Kassiou M (2023) Late-stage functionalization for improving drug-like molecular properties. Chem Rev 123(13):8127–8153. https://doi.org/10.1021/acs.chemrev.2c00797
45. Guillemard L, Kaplaneris N, Ackermann L, Johansson MJ (2021) Late-stage C–H functionalization offers new opportunities in drug discovery. Nat Rev Chem 5(8):522–545. https://doi.org/10.1038/s41570-021-00300-6
46. Pyser JB, Chakrabarty S, Romero EO, Narayan ARH (2021) State-of-the-art biocatalysis. ACS Cent Sci 7(7):1105–1116. https://doi.org/10.1021/acscentsci.1c00273
47. Romero E, Jones BS, Hogg BN, Rué Casamajo A, Hayes MA, Flitsch SL, Turner NJ, Schnepel C (2021) Enzymatic late-stage modifications: better late than never. Angew. Chemie Int. Ed. 60(31):16824–16855. https://doi.org/10.1002/anie.202014931
48. Feng Z, Chakraborty D, Dewell SB, Reddy BVB, Brady SF (2012) Environmental DNA-encoded antibiotics Fasamycins a and B inhibit FabF in type II fatty acid biosynthesis. J Am Chem Soc 134(6):2981–2987. https://doi.org/10.1021/ja207662w
49. Li J, Renata H (2021) Concise chemoenzymatic synthesis of fasamycin A. J Org Chem 86(16): 11206–11211. https://doi.org/10.1021/acs.joc.1c00526
50. Hanif M, Zahoor AF, Saif MJ, Nazeer U, Ali KG, Parveen B, Mansha A, Chaudhry AR, Irfan A (2024) Exploring the synthetic potential of epoxide ring opening reactions toward the synthesis of alkaloids and terpenoids: a review. RSC Adv 14(19):13100–13128. https://doi.org/10.1039/D4RA01834F
51. Morgan KM, Ellis JA, Lee J, Fulton A, Wilson SL, Dupart PS, Dastoori R (2013) Thermochemical studies of epoxides and related compounds. J Org Chem 78(9):4303–4311. https://doi.org/10.1021/jo4002867
52. Pellissier H, Lattanzi A, Dalpozzo R (2017) Asymmetric epoxidation. In: Asymmetric synthesis of three-membered rings, pp 379–538. https://doi.org/10.1002/9783527802029.ch3
53. See WWL, Choo JPS, Du L, Li Z (2024) Engineering styrene oxide isomerase towards enhanced chemoenzymatic synthesis of (S)-flurbiprofen. Adv Synth Catal 366(13): 2967–2975. https://doi.org/10.1002/adsc.202400325
54. Zhang Q, Hua X, Shi H, Liu J, Tian M, Wang M (2015) Enantioselective bioactivity, acute toxicity and dissipation in vegetables of the chiral triazole fungicide flutriafol. J Hazard Mater 284:65–72. https://doi.org/10.1016/j.jhazmat.2014.10.033
55. Calcaterra A, D'Acquarica I (2018) The market of chiral drugs: chiral switches versus de novo enantiomerically pure compounds. J Pharm Biomed Anal 147:323–340. https://doi.org/10.1016/j.jpba.2017.07.008

56. Hu D, Lu Z-Y, Liao X, Jia X-W, Song W-H, Hu Y-Y, He Y-C (2024) Engineering an epoxide hydrolase for chemoenzymatic asymmetric synthesis of chiral triazole fungicide (S)- and (R)-flutriafol. J Agric Food Chem 72(39):21741–21751. https://doi.org/10.1021/acs.jafc.4c07318
57. Lipner SR, Scher RK (2015) Efinaconazole in the treatment of onychomycosis. Infect Drug Resist 8:163–172. https://doi.org/10.2147/IDR.S69596
58. Ju Z, Li Z, Li M, Xu S, Kaliaperumal K, Chen F-E (2023) A chemo-enzymatic approach for preparing efinaconazole with high optical yield. J Org Chem 88(20):14803–14808. https://doi.org/10.1021/acs.joc.3c01641
59. Northrop BH, Frayne SH, Choudhary U (2015) Thiol–Maleimide "click" chemistry: evaluating the influence of solvent, initiator, and thiol on the reaction mechanism, kinetics, and selectivity. Polym Chem 6(18):3415–3430. https://doi.org/10.1039/C5PY00168D
60. Dudchak R, Podolak M, Holota S, Szewczyk-Roszczenko O, Roszczenko P, Bielawska A, Lesyk R, Bielawski K (2024) Click chemistry in the synthesis of antibody-drug conjugates. Bioorg Chem 143:106982. https://doi.org/10.1016/j.bioorg.2023.106982
61. Lauder K, Toscani A, Qi Y, Lim J, Charnock SJ, Korah K, Castagnolo D (2018) Photo-biocatalytic one-pot cascades for the enantioselective synthesis of 1,3-mercaptoalkanol volatile sulfur compounds. Angew Chemie Int Ed 57(20):5803–5807. https://doi.org/10.1002/anie.201802135
62. Lu W-F, Yu Y, Lin R-D, Yao Y, Liu Y, Wu Z-L, Liu Y-H, Wang N (2022) Enantioselective biosynthesis of (R)-γ-hydroxy sulfides via a one-pot approach with ChKRED20. Mol Catal 531:112655. https://doi.org/10.1016/j.mcat.2022.112655

Top Heterocycl Chem (2026) 63: 225–256
https://doi.org/10.1007/7081_2025_95

Published online: 30 December 2025

The Sustainability Aspects of Click-Chemistry Transformations

Filippo Campana, Giulia Brufani, Daniela Lanari, and Luigi Vaccaro

Contents

1 Introduction 226
2 Classification of Click Reactions 229
2.1 Nucleophilic Ring Opening of Epoxides and Aziridines 230
2.2 Copper-Catalyzed and Ruthenium-Catalyzed Alkyne-Azide Cycloaddition 232
2.3 Strain-Promoted Azido-Alkyne Cycloaddition (SPAAC) and Thiol-Ene Click Coupling 234
3 Alternative Solvents in Click Chemistry 234
4 Heterogeneous Catalyzed Click Reactions 236
5 Click Chemistry Under Continuous Flow Conditions 240
6 Alternative Energy Sources 243
6.1 Photo-Click Chemistry 244
6.2 Microwave-Assisted Click Chemistry 246
6.3 Mechanochemical Approaches 246
6.4 Sonochemistry 247

Filippo Campana and Giulia Brufani are co-first authors.
The original version of the chapter has been revised. In page 226, "Prevention Pollution Act" changed to "Pollution Prevention Act"; page 229, "Protecting group reaction" changed to "Protecting group reactions";page 230, "Sn2-type mechanism" changed to "SN2-type mechanism"; page 237, Figure 6 has been replaced with a revised Figure; page 238, "reaction among 2-phenyl-1-tosylaziridine and ethanol" changed to "reaction between 2-phenyl-1-tosylaziridine and ethanol"; page 239, "substituents" changed to "substitutes"; page 239, "Regarding CuAAc and RuAAc" changed to "Regarding CuAAC and RuAAC"; page 245, caption of Figure 10 "b) metal-free photo-triggerd azide cycloaddition" changed to "b) metal-free photo-triggered azide-alkyne cycloaddition"; page 251, "Rostovtsev VV, Green LG, Fokin VV, Sharpless KB (2002) A stepwise huisgen cycloaddition" changed to "Rostovtsev VV, Green LG, Fokin VV, Sharpless KB (2002) A stepwise Huisgen cycloaddition".

F. Campana, G. Brufani, and L. Vaccaro (✉)
Laboratory of Green S.O.C.—Dipartimento di Chimica, Biologia e Biotecnologie Università degli Studi di Perugia, Perugia, Italy
e-mail: luigi.vaccaro@unipg.it

D. Lanari (✉)
Dipartimento di Scienze Farmaceutiche, Università degli Studi di Perugia, Perugia, Italy
e-mail: daniela.lanari@unipg.it

6.5 Electrochemistry 248
7 Conclusion, Future Outlooks, and Trends 249
7.1 Conclusion 249
7.2 Future Outlooks and Trends 250
References 251

Abstract Click chemistry is a synthetic strategy that enables the formation of heteroatom-based systems. It provides both stereoselectivity and regioselectivity, along with high yields, under mild reaction conditions. These features make this chemical strategy inherently characterized by complete atom economy and reduced waste production.

Nevertheless, since its inception, several upgrades have been implemented to further enhance the overall sustainability of this process, such as replacing fossil fuel-derived and hazardous solvents, introducing recoverable heterogeneous catalysts, and establishing continuous flow synthesis and/or using alternative energy sources. In this chapter, we aim to focus on the most significant green achievements that have contributed significantly to greening this useful transformation and reducing its environmental and human health footprint.

1 Introduction

As defined by Sharpless et al., in 2001, a click reaction "*must be modular, must be wide in scope, must give very high yields, must generate only inoffensive byproducts that can be removed by nonchromatographic methods, and must be stereospecific (but not necessarily enantioselective). The required process characteristics include simple reaction conditions (ideally, the process should be insensitive to oxygen and water), readily available starting materials and reagents, use of no solvent or a solvent that is benign (such as water) or easily removed, and simple product isolation. Purification, if required, must be by nonchromatographic methods, such as crystallization or distillation, and the product must be stable under physiological conditions*" [1].

Considering the provided definition, the rationale behind click chemistry features is strictly and deeply linked to the green chemistry principles, a set of 12 milestones that chemists need to follow to design a benign and waste-minimized reaction process [2].

These guiding benchmarks, developed by Anastas and Warner in 1998, shape the challenges highlighted some years earlier by the Pollution Prevention Act, a document delivered by the United States Environmental Protection Agency that defined the transition from a strategy of "command and control" of the environment to one of "prevention and protection" from pollution through a new way to exploit chemistry [3].

In 1991, the same institute had already defined green chemistry as "*the design of chemical products and processes that reduce or eliminate the use or generation of hazardous substances. Green chemistry applies across the life cycle of a chemical product, including its design, manufacture, use, and ultimate disposal. Green chemistry is also known as sustainable chemistry*" [2].

The 12 principles of green chemistry (GCP) represent the key elements in the redesign of chemical processes. They encompass waste prevention (GCP1), reducing hazardous chemical synthesis (GCP3), and derivatives (GCP8). Furthermore, design of safer chemicals (GCP4), real-time analysis of pollutants (GCP11), and inherently safer chemistry for accident prevention (GCP12) play a crucial role in ensuring process safety for both workers and the environment.

Moreover, special emphasis is placed on minimizing energy consumption (GCP6) and using renewable feedstocks (GCP7), leading to more environmentally friendly and energy-efficient chemical processes.

Finally, the atom economy (GCP2) deserves special attention as it focuses on maximizing the number of reactant atoms incorporated into the final product. Achieving a 100% atomic economy guarantees intrinsically limiting waste generation and reduces energy requirements.

When aiming at implementing a chemical route from a sustainable point of view, it is essential to keep in mind a series of tools that can be adopted. Among these strategies, the use of alternative solvents, such as water or those derived from biomass, or the valorization of industrial waste, has gained prominence. However, as Sheldon has stated, "*the best solvent is no solvent*," highlighting that solvents typically account for approximately 80% of the waste generated in chemical processes [4]. Consequently, solvent-free conditions (SolFC) represent the most effective approach to waste minimization [5].

An alternative approach involves employing heterogeneous catalysis, which simplifies the separation and recovery of the insoluble catalyst, thereby eliminating the need for time-consuming and expensive purification steps to isolate the desired product. The combination of the above mentioned tools can often be merged within a continuous flow procedure. Flow chemistry is a recently emerged technology in which the product is continuously produced through tubes, pipes, and reactors that can be prepared or packed with heterogeneous catalytic species.

The last point that can be addressed to improve the process sustainability is the research of alternative energy sources to limit electricity consumption, which, as reported by the European Union, is primarily responsible for greenhouse gas emissions [6].

Once the possible strategies to design a more benign synthetic route have been defined, it is essential to evaluate the actual quantitative and qualitative improvements provided, always relating to the alternative and existing examples. In fact, *a priori* choice, for instance, in the selection of alternative solvents or catalysts, is not always possible and usually not recommended as each process has its peculiarities [7]. In this context, several "green metrics" have been introduced in the last decades to measure green enhancements.

The first green metric that can be considered is the atom economy (AE), discussed in the second principle of green chemistry. It describes the conversion efficiency of a chemical process in terms of all the atoms included in the desired products.

The E-factor, introduced by R. Sheldon in 1992 [8], is one of the most useful and indicative metrics for assessing the process' sustainability. Defined as the ratio of total waste mass to the mass of the final product, it provides a quantitative measure of the waste generated per kilogram of the desired product. However, the E-factor is not

Fig. 1 Life cycle assessment (LCA) stages

without limitations. A key drawback lies in how waste is accounted for, and it does not distinguish between harmless and hazardous waste, failing to provide insights into the toxicological and environmental impact of the generated waste.

To further overcome this issue, the environmental quotient (EQ) was introduced, where Q represents an arbitrarily assigned unfriendliness multiplier or weighting factor that must disclose the nature of the waste, to assess their environmental impact [5].

In this context, a more extensive and complete evaluation, both quantitative and qualitative, is necessary to evaluate the sustainability of chemical transformations efficiently. Life cycle assessment (LCA) (Fig. 1) is a complete and exhaustive method that considers the environmental impact of a product or process, in all stages of its "life," from the raw material extraction or acquisition stage through production and use to the end-of-life treatment and disposal or recycling (cradle-to-grave) [9], also taking into account the energetic consumptions and transportation [10]. Multiple useful pieces of information can be derived from an LCA analysis, depending on the kind of characterization performed. If the midpoint level is analyzed through different impact categories, it is possible to investigate the impact of the different components that are part of the process under investigation on specific sectors in the cause-effect chain. At the same time, the endpoint level represents the final effect on damage areas (human health, ecosystems, and resources) through environmental mechanisms. A clarifying example is represented by the "*global warming potential*" (midpoint) that contributes to human health damage area (endpoint level) through heatwaves, disease spread, and extreme weather events [11].

With this framework established, the primary goal of this chapter is to highlight key advancements in the chemical sustainability of click chemistry. Keeping in mind the green chemistry features, four macro areas that are usually considered to define a green protocol have been discussed. Particularly, we delve into the use of alternative reaction media and the exploitation of heterogeneous catalysts in combination with flow chemistry. Additionally, given the current environmental and climate change concerns, we assess the adoption of alternative energy sources as replacements for conventional heating systems. These innovations aim to improve energy efficiency, reduce electricity consumption, and minimize greenhouse gas emissions.

2 Classification of Click Reactions

Click reactions can be considered nearly "ideal" transformations since they achieve a complete atom economy. Moreover, due to their significant thermodynamic driving force (~20 kcal mol^{-1}), they ensure the preferential and selective formation of the thermodynamically stable product, maximizing energy efficiency [1].

Sharpless originally classified click reactions into two main categories. The first involves proton rearrangements, like the nucleophilic opening of epoxides, aziridines, cyclic sulfates, sulfamidates, aziridinium ions, and episulfonium ions; the second one includes reactions where σ-bonds remain intact, such as cycloadditions like Diels-Alder and Huisgen dipolar cycloaddition.

Protecting group reactions can be described as a third-generation click transformation. This unique class of click transformation is based on the reversible carbonyl reaction, in which hydroxy groups react with acids, aldehydes, and ketones to form cyclic 1,3-dioxolane rings.

In the past 20 years, novel "ideal" click transformations have been developed. Huisgen's thermal 1,3-dipolar cycloaddition of alkynes and azides was an early example, accessing 1,4- and 1,5-disubstituted 1,2,3-triazoles. However, the lack of regioselectivity and the required harsh conditions pushed the development of catalyzed protocols [12].

The copper-catalyzed alkyne-azide cycloaddition (CuAAC), developed by the Nobel Prize winners Sharpless and Meldal, overcomes these challenges, allowing the regioselective synthesis of 1,4-disubstituted 1,2,3-triazoles under mild conditions. The CuAAC fits so well inside the definition of click chemistry that they are almost synonymous. On the other hand, the ruthenium-catalyzed azide-alkyne cycloaddition (RuAAC) enables the formation of 1,5-disubstituted 1,2,3-triazoles with internal alkynes.

Among non-azide click reactions, thiol-ene coupling and copper-catalyzed sydnone-alkyne cycloaddition (CuSAC) are notable for their efficiency under mild conditions. Additionally, photo-click chemistry provides precise timing control, expanding click chemistry's applications (Fig. 2).

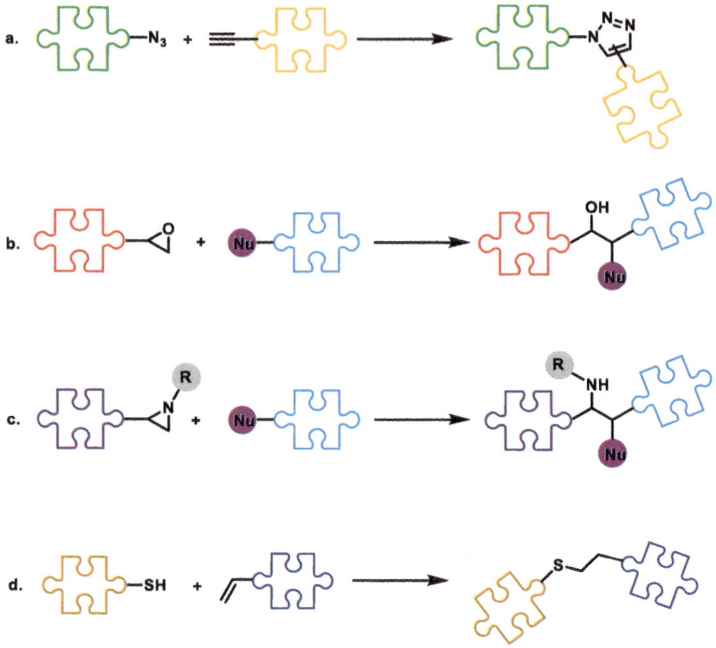

Fig. 2 Different types of click reaction: (**a**) alkyne-azide cycloaddition; (**b**) nucleophilic ring opening of epoxides; (**c**) nucleophilic ring opening of aziridines; (**d**) thiol-ene coupling

2.1 Nucleophilic Ring Opening of Epoxides and Aziridines

Epoxides and aziridines are highly reactive strained-ring systems that serve as crucial intermediates in organic synthesis, metabolic pathways, and pharmaceutical applications. Their ring-opening reactions, facilitated by nucleophiles, offer an efficient route to diverse 1,2-functionalized compounds with high regio- and stereoselectivity [13].

Due to the inherent ring strain, they easily undergo nucleophilic attack to open the ring, making them versatile building blocks in synthetic chemistry. Typically, the reaction follows an S_N2-type mechanism under neutral or basic conditions, leading to inversion of configuration at the less substituted carbon. This process yields 1,2-disubstituted products with *trans*- or *anti*-stereochemistry [14].

The scope of epoxide reactivity has been expanded through transition metal-catalyzed desymmetrization of meso-epoxides and kinetic resolution strategies, enabling enantio- and diastereoselective synthesis. Carbon-based nucleophiles facilitate C–C bond formation, while nitrogen-based nucleophiles provide access to amino alcohols and azido alcohols. On the other hand, their reaction with oxygen-based nucleophiles such as water, alcohols, and phenols results in less efficient reactions. However, it remains crucial to prepare alkoxy alcohols and 1,2-diols.

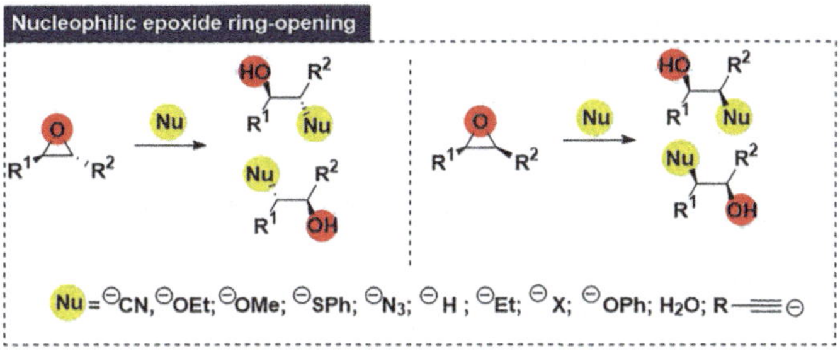

Fig. 3 Epoxide ring-opening reactions with various nucleophiles

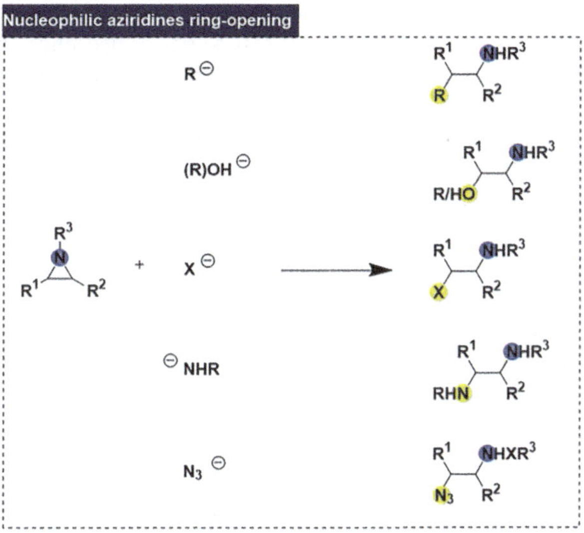

Fig. 4 Nucleophiles employed in aziridine ring-opening reactions

Thiols, often requiring Lewis acid activation, undergo efficient ring opening under milder conditions (Fig. 3).

Aziridines, valuable intermediates in fine chemical and pharmaceutical synthesis, undergo nucleophilic opening, yielding amino alcohols, diamines, and other bioactive molecules. Specifically, activated aziridines (bearing electron-withdrawing groups) undergo nucleophilic attack at the less hindered carbon, while nonactivated aziridines exhibit varied regioselectivity influenced by nucleophile and substituent effects. Common nucleophiles include carbanions, enolates, cyanides, thiols, amines, and halides. Additionally, hydrogenation of aziridines yields amines, while Lewis acid-mediated activation generates reactive aziridinium intermediates, useful for further functionalization (Fig. 4) [15].

2.2 *Copper-Catalyzed and Ruthenium-Catalyzed Alkyne-Azide Cycloaddition*

The copper-catalyzed alkyne-azide cycloaddition (CuAAC) first discovered in 2002, by both Sharpless and Meldal, is a cornerstone of click chemistry [16, 17]. This transformation involves the selective formation of 1,4-disubstituted 1,2,3-triazoles, which have shown promising biological properties, including anti-HIV, antibacterial, and antiviral effects. CuAAC has broad applications across medicinal chemistry, biology, and material science, primarily because of its efficiency, regioselectivity, and tolerance for a plethora of functional groups. The copper (I) catalyst is typically generated *in situ* from copper(II) salts with a reducing agent such as sodium ascorbate, in organic solvents such as acetonitrile or toluene. This method is highly effective because it can be performed under mild conditions and is compatible with a wide range of other chemical transformations, making it ideal for one-pot reactions.

From a mechanistic point of view, CuAACs begin with the formation of a copper acetylide, followed by the coordination of an azide with the copper(I) catalyst. The reaction proceeds with the formation of a copper-containing intermediate, where acetylide and azide are coordinated to different copper atoms. The final rearrangement affords the 1,4-disubstituted triazoles. This reaction is generally unaffected by the steric and electronic properties of the reactants, allowing for high yields of the desired product [18] (Scheme 1).

The reaction's versatility extends to the use of various copper sources, both homogeneous and heterogeneous, and has found synergy with other reactions, such as alkylation, acylation, and palladium-catalyzed arylation. Additionally,

Scheme 1 CuAAC mechanism

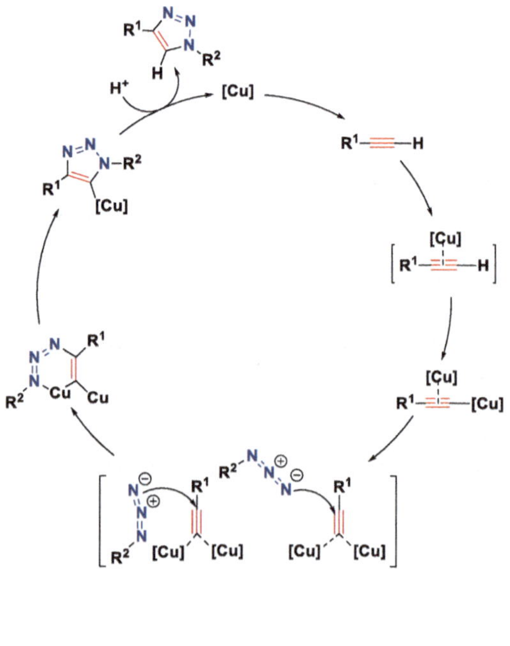

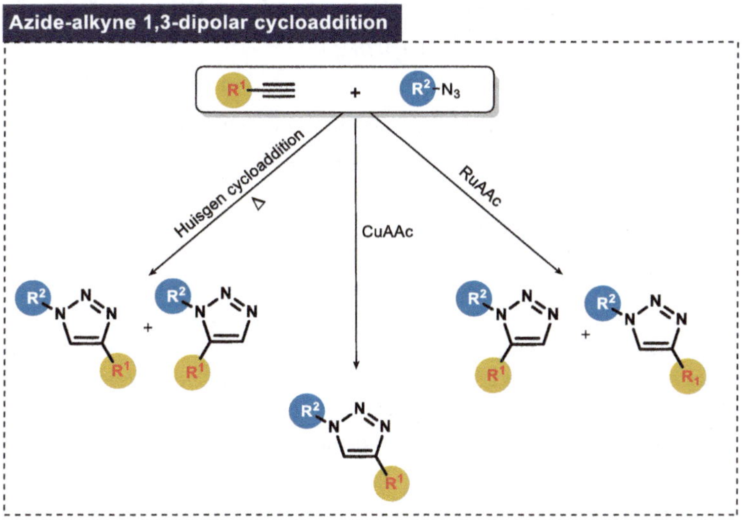

Fig. 5 Catalyzed and uncatalyzed azide-alkyne 1,3-dipolar cycloadditions

CuAAC has been combined with S_N2 reactions, making it highly adaptable in synthetic chemistry.

In comparison, ruthenium-catalyzed alkyne-azide cycloaddition (RuAAC) is highly selective for 1,5-disubstituted 1,2,3-triazole isomers and can offer an alternative that tolerates internal alkynes, providing access to 1,4,5-trisubstituted triazoles. Moreover, as specified by Fokin and colleagues [19], ruthenium catalysts lacking a cyclopentadienyl ligand afforded the 1,4-disubstituted isomer instead of the 1,5-disubstituted one.

RuAAC works similarly to CuAAC but involves different mechanistic steps. The reaction uses a Ru(II) complex, with a pentamethylcyclopentadienyl (Cp*) ligand, to catalyze the cycloaddition. The mechanism involves irreversible oxidative coupling between the alkyne and azide, leading to the formation of a ruthenacycle intermediate. This successively undergoes reductive elimination, providing the desired triazole and regenerating the catalyst [20].

RuAAC also offers a high tolerance for various functional groups, with aromatic and ether solvents like benzene and THF typically used. This versatility has made RuAAC a valuable tool in medicinal chemistry, providing access to a wide range of triazole-containing compounds with important biological activities.

Both CuAAC and RuAAC (Fig. 5) reactions proved to be crucial for synthesizing biologically active molecules and advancing various chemical and medicinal applications.

2.3 Strain-Promoted Azido-Alkyne Cycloaddition (SPAAC) and Thiol-Ene Click Coupling

In contrast to CuAAC and RuAAC, strain-promoted azido-alkyne cycloaddition (SPAAC) is a metal-free and highly efficient alternative to access triazole moieties. Unlike CuAAC, which is limited to terminal alkynes and requires copper, SPAAC uses strained cycloalkynes, such as cyclooctynes, to undergo a [3 + 2] dipolar cycloaddition with azides, producing triazoles without the need for a catalyst. Developed by Bertozzi and others [21], SPAAC eliminates metal toxicity concerns as those associated with copper and offers a simple, high-yield process with excellent atom economy to be employed in biological systems. Variants of cyclooctynes, such as fluorinated or thiacycloalkynes, have been used to accelerate the reaction. SPAAC is compatible with various solvents and operates under mild conditions, fitting well with the click chemistry principles of efficiency, selectivity, and versatility.

In the context of metal-free click reactions, thiol-ene chemistry has also gained attention for its simplicity and broad application potential, yielding high-quality products under mild reaction conditions and with reduced purification steps. Thiol-ene reactions can be categorized into two types: free-radical thiol-ene addition and thiol-Michael addition [22].

In the free-radical variant, a thiol adds to an alkene to form a thioether, following either a step-growth or chain-growth mechanism. In the thiol-Michael reaction, a thiol adds to an electron-deficient olefin, such as acrylates or maleimides, often under mild conditions using nontoxic catalysts. Thiol-ene click reactions are widely used in polymer chemistry [23], biomedical applications, and bioorganic modification due to their efficiency, selectivity, and environmental compatibility. Both SPAAC and thiol-ene click reactions exemplify the growing trend of metal-free, highly adaptable click chemistry methodologies, thus making a fundamental step toward sustainability.

3 Alternative Solvents in Click Chemistry

Since its inception, the chemical sector, encompassing both academia and industry, has heavily relied on the use of solvents. These substances are essential in chemical processes, facilitating the solubilization of reagents, stabilizing transition states, enabling heat transfer, and, most importantly, serving as extraction and purification agents. Beyond their role in synthetic chemistry, solvents are also key components in cosmetic, paint, and cleaning product formulations, with an estimated market value projected to reach approximately 61 billion dollars by 2034 [24].

Due to their petroleum-based origin and the great quantities employed, solvents contribute significantly to waste production, posing serious risks to both human health and the environment. As a result, greener alternatives offer a promising path to reducing net waste and emissions, supporting a transition toward more sustainable chemical practices. In this regard, solvents derived from biomass or industrial byproducts or those with a favorable ecotoxicological profile are increasingly recognized as viable substitutes.

The quest for greener solvents, a cornerstone of green chemistry, has been the focus of extensive research over recent decades. Many chemical processes have been refined by replacing toxic and hazardous solvents with safer, more environmentally friendly options. Notable advancements have been made in material chemistry, such as the synthesis of organic thin-film transistors, metal-organic frameworks (MOFs) [25], and organic borazines [26], as well as in biomass valorization [27] and development of crucial synthetic methodologies, including C-H activation [28], cross-coupling reactions [29, 30], and a broad spectrum of other chemical transformations [31, 32].

In this scenario, click chemistry also played a pivotal role, representing a fundamental actor in the pursuit of safer solvents. Epoxide ring opening is among those click chemistry transformations where the classical organic solvents have been successfully replaced by safer alternatives, especially water. Thanks to its unique chemical-physical properties and structure, including pH control, ability to establish hydrogen bonds, and ease of activating ionic nucleophiles, this medium may serve as an ideal reaction solvent for achieving optimal reaction rates and yields with a variety of *O*-, *N*-, and *S*-nucleophiles [14]. Notably, the thiolysis of epoxides, traditionally carried out under harsh reaction conditions using toxic and hazardous solvents like tetrahydrofuran (THF), dichloromethane (DCM), and methanol, has been shown to proceed effectively in aqueous media when using arene- and alkanethiols. In aqueous basic conditions, these thiols are readily deprotonated by hydroxide ions, forming the corresponding thiolate ions, which makes water a valuable and sustainable medium for this transformation [33–35].

To confirm the advantages of using water as a reaction medium for epoxide opening, not only from an environmental perspective but also in terms of enhanced chemical reactivity, it is worth mentioning the work of Hou and colleagues [36] who reported the reaction involving cyclohexene oxide and aniline (or benzyl-amine) catalyzed by tributylphosphane, where water allowed an enhanced reaction efficiency compared to that obtained when using acetonitrile.

As observed with epoxide rings, aziridines have also been properly opened in water, deep eutectic solvents, and solvent-free conditions, sensitively diminishing the amount of waste produced [37, 38].

The classic reaction solvents used for 1,3-dipolar cycloaddition include toxic and hazardous benzene, *N*,*N*-dimethylformamide, and dichloromethane. With the growing awareness of their intrinsic hazardousness, such that they have been recognized as "substances of very high concern" by the European Union [39], several efforts have been made to develop alternative procedures employing more benign media. Sharpless made a fundamental step toward a more sustainable protocol by selectively synthesizing 1,4-triazoles through a Huisgen-type 1,3-dipolar cycloaddition using H_2O/alcohol as reaction solvent. Particularly, the use of H_2O/*t*-butanol significantly greened the overall protocol, allowing a proper solubilization of both the organic and inorganic reactants, catalyst, and additives, even if the recovery of the reaction media proved tedious [17].

Further improvements were additionally provided by Vaccaro and colleagues, who introduced the use of two novel solvent blends, such as the azeotrope mixture composed of furfuryl alcohol/water [40] and the mixture Polarclean/water 4:1 [41]. Furfuryl alcohol is a bio-based alcohol deriving from the hydrogenation of furfural, a building block originated from the valorization of hemicellulose [42], while Polarclean is a 20:1 mixture of an ester-amide and a diester, originating from the valorization of Nylon 6,6 industrial waste [43]. The use of these solvent systems not only serves as an outstanding example of circular economy applied to chemical synthesis but also significantly reduces waste production, as the solvent mixtures and the catalyst as well, when using Polarclean/water, can be both recovered and reused. These eco-friendly improvements were further validated by the calculation of the E-factor, which was considerably lower (~3–10 kg waste/kg product) compared to methods where the reaction solvent is not recycled, showing an average E-factor of 150 for the H_2O/*t*-butanol system and 50 for the CH_2Cl_2/H_2O system [40].

Neat water was also adopted as a solvent to produce 1,4-disubstituted 1,2,3-triazoles [44], as well as ethylene glycol [45], and glycerol, the side product of biodiesel production by Alvarez et al. (Fig. 6) [46].

Recently, dihydrolevoglucosenone (marketed as Cyrene™), a biodegradable and nontoxic solvent, has also been shown to be effective in CuAAC, enabling the production of a plethora of 1,4-disubstituted triazoles under mild reaction conditions [47].

4 Heterogeneous Catalyzed Click Reactions

The use of catalysts is essential for establishing environmentally friendly procedures, as they significantly enhance the efficiency of chemical processes by increasing reaction rates and selectivity while simultaneously reducing waste formation and energy consumption and minimizing the formation of undesired byproducts.

Compared to stoichiometric reagents, catalysts offer a major advantage in terms of waste reduction. In fact, sensitively reducing activation energy enhances the reaction rate selectively toward the desired products, limiting secondary pathways. Moreover, by establishing a catalytic cycle, they can be used in sub-stoichiometric amounts, limiting the amount of materials required and so reducing waste formation. Furthermore, when working with heterogeneous ones, this waste limitation can be further increased, thanks to the capability of the catalysts to remain unaltered at the end of the catalytic cycle, and so to be recovered and reused. Although over 80% of chemicals today are produced through various catalytic systems [48], many industrial processes still favor homogeneous catalysts despite their drawbacks, such as time-consuming and costly removal from the reaction mixture and the need for extensive product purification.

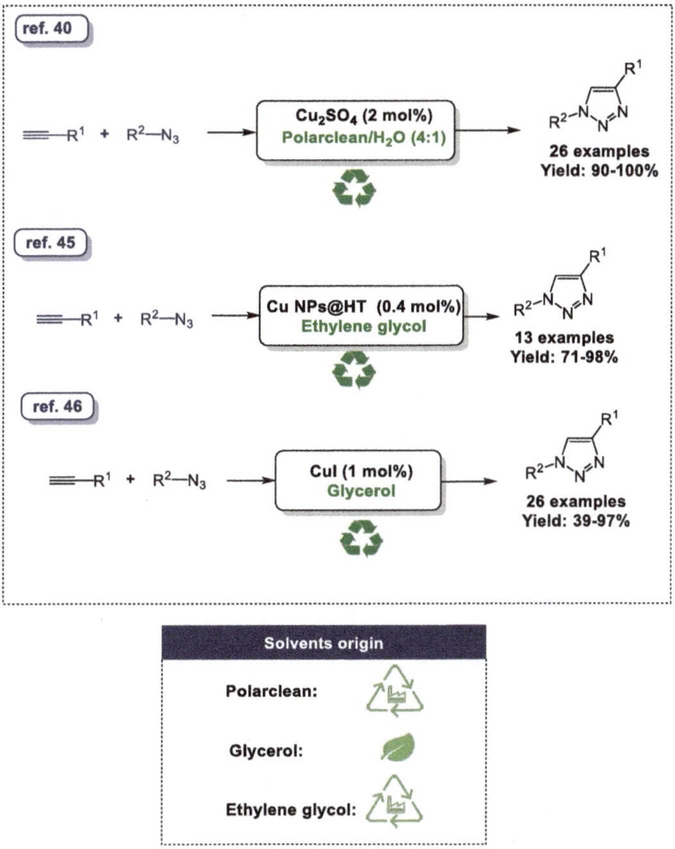

Fig. 6 Catalyst and reaction solvent recycling in CuAAC

By contrast, heterogeneous catalysts, so named because they exist in a different physical phase from the reaction mixture, are gaining increasing attention due to their technical advantages. Their easy separation and recovery, along with the ability to be reused multiple times, make them ideal for developing waste-minimized protocols, thereby aligning perfectly with the principles of sustainable synthesis [49].

Various heterogeneous catalytic systems have been designed, developed, and widely employed in click chemistry, often paired with the choice of a less hazardous solvent. For example, silica is capable of activating aziridine rings and facilitating their opening due to its weakly acidic nature and ability to adsorb organic molecules through hydrophobic interactions. The combination of the silica gel with solvent-free conditions, as reported by Singh et al., presents an intriguing method for waste minimization (Table 1, entry 1) [50].

More recently, in 2020, Chatterjee et al. employed CuO nanoparticles to ring-open aziridines, enabling their reaction with a wide range of nucleophiles, such as

Table 1 Examples of heterogeneous catalyzed aziridine and epoxide ring opening

Entry	Ring	Nucleophile	Catalyst	Solvent	Examples (yields)	Catalyst recycling	Ref
1	Aziridines	Aromatic diamines	Silica gel	–	20 (35–95%)	–	[50]
2	Aziridines	– Aliphatic and aromatic alcohols – Indoles – Thiols	CuO NPs	–	30 (57–96%)	5 runs	[51]
3	Epoxides	– Morpholine – Piperidine	– APTES@RH-SiO_2 – MPTES@RH-SiO_2	–	12 (43–93%)[a]	4 runs	[52]
4	Epoxides	Aromatic and aliphatic amines	MOF-Cu-1	–	7 (98–85%)	4 runs	[53]
5	Epoxides	– Aliphatic alcohols – Anilines	– MIP-202(Zr) – UiO-66-$(CO_2H)_2$	–	26 (0–99%)[a]	5 runs	[54]

[a]Conversion

aliphatic and aromatic alcohols, indoles, and thiols. Specifically, in reference to the reaction between 2-phenyl-1-tosylaziridine and ethanol, the nanocatalyst was recovered and reused for five consecutive opening reactions without losing its catalytic effectiveness (Table 1, entry 2) [51].

The positive impact of catalyst recovery in this case can be demonstrated through the E-factor evaluation, which revealed that recovering the catalyst reduces waste by 30 g for every kilogram of product produced. Even if this value can appear negligible, we need to consider if a comparable reaction using a similar nanocatalyst is performed on a larger or even an industrial scale. In that case, we will be facing a massive amount of CuO waste, which must be properly disposed of from the reaction mixture and, once landfilled, may be dispersed in the atmosphere as fine particulate that can impact both our health, causing lung diseases, and the environment, favoring global warming.

As with the aziridine ring opening, heterogeneous catalysis represents a key green tool in the epoxide ring-opening process as well. In 2022, Pandey et al. compared two functional groups, a base ((3-aminopropyl)triethoxysilane (APTES)) and an acid ((3-mercaptopropyl)triethoxysilane (MPTES)), for modifying an MCM mesoporous silica support. They found that the base-functionalized catalyst (APTES@RH-SiO_2) performed better than the acidic one (MPTES@RH-SiO_2) in ring-opening reactions of various oxiranes with N-heterocycles. Specifically, in the reactions with morpholine and piperidine, APTES@RH-SiO_2 achieved conversion rates of 93% and 91%, while MPTES@RH-SiO_2 only reached 52% and 57%. The catalyst containing amine moieties was recycled and reused in the benchmark reaction involving propylene oxide and morpholine, highlighting their high recyclability and efficiency, which slightly decreased only after four runs (Table 1, entry 3) [52]. Huang and colleagues took a similar approach by synthesizing a 3D copper-based metal-organic framework (MOF) featuring unsaturated metal sites and an amino tridentate *N*-donor ligand, (*N*1-(4-(1H-imidazol-1-yl)benzyl)-*N*1-(2-aminoethyl)ethane-1,2-diamine). This catalyst demonstrated outstanding performance in the ring opening of epoxides with various amines under solvent-free conditions and at room temperature. Its efficiency was further highlighted by its straightforward recoverability, allowing reuse in four consecutive reactions without significant loss of yield or structural integrity (Table 1, entry 4) [53].

Besides their chemical effectiveness, the present catalyst still shows some environmental concerns since they are prepared in *N,N*-dimethylformamide. In fact, as stated above, its toxicity has been extensively discussed and ascertained, and proper substitutes, even if for UiO-66, have already been disclosed [25].

Once feasibility is ensured, such a substitution will further enhance the overall sustainability of the process, sensitively while reducing the impact induced by the adopted solvent. Furthermore, this is not the only example of MOF catalysts employed to convert epoxides into alcohols or amino-alcohols. Bagherzadeh et al. [54] successfully utilized two different zirconium-based MOFs, MIP-202(Zr) and UiO-66-$(CO_2H)_2$, using both amines and alcohols as nucleophiles. These hybrid materials were also easily recovered and reused, further demonstrating their efficiency and stability (Table 1, entry 5).

Regarding CuAAC and RuAAC, several heterogeneous catalysts have been designed and developed to perform azide-alkyne cycloaddition. Recently, in the Vaccaro research group, Cu(0) rods were used to catalyze the synthesis of a wide range of 1,4-disubstituted 1,2,3-triazoles starting from β-azido-ketones (Table 2, entry 1). The oxidative process induced by the air continuously forms Cu_2O, a Cu (I) species that carries out the catalytic activity [55]. The developed methodology allowed easy solvent and catalyst recovery, minimizing waste production. This green enhancement was further corroborated by the E-factor calculation, which highlighted values ranging from 1.88 to 0.84.

Another example of a heterogeneous catalyst is a copper-functionalized covalent organic framework (COF), Cu-TFP-PyH, reported by Justin Thomas (Table 2, entry 2) [56]. This catalyst is suitable for catalyzing the azide-alkyne cycloaddition in water, under mild reaction conditions (9 h, rt).

Many other copper catalysts have been employed for the same transformations, such as Cu nanoparticles on carbon nanotubes (Table 2, entry 3) [57], copper on graphene oxide (Table 2, entry 4) [58], copper on cellulose acetate (Table 2, entry 5) [59], copper/cobalt magnetic nanoparticles (Table 2, entry 6) [60], and copper(II)-coated magnetic core-shell nanoparticles $Fe_3O_4@SiO_2$ modified by isatoic anhydride (Table 2, entry 7) [61].

Sasson and his team [62] reported a heterogeneous Ru(II) catalysis method for synthesizing 1,4-disubstituted 1,2,3-triazoles, utilizing a Ru catalyst where a ligand is covalently attached to SBA-15, a mesoporous material valued for its hydrothermal

Table 2 Examples of heterogeneous catalyzed CuAAC and RuAAC

Entry	Cycloaddition	Catalyst	Solvent	Examples (yields)	Ref.
1	CuAAC	Cu(0) rods	CH_3CN/ H_2O az.	23 (65–99%)	[55]
2	CuAAC	Cu-TFP-PyH	H_2O	6 (98–83%)	[56]
3	CuAAC	Cu@PCT@CNT	H_2O	18 (45–97%)	[57]
4	CuAAC	GO-M-Cu	H_2O	20 (86–98%)	[58]
5	CuAAC	Cu(II)-CA	H_2O	9 (90–97%)	[59]
6	CuAAC	Cu/Co@NPC	H_2O	12 (62–99%)	[60]
7	CuAAC	Copper(II)-coated Fe_3O_4@SiO nanoparticles	H_2O	15 (85–98%)	[61]
8	RuAAC	SBA-15-Tz-Ru(II)TPP	H_2O	14 (70–92%)	[62]
9	RuAAC	Metalloprotein-inspired ruthenium polymeric complex	H_2O	19 (82–98%)	[63]

stability, thick pore walls, and large pore size. The free hydroxyl groups of SBA-15 were functionalized with an azido linker, which was subsequently combined with propargylamine through a click reaction. The resulting ligand was then mixed with $[RuCl_2(PPh_3)_3]$ to form the definitive catalyst. This Ru complex proved to be highly effective in a multicomponent cycloaddition reaction involving an alkyne, sodium azide, and benzyl bromide. Optimal conditions were achieved using 0.5 mol% of the catalyst at 90 °C in water, affording functionalized 1,4-disubstituted triazoles with yields from 70% to 92% (Table 2, entry 8).

Another notable example is the work of Gupta and co-workers [63], who developed an efficient Ru-based metalloprotein capable of regioselectively catalyzing the synthesis of 1,4,5-trisubstituted 1,2,3-triazoles in water at room temperature with remarkably short reaction times. The catalyst was easily separated from the reaction crude and recycled, maintaining its efficiency over six consecutive runs without any loss of performance (Table 2, entry 9).

5 Click Chemistry Under Continuous Flow Conditions

The application of click chemistry in flow systems minimizes exposure and handling of toxic reactants and intermediates, significantly enhancing process safety. Moreover, the flow chemistry ensures easy scaling up of the protocols and facilitates the workup procedure, contributing to reducing the environmental footprint of the protocols involved.

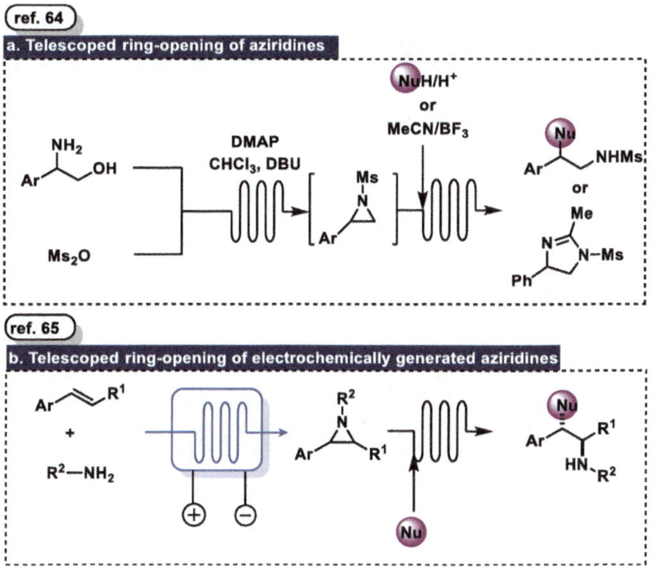

Fig. 7 Telescoped ring opening under continuous flow of **a)** aziridines, **b)** electrochemically generated aziridines

Notably, organic azides and aziridines present substantial hazards. Aziridine, a potent alkylating agent, is classified as a possible human carcinogen due to its strong mutagenic and genotoxic properties. Meanwhile, organic azides undergo highly exothermic decomposition, posing a serious explosion risk, particularly in protocols that involve the use or in situ generation of hydrazoic acid, a highly toxic and explosive compound. Implementing flow conditions in these protocols can enhance safety by minimizing operator exposure and improving process control.

Given the protocols involving aziridines, telescoped continuous flow reactions that enable their in-line formation have been extensively studied. Shipman et al. reported the continuous flow synthesis of *N*-sulfonyl aziridines from 1,2-amino alcohols, along with their subsequent ring opening to amines and ring expansion to imidazolines [64].

Ošeka et al. developed a telescoped approach for vicinal diamine synthesis, featuring the electrochemical aziridination of alkenes with primary amines, followed by strain release-driven ring opening using various nitrogen nucleophiles (Fig. 7) [65].

The safe delivery of azides, including their potential in-line generation, combined with the inherent scalability of continuous flow processing, makes CuAAC particularly advantageous. However, a key concern is copper contamination in the final products. Flow-based applications address this issue by incorporating an in-line copper-scavenging unit, effectively reducing contamination associated with homogeneous Cu catalysis [66].

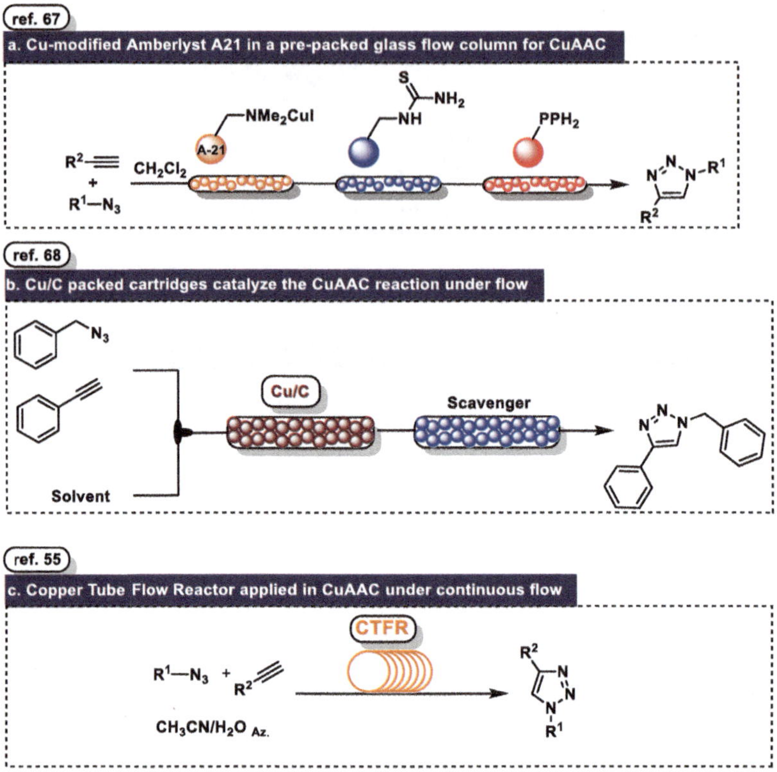

Fig. 8 Examples of different configurations under continuous flow: **a)** Cu-modified Amberlyst A21 in a pre-packed glass flow column for CuAAC, **b)** Cu/C packed cartridge catalyze the CuAAC reaction under flow, **c)** Copper Tube Flow Reactor applied in CuAAC

The integration of flow technologies with heterogeneous Cu-based catalysts has demonstrated their effectiveness in improving catalyst recovery, reducing metal contamination, and enhancing process efficiency.

In 2007, Ley et al. [67] reported a continuous flow gram-scale CuAAC reaction using a combination of immobilized catalysts and scavenger resins in a mesofluidic flow system. A Cu-modified Amberlyst A21 resin in a prepacked glass flow column was used as a reactor. Since the Cu catalyst is only attached to the resin by weak coordination, the Cu species is expected to leach into the solution phase. A subsequent in-line scavenging step using a thiourea-based metal resin removes all Cu in the solution. Ultimately, unreacted azide used in excess and contaminating the triazole product could be removed by flowing the reaction solution through a column of phosphine resin, thus capturing the azide onto the solid phase *via* a Staudinger reaction (Fig. 8a).

Copper powder or Cu/C, loaded into catalyst cartridges, is widely used in flow chemistry. In these systems, the catalytically active species originates from a surface layer of Cu_2O formed on metallic copper. Consequently, when using Cu turnings, powders, or coils, the addition of a reducing agent may not be necessary under appropriate reaction conditions. However, with Cu/C catalysts, the reaction mechanism may involve homogeneous catalysis, where Cu/C serves as a reservoir in a dynamic release and catch cycle. Notably, catalyst leaching is a critical factor in flow chemistry and should be carefully monitored (Fig. 8b) [68].

An alternative approach utilizes a copper tube flow reactor (CTFR), a tubular reactor made entirely of copper that serves as the catalyst, eliminating the need for a catalyst cartridge or any additional homogeneous or heterogeneous copper-based catalysts. The CTFR effectively combines the benefits of a heterogeneous copper source with the advantages of easy incorporation into the line and rapid heating or cooling of the reactor system [69]. This technology has been efficiently applied in the synthesis of rufinamide, cyclic peptides, and drug-like macrocycles, resulting in excellent yields and control of reaction parameters. A typical drawback of using CTFR is the potential metal contamination of the products [70].

Vaccaro et al. [55] reported the optimization of flow CuAAC using CTFR in combination with $CH_3CN:H_2O$ azeotrope, achieving a significant reduction in copper contamination from 362.54 μg/g in batch approaches to 50.5 μg/g in flow processes. This improvement enhanced the sustainability of the CuAAC reaction, as highlighted by a lower E-factor, with values ranging from 0.11 to 0.25, compared to 0.84 to 1.88 in batch processes (Fig. 8c).

Vaccaro et al. reported a flow-based approach that integrates the in-line generation of azides with CTFR to synthesize 1,4-disubstituted 1,2,3-triazoles *via* CuAAC. The multistep flow protocol begins with the synthesis of β-azidation starting from α,β-unsaturated carbonyls. This first step is achieved using a packed-bed reactor equipped with a tailor-made POLITAG-F organocatalyst [71]. The resulting β-azido intermediates are then directly streamed into a CTFR for β-keto triazole formation, eliminating the need for intermediate purification. Process optimization was achieved using a $CH_3CN:H_2O$ azeotrope, which enhanced catalytic performance across both reaction steps. Notably, the aqueous acetonitrile azeotrope was largely recovered and reused, contributing to the protocol's sustainability. The method afforded 1,4-disubstituted β-keto 1,2,3-triazoles in high yields, with impressively low E-factor values ranging from 0.9 to 1.6 (Fig. 9) [72].

6 Alternative Energy Sources

The application of nontraditional activation methods in green chemistry, such as microwaves, ultrasounds, electrochemistry, photochemistry, and mechanochemistry, offers an energy-efficient alternative to conventional heating systems. These approaches enhance energetic efficiency, reduce electricity consumption, and minimize greenhouse gas emissions while maintaining excellent performance and effectiveness [73].

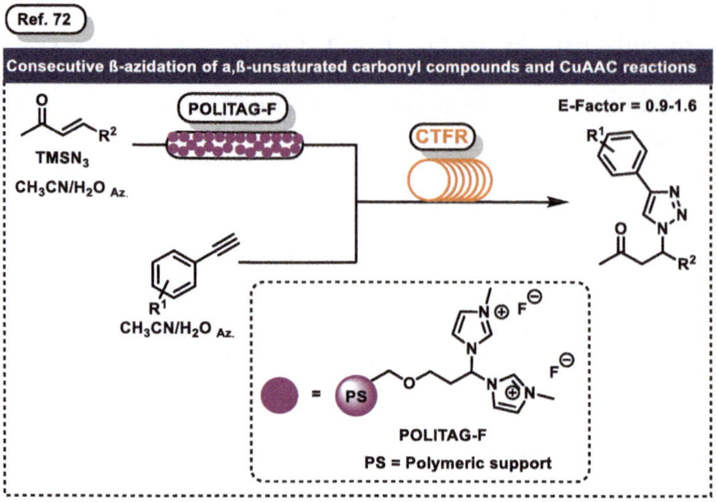

Fig. 9 Continuous flow synthesis of 1,4-disubstituted 1,2,3-triazoles *via* consecutive β-azidation of α,β-unsaturated carbonyl compounds and CuAAC reactions

6.1 Photo-Click Chemistry

The integration of click chemistry with photochemically driven processes has led to the development of light-triggered click reactions, known as photo-click chemistry, which strictly adhere to the principles of click chemistry. Notably, light as an external stimulus, through precise control of wavelength, intensity, and irradiation time, enhances their efficiency while ensuring mild reaction conditions [74].

Photo-click chemistry was initially linked to the tetrazole-ene reaction. Over the past decade, a growing number of photo-click reactions have been developed, which can be classified into three categories. The first category includes light-triggered tetrazole−alkene cycloaddition, sydnone−alkene/alkyne cycloadditions, oxime ligation, and azirine−alkene cycloaddition, all of which involve the irreversible photogeneration of a reactive intermediate with varying stability. The second category includes photoinduced hetero-Diels-Alder reactions based on *o*-naphthoquinone methide and *o*-quinodimethane; these reactions rely on photoisomerization to produce a highly unstable reactive intermediate that can either revert to the starting materials or react with a partner to form a covalent adduct. The third category involves reactions that require a light-activated catalyst, such as light-triggered azide-alkyne cycloaddition (photo-CuAAC) and tetrazine ligation. In photo-CuAACs, active Cu(I) catalysts are generated *in situ* from Cu(II) salts through photochemical approaches. This photochemical generation of Cu(I) catalysts enables precise spatial and temporal control over the reaction [74].

In 2017, Heredia and Argüello reported the synthesis of 1,4-disubstituted 1,2,3-triazoles *via* the efficient reduction of Cu(II), facilitated by organic dyes and promoted by visible light. This light-driven approach enables the reaction to proceed

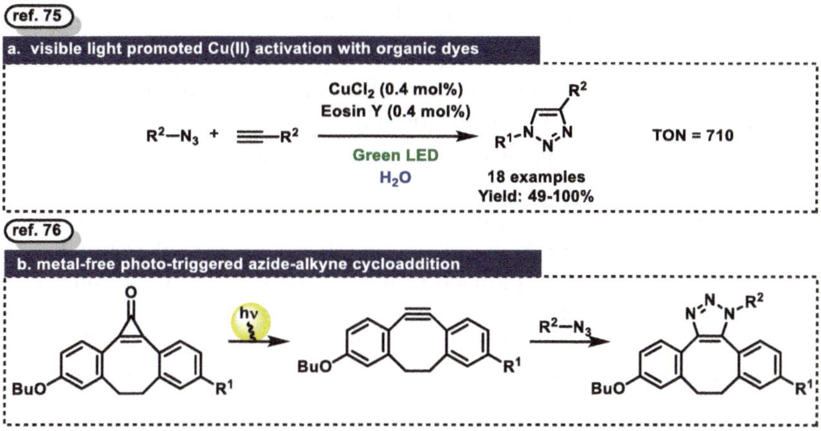

Fig. 10 Examples of light-triggered click reactions: **a)** visible light promoted Cu(II) activation with organic dyes, **b)** metal-free photo-triggered azide-alkyne cycloaddition

under exceptionally mild conditions while adhering to several green chemistry principles. Water serves as a nontoxic solvent, eliminating the need for organic solvents. Additionally, the catalyst precursor and organic dye, acting as a reducing agent, are required in reduced amounts compared to conventional methods (TON = 710). Notably, the copper catalyst and solvent were successfully recycled at least three times without any loss of efficiency (Fig. 10a) [75].

In 2009, Popik and Boons et al. reported a photo-triggered metal-free azide-alkyne cycloaddition achieved by masking the triple bond of dibenzocyclooctynes as cyclopropenones. Irradiation of cyclopropenones results in the efficient and clean regeneration of the corresponding dibenzocyclooctynes, which then undergo facile catalyst-free cycloadditions with azides to give corresponding triazoles under ambient conditions. Following this pioneering work, various strained alkynes, such as fluorinated cyclooctynes, dibenzocyclooctynes, and thiacycloalkynes, have been specifically crafted to facilitate the accelerated copper-free azide-alkyne cycloaddition (Fig. 10b) [76].

In Diels-Alder reactions, light can be exploited to generate reactive intermediates as made in 2009 by Popik et al., who reported that upon irradiation, 3-hydroxy-2-naphthalenemethanol and 2-hydroxy-1-naphthalenemethanol undergo dehydration, generating isomeric naphthoquinone methides. In aqueous solution, these intermediates rapidly rehydrate, regenerating the starting materials. However, reactive naphthoquinone methides can also be intercepted by nucleophiles such as azide ions or thiols. Notably, ethyl vinyl ether readily reacts with these intermediates, leading to the formation of Diels-Alder adducts [77].

6.2 Microwave-Assisted Click Chemistry

The application of microwaves as an alternative energy source in click transformations significantly reduces synthetic time (and sometimes the volume of solvent used) while enhancing yields. Moreover, the integration of microwave techniques into organic methodologies has advanced green and sustainable practices.

Fokin et al. [78] reported the microwave-assisted synthesis of 1,5-disubstituted 1,2,3-triazoles from aryl azides and alkynes *via* RuAAC using a $[Cp^*RuCl]_4$ catalyst in dimethylformamide. This approach resulted in higher yields, a cleaner product, and significantly shorter reaction times.

Numerous examples of CuAAC demonstrate the effectiveness of microwave irradiation in aqueous media, often in combination with heterogeneous Cu-based catalysts [79].

Xiong et al. [80] reported the synthesis of 1,4-disubstituted 1,2,3-triazoles using a superparamagnetic graphene oxide (GO)/Fe_3O_4–CuBr catalyst. This heterogeneous magnetic system facilitated the in situ reaction of a primary halide, NaN_3, and a terminal alkyne under microwave irradiation, exhibiting high catalytic efficiency and easy recyclability.

Furthermore, microwave-assisted CuAAC reactions have been reported using carbon-supported Cu(I)-phosphine catalysts and NiO/Cu_2O nanocatalysts, both of which have been successfully employed in the synthesis of 1,4-disubstituted 1,2,3-triazoles. These reactions demonstrated impressive TON of 64–101 and 890–960, respectively, in aqueous media (Fig. 11a) [81, 82].

In 2022, Hernaiz et al. reported a combination of ball milling and microwave-assisted unconventional heating to achieve the sustainable synthesis of glucuronic acid glycodendrimers via CuAAC. This mechanochemical method enables solvent-free reactions, yielding the desired product with excellent conversion rates, while also simplifying the purification process (Fig. 11b) [83].

6.3 Mechanochemical Approaches

Mechanochemical activation of metallic copper powder is well known to enable solvent-free CuAAC, providing a straightforward and eco-friendly approach. Cravotto et al. reported a mechanochemical CuAAC protocol that utilizes copper powder as a catalyst, achieving high reaction efficiency and scalability across various substrates. The mechanochemical activation of metallic copper powder in a ball mill enables solvent-free alkyne-azide click reactions. The desired products were obtained in high yields through simple filtration, making the process both rapid and highly practical (Fig. 12a) [84].

Liu et al. reported a solvent-free lignin modification *via* CuAAC, utilizing ball-milling mechanochemistry. Initially, the hydroxyl groups of lignin were functionalized with alkyne end groups. Subsequently, the lignin-alkyne was ball-milled with ethyl iodoacetate and $Cu(OAc)_2$ as a catalyst under solvent-free conditions, preserving the integrity of the lignin structure (Fig. 12b) [85].

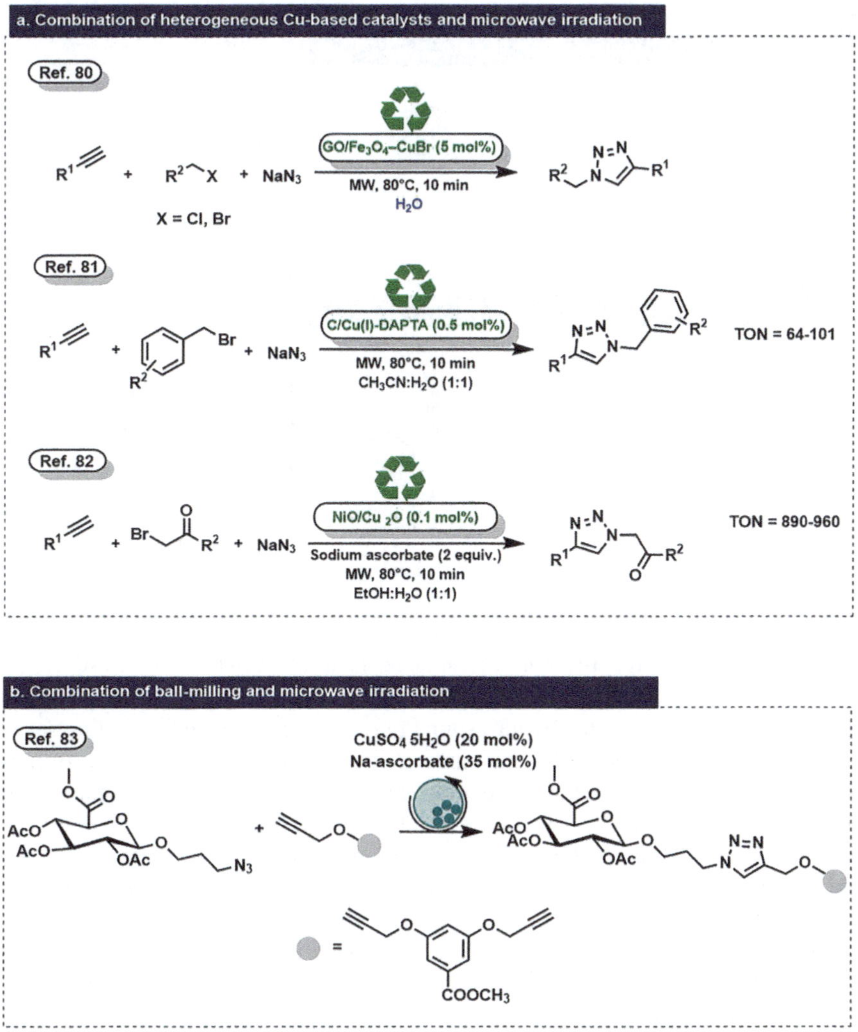

Fig. 11 Application of microwaves as an alternative energy source in CuAAC **a)** combination of heterogeneous Cu-based catalysts and microwave irradiation, **b)** combination of ball milling and microwave irradiation

6.4 Sonochemistry

Another unconventional heating method successfully applied in click transformations is ultrasound irradiation. Utilizing water as the reaction medium alongside a photocatalytic system, ultrasound irradiation enhances the activity of Cu(I) salts, significantly improving both energy efficiency and reaction mildness. The

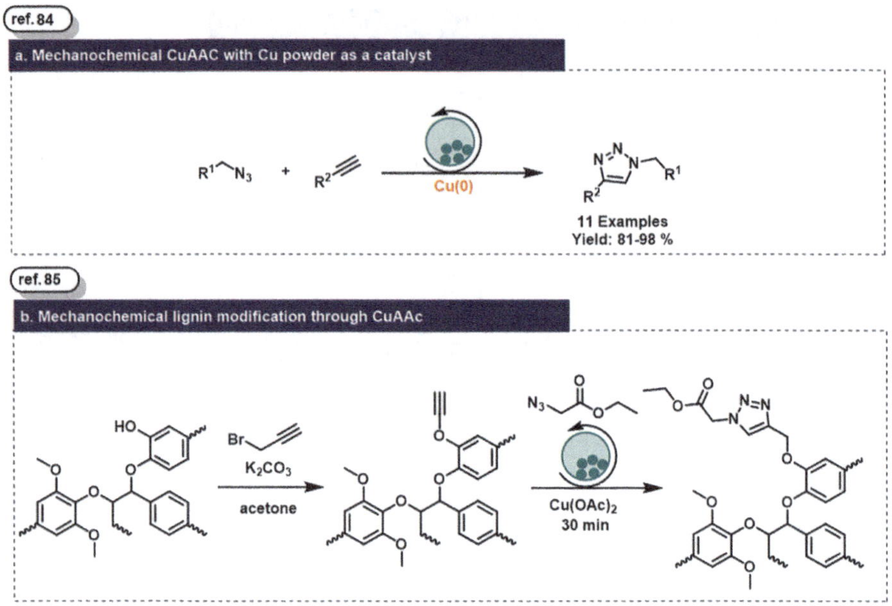

Fig. 12 Application of mechanochemical activation of CuAAC: **a)** mechanochemical CuAAC with Cu powder as a catalyst, **b)** mechanochemical lignin modification through CuAAC

cavitational energy generated by ultrasound plays a crucial role in facilitating several key processes: activating hard metal surfaces, promoting reagent diffusion from the solution to the metal, enabling electron transfer from the active surface to the reducible site of the organic substrate, and ultimately extracting ions from the surface to form a soluble product. Cravotto et al. reported an ultrasound-assisted CuAAC starting both from small molecules and oligomers such as cyclodextrins (CDs) significantly reducing the reaction time (Fig. 13a) [86].

The use of water as a reaction medium, combined with ultrasound irradiation, has been shown to enhance the activity of Cu(I) salts, improving both the energy efficiency and mildness of the process.

Chen, Qu et al. reported a Cu(I)-catalyzed CuAAC reaction that efficiently targets a wide range of azides and alkynes at room temperature (Fig. 13b) [87].

Lunkov et al. reported an ultrasound-assisted CuAAC that promotes the *in situ* formation of Cu(I) under ultrasound irradiation in aerobic conditions, with acetic acid and copper turnings as reagents. This method enables the easy and efficient synthesis of novel chitosan derivatives [88].

6.5 *Electrochemistry*

The last activation method to consider is the electrochemical technique applied to the click reaction. Seavill and Wilden [89] reported the electrochemical synthesis of Cu

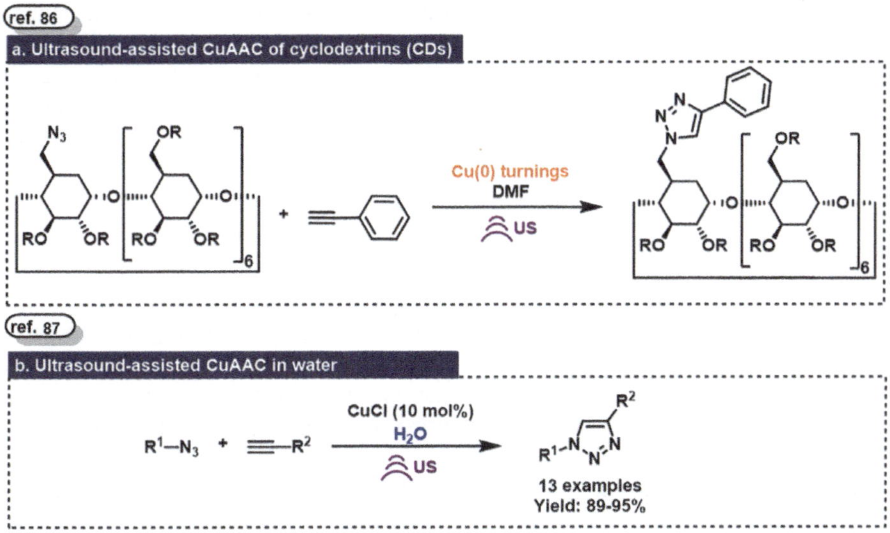

Fig. 13 Ultrasound-assisted CuAAC **a)** of cyclodextrins(CDs), **b)** in water

(I)-acetylides through the simultaneous copper oxidation and Hofmann elimination of quaternary ammonium salts. The electrochemically generated base was also regenerated electrochemically, making the process catalytic. The electrochemically promoted CuAAC reaction demonstrated the potential application of this approach in pharmaceutically relevant reactions.

7 Conclusion, Future Outlooks, and Trends

7.1 Conclusion

In recent decades, spurred by the Nobel Prize award to Sharpless, Meldal, and Bertozzi, click chemistry has emerged as a valuable tool for promoting more benign organic synthesis, maximizing product yields while simultaneously reducing waste generation. Over the years, its intrinsic sustainability aspects have been further amplified by growing awareness of the need to develop cleaner production routes.

Encouraged by the rise of green chemistry concepts, several additional factors have been considered to improve overall sustainability. The use of alternative reaction media, such as those derived from residual biomass or the valorization of industrial waste, and the adoption of recoverable and reusable heterogeneous catalysts are now fundamental prerequisites for establishing more environmentally respectful synthesis. These approaches become even more effective when combined with the development of continuous flow procedures or with technological setups that minimize energy consumption, such as mechanochemical, ultrasound, microwave, and electrochemical reactions.

In summary, this chapter discusses and comments on some of the most recent examples of the application of these green chemistry tools, with a particular focus on how the improvements achieved have been evaluated using green chemistry metrics.

7.2 Future Outlooks and Trends

Since its inception, click chemistry has been recognized as a powerful green chemistry tool, enabling high product yields and excellent atom economy through selective, catalyzed reactions that minimize both waste generation and purification steps. These transformations inherently embody the principles of safe and sustainable by design (SSbD) [90]. The integration of green chemistry, circular chemistry, and SSbD creates a holistic approach to click chemistry that enables the increased sustainability of this crucial transformation for industrial and medicinal chemistry.

Continuous efforts are being made to enhance the sustainability of click transformations. As discussed in this chapter, various solvent systems have been explored and successfully applied in 1,3-dipolar cycloadditions. Additionally, solvent-free conditions, especially when combined with mechanochemistry, offer a promising green alternative. Research for alternative catalysts represents one of the hottest topics to increase overall sustainability.

One aspect that must urgently be addressed is the use of copper and ruthenium, whose toxicity to human health and ecosystems is well known. In this context, research for alternative metals is a key step to enhance the reaction greenness. An example is the use of zinc, a nontoxic, cheaper, and more abundant metal, which extraction is even less energy demanding than copper. As an example, it has been recently used in the synthesis of 1,2,3-triazoles, affording excellent yields, even if slightly lower compared to the use of copper [91]. In fact, as documented, although it seems to work with a mechanism like that of Cu, its selectivity is still lower, even if further catalyst design could mitigate this difference.

The integration of metal catalysts with ferromagnetic nanoparticles offers a promising strategy to promote click transformations through magnetocatalysis. Among the few examples reported, the use of Cu(I)- or Cu(II)-decorated Fe_3O_4 nanoparticles combines high catalytic efficiency with straightforward recoverability [92].

By combining innovative energy sources with flow technology, click transformations can be efficiently scaled up with minimal energy input.

Plasmon assistance can induce the azide-alkyne cycloaddition, enabling low-temperature reactions that are cleaner and more accessible to more delicate substrates [93].

Machine learning can be used to guide the discovery of new bio-orthogonal click reactions using models capable of predicting activation and reaction energies, allowing the in silico screening of thousands of candidates for greener reactions.

Acknowledgments This work was supported by the funding from the European Union—NextGenerationEU—under the Italian Ministry of University and Research (MUR) National Innovation Ecosystem grant ECS00000041——VITALITY—and within the program Circular and Sustainable Made in Italy Extended Partnership (MICS) for supporting the project CIRCLE MICS_ODR12_2024. We also thank the University of Perugia and MUR for financial support within the project PRIN-2022 "20223ARWAY—REWIND." E.B. thanks the National PhD Program in Catalysis coordinated by the University of Perugia.

References

1. Kolb HC, Finn MG, Sharpless KB (2001) Angew Chem Int Ed Engl 40:2004–2021
2. Anastas PT, Warner JC (2000) Green chemistry: theory and practice. Oxford University Press, Oxford
3. https://www.epa.gov/p2/pollutionpreventionact1990#:~:text=The%20Congress%20hereby%20declares%20it,be%20prevented%20or%20recycled%20should
4. Sheldon RA (2005) Green solvents for sustainable organic synthesis: state of the art. Green Chem 7:267–278
5. Sheldon RA (2023) The E factor at 30: a passion for pollution prevention. Green Chem 25: 1704–1728
6. https://commission.europa.eu/strategy-and-policy/priorities-2019-2024/european-green-deal/energy-and-green-deal_en
7. Bardow A, Perez-Ramírez J, Sala S, Vaccaro L (2024) Measuring green chemistry: methods, models, and metrics. Green Chem 26:11016–11018
8. Sheldon RA (1994) ChemTech:38–47
9. Fava JA, Denison R, Jones B, Curran MA, Vigon B, Selke S, Barnum J (1991) SETAC workshop report: a technical framework for life-cycle assessment. August 18–23, 1990, Smugglers Notch, Vermont. SETAC, Washington, DC
10. Campana F, Zhou K, Yunda JA, Khodadadi AN, Bonifazi D, Melinte S, Vaccaro L (2025) Assessing the energetic and environmental sustainability of organic borazines preparation: a comprehensive life cycle assessment and uncertainty analysis. Chem Eng J 504:158822
11. Goedkoop M, Heijungs R, Huijbregts M, De Schryver A, Struijs J, Van Zelm R (2009) ReCiPe 2008: a life cycle impact assessment method which comprises harmonised category indicators at the midpoint and the endpoint level: report 1: characterisation. Ministry of Housing, Spatial Planning and Environment (VROM)
12. Huisgen R (1963) 1,3-dipolar cycloadditions. past and future. Angew Chem Int Ed Engl 2:565–598
13. Sun WN, Mei GJ, Jia SK (2024) ChemCatChem 16:1–9
14. Bonollo S, Lanari D, Vaccaro L (2011) Ring-opening of epoxides in water. Eur J Org Chem 2011:2587–2598
15. Hu XE (2004) Nucleophilic ring opening of aziridines. Tetrahedron 60:2701–2743
16. Tornøe CW, Christensen C, Meldal M (2002) Peptidotriazoles on solid phase: [1,2,3]-triazoles by regiospecific copper(I)-catalyzed 1,3-dipolar cycloadditions of terminal alkynes to azides. J Org Chem 67:3057–3064
17. Rostovtsev VV, Green LG, Fokin VV, Sharpless KB (2002) A stepwise Huisgen cycloaddition process: copper(I)-catalyzed regioselective "ligation" of azides and terminal alkynes. Angew Chem Int Ed 41:2596–2599
18. Hein JE, Fokin VV (2010) Copper-catalyzed azide–alkyne cycloaddition (CuAAC) and beyond: new reactivity of copper(I) acetylides. Chem Soc Rev 39:1302–1315

19. Johansson JR, Beke-Somfai T, Said Stålsmeden A, Kann N (2016) Ruthenium-catalyzed azide alkyne cycloaddition reaction: scope, mechanism, and applications. Chem Rev 116:14726–14768
20. Boren BC, Narayan S, Rasmussen LK, Zhang L, Zhao H, Lin Z, Jia G, Fokin VV (2008) Ruthenium-catalyzed azide−alkyne cycloaddition: scope and mechanism. J Am Chem Soc 130: 8923–8930
21. Agard NJ, Prescher JA, Bertozzi CR (2004) A strain-promoted [3 + 2] azide−alkyne cycloaddition for covalent modification of biomolecules in living systems. J Am Chem Soc 126:15046–15047
22. Hoyle CE, Bowman CN (2010) Thiol–ene click chemistry. Angew Chem Int Ed 49:1540–1573
23. Campana F, Brufani G, Mauriello F, Luque R, Vaccaro L (2024) Green Synth Catal. In press
24. https://www.precedenceresearch.com/solvents-market
25. Venturi DM, Campana F, Marmottini F, Costantino F, Vaccaro L (2020) ACS Sustain Chem Eng 8:17154–17164
26. Nazari Khodadadi A, Cela E, Marchionni D, Huang F, Ferlin F, Vaccaro L (2024) Efficient access to hexaaryl-substituted borazines in batch and continuous-flow. Green Chem 26:7059–7066
27. Pateromichelakis A, Psycha M, Pyrgakis K, Maréchal F, Kokossis A (2022) The use of GVL for holistic valorization of biomass. Comput Chem Eng 164:107849
28. Gandeepan P, Kaplaneris N, Santoro S, Vaccaro L, Ackermann L (2019) ACS Sustain Chem Eng 7:8023–8040
29. Valentini F, Ferlin F, Lilli S, Marrocchi A, Ping L, Gu Y, Vaccaro L (2021) Valorisation of urban waste to access low-cost heterogeneous palladium catalysts for cross-coupling reactions in biomass-derived γ-valerolactone. Green Chem 23:5887–5895
30. Valentini F, Di Erasmo B, Ciani M, Chen S, Gu Y, Vaccaro L (2024) Microwave assisted batch and continuous flow Suzuki–Miyaura reactions in GVL using a Pd/PiNe biowaste-derived heterogeneous catalyst. Green Chem 26:4871–4879
31. Jordan A, Hall CGJ, Thorp LR, Sneddon HF (2022) Replacement of less-preferred dipolar aprotic and ethereal solvents in synthetic organic chemistry with more sustainable alternatives. Chem Rev 122:6749–6794
32. Quaglia G, Campana F, Latterini L, Vaccaro L (2022) ACS Sustain Chem Eng 10:9123–9130
33. Fringuelli F, Pizzo F, Vaccaro L (2004) NaOH-catalyzed thiolysis of α,β-epoxyketones in water. A key step in the synthesis of target molecules starting from α,β-unsaturated ketones. J Org Chem 69:2315–2321
34. Fringuelli F, Pizzo F, Tortoioli S, Vaccaro L (2003) Easy and environmentally friendly uncatalyzed synthesis of β-hydroxy arylsulfides by thiolysis of 1,2-epoxides in water. Green Chem 5:436–440
35. Amantini D, Fringuelli F, Pizzo F, Vaccaro L (2001) Bromolysis and Iodolysis of α,-β-epoxycarboxylic acids in water catalyzed by indium halides. J Org Chem 66:4463–4467
36. Fan R-H, Hou X-L (2003) Efficient ring-opening reaction of epoxides and aziridines promoted by tributylphosphine in water. J Org Chem 68:726–730
37. Sabir S, Kumar G, Verma VP, Jat JL (2018) Aziridine ring opening: an overview of sustainable methods. ChemistrySelect 3:3702–3711
38. Huang Y-Y, Lv Z-C, Yang X, Wang Z-L, Zou X-X, Zhao Z-N, Chen F (2017) Nucleophilic ring opening of aziridines with amines under catalyst- and solvent-free conditions. Green Chem 19:924–927
39. EUR-Lex. https://eur-lex.europa.eu/legal-content/EN/TXT/PDF/?uri=CELEX:02006R1907-20180509&from=EN
40. Rasina D, Lombi A, Ferlin F, Santoro S, Vaccaro L (2016) Searching for novel reusable biomass-derived solvents: furfuryl alcohol/water azeotrope as a medium for waste-minimised copper-catalysed azide–alkyne cycloaddition. Green Chem 18:6380–6386

41. Luciani L, Goff E, Lanari D, Santoro S, Vaccaro L (2018) Waste-minimised copper-catalysed azide–alkyne cycloaddition in Polarclean as a reusable and safe reaction medium. Green Chem 20:183–187
42. Paniagua M, Melero JA, Iglesias J, Morales G, Hernándes B, López-Aguado C (2017) Appl Catal A Gen 537:74–82
43. Campana F, Maselli A, Falcini C, Selvi A, Piermatti O, Vaccaro L (2024) Waste minimized one-pot synthesis of azido- or triazole-pyrazolines using polarclean as reaction medium. Sustain Chem Pharm 40:101632
44. Barman MK, Sinha AK, Nembenna S (2016) An efficient and recyclable thiourea-supported copper(i) chloride catalyst for azide–alkyne cycloaddition reactions. Green Chem 18:2534–2541
45. Chetia M, Singh Gehlot P, Kumar A, Sarma D (2018) A recyclable/reusable hydrotalcite supported copper nano catalyst for 1,4-disubstituted-1,2,3-triazole synthesis via click chemistry approach. Tetrahedron Lett 59:397–401
46. Vidal C, García-Álvarez J (2014) Glycerol: a biorenewable solvent for base-free Cu(i)-catalyzed 1,3-dipolar cycloaddition of azides with terminal and 1-iodoalkynes. Highly efficient transformations and catalyst recycling. Green Chem 16:3515–3521
47. Citarella A, Fiori A, Silvani A, Passarella D, Fasano V (2025) Sustainable synthesis of 1,2,3-triazoles using cyrene as a biodegradable solvent in click chemistry. ChemSusChem 18: e202402538
48. Noyori R (2010) Insight: green chemistry: the key to our future. Tetrahedron 66:1028
49. De S, Dutta S, Saha B (2016) Critical design of heterogeneous catalysts for biomass valorization: current thrust and emerging prospects. Catal Sci Technol 6:7364–7385
50. Anand RV, Pandey G, Singh VK (2002) Silica gel induced cleavage of aziridines by aromatic amines under solvent free conditions. Tetrahedron Lett 43:3975–3976
51. Chatterjee R, Santra S, Chakraborty Ghosal N, Giri K, Zyryanov GV, Majee A (2020) CuO nanoparticles as a simple and efficient green catalyst for the aziridine ring-opening: examination of a broad range of nucleophiles. ChemistrySelect 5:4525–4529
52. Pandey M, Tsunoji N, Kubota Y, Bandyopadhyay M (2022) Amine and sulfonic acid anchored porous silica as recyclable heterogeneous catalysts for ring-opening of oxiranes. ChemistrySelect 7:e202201756
53. Huang ZQ, Xu ZH, Liu XH, Zhao Y, Wang P, Sun WY (2021) Appl Organomet Chem. 35: e6262
54. Bagherzadeh M, Chegeni M, Bayrami A, Amini M (2024) Sci Rep 14:1–19
55. Brufani G, Valentini F, Rossini G, Rosignoli L, Gu Y, Liu P, Vaccaro L (2023) Waste-minimized continuous flow copper-catalyzed azide-alkyne cycloaddition with low metal contamination. Green Synth Catal 4:154–159
56. Mohit M, Thomas KRJ (2024) ChemCatChem 16:e202401378
57. Kwak SJ, Shin US, Kim SH (2023) A heterogeneous Cu-catalyst immobilized on poly (3-carboxythiophene)-modified multi-walled carbon nanotubes for click reaction. J Chem Sci 135:10
58. Eftekhar M, Raoufi F (2022) Synthesis, characterization and first application of graphene oxide functionalized Cu(II) complex for the synthesis of 1,2,3-triazole derivatives. Polycycl Aromat Compd 42:4780–4792
59. Stiriba SE, Bahsis L, Benhadria E, Oudghiri K, Taourirte M, Julve M (2023) Cellulose acetate-supported copper as an efficient sustainable heterogenous catalyst for azide-alkyne cycloaddition click reactions in water. Int J Mol Sci 24:9301
60. Buğday N, Yaşar S, Altin S, Duygulu O, Önal Y (2023) Magnetic Cu/Co nanoparticles supported on nitrogen-doped porous carbon derived from Cu/Co@aZIF: investigation of catalytic activity and structural properties. Appl Organomet Chem 37:e7017
61. Rajabi-Moghaddam H, Naimi-Jamal MR, Tajbakhsh M (2021) Fabrication of copper(II)-coated magnetic core-shell nanoparticles Fe3O4@SiO2-2-aminobenzohydrazide and investigation of its catalytic application in the synthesis of 1,2,3-triazole compounds. Sci Rep 11:2073

62. Sharma P, Rathod J, Singh A, Kumar P, Sasson Y (2018) Synthesis of heterogeneous Ru(ii)-1,2,3-triazole catalyst supported over SBA-15: application to the hydrogen transfer reaction and unusual highly selective 1,4-disubstituted triazole formation via multicomponent click reaction. Catal Sci Technol 8:3246–3259
63. Gupta A, Jamatia R, Mahato M, Pal AK (2017) Metalloprotein-inspired ruthenium polymeric complex: a highly efficient catalyst in parts per million level for 1,3-dipolar Huisgen's reaction in aqueous medium at room temperature. Ind Eng Chem Res 56:2375–2382
64. Hsueh N, Clarkson GJ, Shipman M (2015) Generation and ring opening of aziridines in telescoped continuous flow processes. Org Lett 17:3632–3635
65. Laktsevich-Iskryk M, Krech A, Fokin M, Kimm M, Jarg T, Noël T, Ošeka M (2024) Telescoped synthesis of vicinal diamines via ring-opening of electrochemically generated aziridines in flow. J Flow Chem 14:139–147
66. Vural Gürsel I, Aldiansyah F, Wang Q, Noël T, Hessel V (2015) Continuous metal scavenging and coupling to one-pot copper-catalyzed azide-alkyne cycloaddition click reaction in flow. Chem Eng J 270:468–475
67. Smith CD, Baxendale IR, Lanners S, Hayward JJ, Smith SC, Ley SV (2007) [3 + 2] Cycloaddition of acetylenes with azides to give 1,4-disubstituted 1,2,3-triazoles in a modular flow reactor. Org Biomol Chem 5:1559–1561
68. Fuchs M, Goessler W, Pilger C, Kappe CO (2010) Mechanistic insights into copper(I)-catalyzed azide-alkyne cycloadditions using continuous flow conditions. Adv Synth Catal 352:323–328
69. Bogdan AR, Sach NW (2009) The use of copper flow reactor technology for the continuous synthesis of 1,4-disubstituted 1,2,3-triazoles. Adv Synth Catal 351:849–854
70. Bogdan AR, James K (2011) Synthesis of 5-Iodo-1,2,3-triazole-containing macrocycles using copper flow reactor technology. Org Lett 13:4060–4063
71. Valentini F, Brufani G, Rossini G, Campana F, Lanari D, Vaccaro L (2023) ACS Sustain Chem Eng 11:3074–3084
72. Brufani G, Valentini F, Rossini G, Carpisassi L, Lanari D, Vaccaro L (2023) Continuous flow synthesis of 1,4-disubstituted 1,2,3-triazolesviaconsecutive β-azidation of α,β-unsaturated carbonyl compounds and CuAAC reactions. Green Chem 25:2438–2445
73. C. Schäfer, B. Török, Nontraditional activation methods in Green and sustainable applications, Ed. C. Schäfer, B. Török, Elsevier, Amsterdam, 2021, ch. 1, pp. 1–26
74. Fairbanks BD, Macdougall LJ, Mavila S, Sinha J, Kirkpatrick BE, Anseth KS, Bowman CN (2021) Photoclick chemistry: a bright idea. Chem Rev 121:6915–6990
75. Castro-Godoy WD, Heredia AA, Schmidt LC, Argüello JE (2017) RSC Adv 7:33967–33976
76. Poloukhtine AA, Mbua NE, Wolfert MA, Boons GJ, Popik VV (2009) Selective labeling of living cells by a photo-triggered click reaction. J Am Chem Soc 131:15769–15776
77. Arumugam S, Popik VV (2009) Photochemical generation and the reactivity ofo-naphthoquinone methides in aqueous solutions. J Am Chem Soc 131:11892–11899
78. Rasmussen LK, Boren BC, Fokin VV (2007) Ruthenium-catalyzed cycloaddition of aryl azides and alkynes. Org Lett 9:5337–5339
79. George N, Singh G, Singh R, Singh G, Devi A, Singh H, Kaur G, Singh J (2022) Sustain Chem Pharm 30:100824
80. Xiong X, Chen H, Tang Z, Jiang Y (2014) Supported CuBr on graphene oxide/Fe3O4: a highly efficient, magnetically separable catalyst for the multi-gram scale synthesis of 1,2,3-triazoles. RSC Adv 4:9830–9837
81. Librando IL, Mahmoud AG, Carabineiro SAC, Guedes da Silva MFC, Geraldes CFGC, Pombeiro AJL (2021) The catalytic activity of carbon-supported cu(I)-phosphine complexes for the microwave-assisted synthesis of 1,2,3-triazoles. Catalysts 11:185
82. Attia YA, Mohamed YMA (2024) J Chem Technol Biotechnol 99:2311–2231
83. García-Oliva C, Merchán A, Perona A, Hoyos P, Rumbero Á, Hernáiz MJ (2022) Development of sustainable synthesis of glucuronic acid glycodendrimers using ball milling and microwave-assisted CuAAC reaction. New J Chem 46:6389–6393

84. Rinaldi L, Martina K, Baricco F, Rotolo L, Cravotto G (2015) Solvent-free copper-catalyzed azide-alkyne cycloaddition under mechanochemical activation. Molecules 20:2837–2849
85. Sang Y, Guan X, Liu H, Yang J, Li S, Wang Q, Huang X, Wang Y, Yu Q, Li M (2024) ACS Sustain Chem Eng 12:13700–11371
86. Cintas P, Barge A, Tagliaprieta S, Boffa L, Cravotto G (2010) Nat Protoc 5:607–616
87. Li X, Chen X, Jiang Y, Chen S, Qu L, Qu Z, Yuan J, Shi H (2016) J Heterocycl Chem 53:1402–1141
88. Lunkov A, Shagdarova B, Lyalina T, Dubinnyi MA, Karpova N, Lopatin S, Il'ina A, Varlamov V (2022) Carbohydr Polym 282:119109
89. Seavill PW, Holt KB, Wilden JD (2019) Electrochemical synthesis of copper (i) acetylidesviasimultaneous copper ion and catalytic base electrogeneration for use in click chemistry. RSC Adv 9:29300–29304
90. Slootweg JC (2024) Sustainable chemistry: green, circular, and safe-by-design. One Earth 7: 754–758
91. Morozova MA, Yusubov MS, Kratochvil B, Eigner V, Bondarev AA, Yoshimura A, Saito A, Zhdankin VV, Trusova ME, Postnikov PS (2017) Regioselective Zn(OAc)2-catalyzed azide–alkyne cycloaddition in water: the green click-chemistry. Org Chem Front 4:978–985
92. Guselnikova O, Vana J, Phuong LT, Panov I, Rulisek L, Trelin A, Postnikov P, Svorcik V, Andris E, Lyutakov O (2021) Plasmon-assisted click chemistry at low temperature: an inverse temperature effect on the reaction rate. Chem Sci 12:5591–5598
93. Kurdtabar M, Soleimani-Amiri S, Alidad M, Bagheri Marandi G (2025) Cu-decorated CMC-g-PAA/Fe3O4 hydrogel nanocomposite: high-efficiency and reusable catalyst for synthesis of 1,2,3-triazoles via click reaction in water. J Polym Environ 33:2334–2350

Top Heterocycl Chem (2026) 63: 257–292
https://doi.org/10.1007/7081_2025_100

Published online: 15 February 2026

Click Immobilization of (Bio)Catalysts in Batch and Continuous Flow Chemistry Systems

Ross D. Jansen-van Vuuren, L. Jedlovčnik, and J. Cerkovnik

Contents

1 General Introduction: Supports for Batch and Continuous Flow Chemistry Systems 258
2 Click Chemistry for the Immobilization of Metal Complex Catalysts 261
3 Click Chemistry for Immobilization of Organocatalysts 262
3.1 Application of Click-Supported Catalysts in Continuous Flow Kinetic Resolution 272
4 Click Chemistry for Immobilization of Biocatalysts 277
5 Summary and Future Implications 279
References 282

Abstract The immobilization of catalysts and enzymes (biocatalysts) on soluble and insoluble supports offers enhanced stability, recoverability, and recyclability of the catalysts/biocatalysts (singularly abbreviated (bio)catalysts) as well as the capacity for incorporating (bio)catalysts into continuous flow chemistry (CFC) systems. There are many approaches to (bio)catalyst immobilization, ranging from physical methods such as adsorption, entrapment, and encapsulation to cross-linking and covalent binding of (bio)catalysts to supports. Each approach offers pros and cons, and the selection of which method to use depends on the specific situation. The use of click chemistry is a desirable approach to covalently bind (bio)catalysts to supports as it offers high specificity and mild reaction conditions, resulting in a stable covalent attachment. However, several challenges are posed by this approach, such as contamination by residual copper catalyst (for copper-catalyzed azide-alkyne cycloaddition reactions), modification of the catalyst during immobilization (may alter the catalyst activity and selectivity), and limited dynamic reversibility. This chapter focuses on the types of catalysts that have been immobilized via click chemistry, as well as the class of click chemistry used. We hope that it will be helpful for practitioners of (bio)catalysis as well as engineers trying to understand

R. D. Jansen-van Vuuren (✉), L. Jedlovčnik, and J. Cerkovnik
Faculty of Chemistry and Chemical Technology, University of Ljubljana, Ljubljana, Slovenia

this aspect of CFC systems, in seeing what has been tried, what works, and promising future directions in the field.

Keywords Click chemistry · Catalyst immobilization · Enzyme immobilization · Recyclable catalyst · Catalysis · Biocatalysis · Flow chemistry · CuAAC · Heterogenization · Organocatalysis · Enzymes · Recyclability · Recoverability · Stability

1 General Introduction: Supports for Batch and Continuous Flow Chemistry Systems

Traditionally, most chemistry is performed in batch systems where a fixed number of reactants are processed in a single vessel at a time, followed by workup and purification to isolate the desired product. In many cases, homogeneous catalysts are used to enable milder conditions for these reactions, and these can take the form of metal catalysts, organocatalysts, or biocatalysts (as well as a list of exceptional cases, e.g., plasmonic catalysts). To be able to easily recover valuable catalysts and/or prevent their contamination of the product, such catalysts can be immobilized on a range of soluble [1–3] and insoluble supports [4–6]. Various approaches can be employed for catalyst immobilization [5] including physical processes such as adsorption [7, 8], electrostatic interactions [9, 10], and encapsulation/entrapment [11–14], as well as covalent bonding of the (bio)catalyst to the support (Fig. 1) [15–17].

Each option has its pros and cons. For instance, while physical adsorption is typically easy and reversible, also allowing the (bio)catalyst to retain its activity, it is susceptible to leaching [18, 19]. Methods involving covalent bonding of (bio)-catalysts to supports offer strong attachment with minimal leaching risk but can lead to potential deactivation of the (bio)catalyst [20, 21]. One way to covalently attach the (bio)catalyst to a surface is through click chemistry. This chapter reviews various methods used with different (bio)catalysts, in batch systems and, more recently, in CFC systems. To clarify, we are *not* discussing (bio)catalyst immobilization in CFC systems for click reactions [22–24], but rather using click chemistry to immobilize (bio)catalysts onto supports. This can facilitate transformations in CFC systems, similar to how click chemistry is used to attach functional molecules onto stationary phases in chromatographic columns or enzymatic reactors [25–28].

CFC systems involve continuously passing reactants through a reactor, collecting the product as it elutes (ideally in a pure form). These systems offer advantages over batch systems such as improved reaction control, increased safety, higher efficiency and productivity, scalability, sustainability, waste reduction, and improved process integration and automation [29–31]. CFC systems are already used for preparing active pharmaceutical ingredients (APIs) [32], polymers [33], and fine chemicals [34, 35], and have been studied for synthesizing a variety of materials, e.g.,

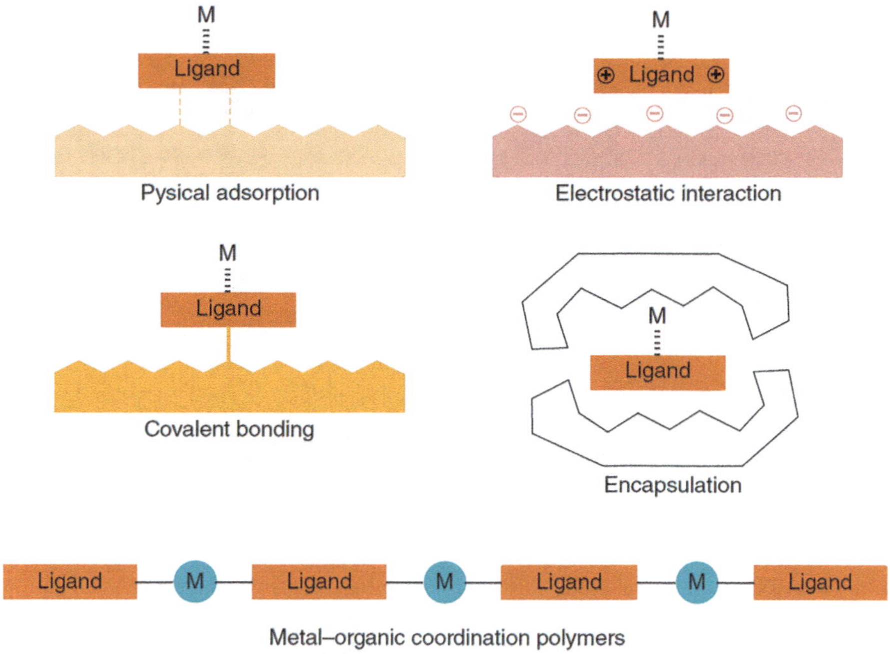

Fig. 1 Methods for the immobilization of catalysts and biocatalysts onto soluble and solid supports: physical adsorption, electrostatic interaction, covalent bonding, encapsulation, and via metal-organic coordination polymers (Figure taken from Piermatti et al. [5])

deuterated compounds [4, 36], nanomaterials [37, 38], proteins and peptides [39, 40], and even for performing click chemistry reactions [22].

Different CFC systems/configurations can be used (Fig. 2). The choice of which type of system to use depends on several factors including the type of chemical transformation, the reaction phase, the reagents and catalysts used (if any), and safety aspects of the reaction [41]. The simplest flow systems include those for reactions without a catalyst (type 1, Fig. 2), while other setups involve the use of a supported reagent in the synthesis (type II, Fig. 2) [42]; inclusion of a homogeneous catalyst, necessitating separation of the product and the catalyst from the output flow stream (type III, Fig. 2) [43]; and the use of immobilized catalysts (and biocatalysts) within the system, enabling efficient heterogeneous catalysis (type IV, Fig. 2) [44–46].

For option (IV), (bio)catalyst immobilization within CFC systems can be achieved using several approaches (Fig. 3): (1) *carrier-bound immobilization* in which the catalysts are attached to solid carriers via adsorption, covalent bonding, or affinity interactions (these are normally placed in a packed-bed reactor or cartridge through which the reaction mixture passes) (Fig. 3a); (2) *carrier-free immobilization*: catalysts are cross-linked or polymerized to form aggregates or crystals without external carriers (normally forming a monolith, as shown in Fig. 3b); (3) *encapsulation*, in which catalysts are enclosed within polymeric matrices, hydrogels, or

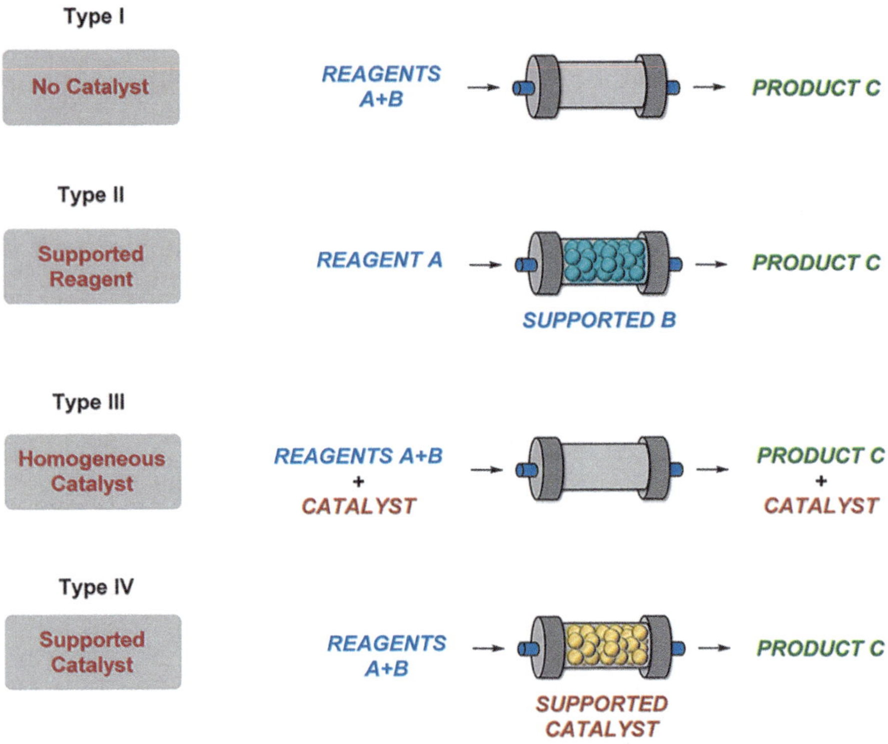

Fig. 2 Types of continuous flow systems. (Figure taken from Porta et al. [35])

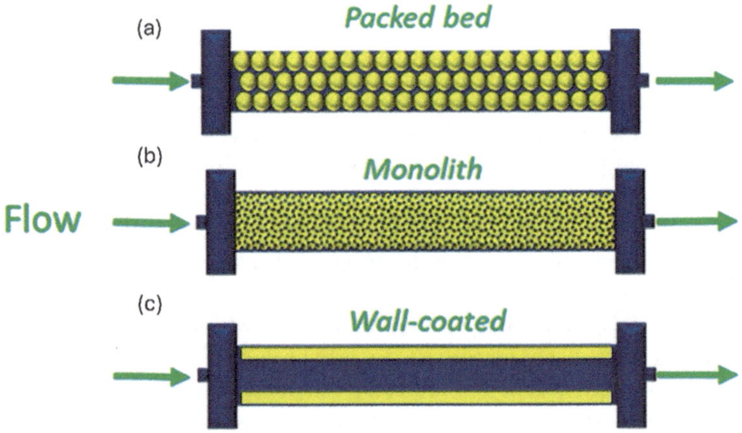

Fig. 3 Schematic representation of three common types of flow reactors: (**a**) packed-bed, consisting of densely packed catalyst particles; (**b**) monolith, with a structured porous matrix; and (**c**) wall-coated, where the catalyst is coated onto the inner walls of the channel. (Adapted from Colella et al. [47])

membranes, allowing selective substrate and product diffusion while retaining the catalyst (also Fig. 3b); and (4) *surface immobilization*, which involves the attachment of catalysts to the walls of the tubing or reactor (Fig. 3c).

For all approaches, a diverse range of supports have been studied [21]. For (a), immobilized catalysts are typically packed into a catalyst cartridge, through which the substrates and reagents flow in a solvent mixture. In (a)–(c), click chemistry can be used for the immobilization process. We have reviewed the literature around this concept in this chapter.

We have divided the chapter into sections based on the *type of catalyst* being immobilized via click chemistry. First, we will review metal complex catalysts (precious metal and earth-abundant metals) immobilized using click chemistry, followed by organocatalysts, and finally, we will highlight significant examples of biocatalysts (enzymes) that have been immobilized in this way.

2 Click Chemistry for the Immobilization of Metal Complex Catalysts

Various strategies have been described for immobilizing metal catalysts, including both covalent and non-covalent bonding to a variety of organic and inorganic supports. Covalent approaches are generally preferred due to their higher stability and applicability. However, they often require further synthetic efforts to customize both the support and the catalyst. Some of the most important factors to consider when immobilizing catalysts are the careful positioning of the anchoring point in the catalyst, the selection of a suitable support, and the proper choice of binding scheme [48]. As a rule, binding is not performed at the active site to avoid or minimize adverse interactions with the carrier. The choice of support is crucial as it can influence the reactivity of the catalyst [49, 50]. Heterogeneous supports are known to facilitate effective recovery and recycling of catalysts, but often alter the performance of supported catalysts. The choice of anchoring chemistry, the type and length of the resulting tether [51–54], and, where possible, the control of catalyst graft density may prove crucial for the successful construction and operation of immobilized catalysts [55–59].

In addition to the undeniable advantages of easier product separation, simpler recovery, and reuse of the often expensive homogeneous catalysts, heterogenization strategies can enable the detailed elaboration of catalytic sites and their environments. In the simplest case, the specific microenvironment created by the support can positively impact the performance of the supported catalyst, both in terms of activity and selectivity. A more versatile, potentially modular tactic that involves the immobilization and positioning of both catalytic sites and cooperative secondary functions has attracted considerable interest in recent years as we move towards tailored catalytic materials. Ligation processes that have high functional tolerance are highly efficient, specific, and versatile, and require mild conditions that are particularly suitable. Copper(I)-catalyzed azide-alkyne cycloaddition (CuAAC) is probably the most popular reaction in the toolbox of "click" chemistry for the heterogenization of homogeneous catalysts. There are many approaches for tethering metal catalysts; we recommend some parts of the reviews by Astruc et al. [60], Fernandes et al. [17], Toy et al. [61], Itsuno et al. [62], Liu et al. [63], Dinică et al. [64], Zuilhof et al. [65], Heravi et al. [66], Garcia et al. [67], Campisciano et al. [68], Kolotilov et al. [69], Motokura et al. [70], Benaglia and Puglisi [5], and Zaera [59] for further reading.

Metal catalysts were attached to various supports, including dendrimers, poly (ethylene glycol) (PEG), polymers and coordination polymers, silica particles, and magnetic nanoparticles, using click chemistry, in batch as well as CFC systems. The studies are summarized in Table 1, followed by a brief commentary about select case studies.

Among the examples, only six metal catalysts have been tested or applied in continuous flow systems (Table 1, entries 17, 30, 32, 41, 85, and 88). In three studies, a polymer-supported metal catalyst for heterogeneous batch reactions was developed and further adapted for continuous flow operation (Table 1, entries 17, 30, and 32; [87, 100, 102]). In another study, a ruthenium-based metathesis catalyst was efficiently immobilized on nanoporous silica via click chemistry (Table 1, entry 41 [108]). The heterogenized catalyst showed good activity and recyclability for various substrates, as demonstrated in a continuous process using a circulating flow reactor.

A specific example of a continuous flow reactor is the biocompatible and biodegradable polymersome nanoreactor platform, which can serve as an artificial organelle within living cells (Table 1, entry 85, and Fig. 4 [151]).

The artificial organelles demonstrated the ability to covalently and non-covalently incorporate tris(triazolylmethyl)-amine-Cu(I) complexes into their membrane. These artificial organelles can effectively catalyze a CuAAC reaction intracellularly without compromising cell integrity. The platform represents a step forward in the application of polymersome-based nanoreactors as artificial organelles.

In another study, carbon-coated cobalt nanoparticles labeled with aza-bis (oxazoline)–Co(II) complexes were used (Table 1, entry 89; [154]). Here, a CuAAC reaction was tested in the kinetic resolution of racemic 1,2-diphenylethane-1,2-diol by asymmetric monobenzoylation. The catalyst was investigated under batch and flow reactor conditions. The extremely high ferromagnetism of the cobalt cores enables a novel design of a continuous flow reactor. The high saturation magnetization, along with the impermeability, chemical inertness, and thermal stability of the carbon coating qualifies the nanopowder as an excellent scaffold for recyclable catalysts.

3 Click Chemistry for Immobilization of Organocatalysts

The integration of organocatalysis with continuous flow technology has become a powerful strategy for developing sustainable, scalable, and (enantio)selective chemical processes [168]. Organocatalysts - small organic molecules capable of

Table 1 Select examples of metal catalysts tethered using click chemistry in batch and CFC systems

Entry	Process	Metal tethering strategy	Click chemistry	Reference
Dendrimer-supported tethering				
1	Cu(II)-catalyzed asymmetric benzoylation	Cu(II) attached to phosphorus dendrimer immobilized azabis (oxazoline) ligands up to the third generation	CuAAC	[71]
2	Pd-catalyzed Suzuki cross-coupling reactions	Dendritic polyglycerol (PG)-supported N-heterocyclic carbene–palladium complexes	CuAAC	[72]
3	Pd-catalyzed Suzuki–Miyaura coupling of aryl halides and phenyl boronic acid	Dendrimerized phosphine ligands precomplexed with Pd $(OAc)_2$	CuAAC	[73]
4	Pd-catalyzed Suzuki–Miyaura, Sonogashira, and Heck cross-coupling reactions	Dendrimers with 2-pyridyl-1,2,3-triazole (pyta) ligands form complex with $Pd(OAc)_2$ units	CuAAC	[74]
5	Selective oxidation of various organic compounds with dioxygen	CNT-appended polyamidoamine dendrimers bearing Fe(II) α-keto acid complexes	CuAAC	[75]
6	Olefin hydrogenation	Pd(II) coordinated to click dendrimers	CuAAC	[76]
7	C–C cross-coupling reactions and reduction of 4-nitrophenol	Pd(II) NPs stabilized by "click" dendrimers	CuAAC	[77]
8	Hydrolysis of ammonia borane	CoPt NPs stabilized by "click" dendrimers	CuAAC	[78]
9	Suzuki–Miyaura cross-coupling in water	Glycerol-dendron-supported NHC–Pd complex	CuAAC	[79]
PEG-supported tethering				
10	Cu(II)-catalyzed asymmetric benzoylation of 1,2-diols	Bisbenzyl-substituted aza (bisoxazoline) ligand complexed with $CuCl_2$ and immobilized on $MeOPEG_{5000}$	CuAAC	[80]
11	Cu(II)-catalyzed asymmetric benzoylations, nitroaldol, and Michael reactions	Perfluoroalkyl-tagged azabis (oxazoline) ligands complexed with $CuCl_2$	CuAAC	[81]
12	Oxidative dehydrogenation of secondary amines using TBHP as oxidant	Co(II) Schiff base attached to the azido-functionalized $MeOPEG_{5000}$	CuAAC	[82]
13	Olefination of aldehydes	Polyethylene glycol (PEG)-immobilized iron(II) porphyrin	CuAAC	[83]

(continued)

Table 1 (continued)

Entry	Process	Metal tethering strategy	Click chemistry	Reference
Polymer-supported tethering				
14	Oxidation of alcohols to carbonyl compounds using molecular oxygen as oxidant	Co(II) Schiff base complexes attached to the polystyrene resin	CuAAC	[84]
15	Pd-catalyzed allylic alkylation of cinnamyl acetate	P,N-type ligand family (ClickPhine) complexed with $[Pd(allyl)Cl]_2$ attached to Merrifield resin	CuAAC	[85]
16	Ring opening of cyclohexene oxide, silylcyanation of benzaldehyde and alkynylation of imines	Pyridine-bis(oxazoline) ligand, immobilized onto a polystyrene resin and precomplexed with various metal salts ($YbCl_3$, $LuCl_3$, and CuOTf) as catalysts	CuAAC	[86]
17*	Pd-catalyzed asymmetric allylic amination	Polystyrene-supported diphenylphosphinooxazoline (PHOX)-Pd complexes (*The system has been adapted to continuous flow operation*)	CuAAC	[87]
18	Addition of arylboronic acids to aldehydes	*N*-heterocyclic carbene (NHC)-rhodium complex anchored on polystyrene resins	CuAAC	[88]
19	Asymmetric, aerobic oxidation of α-hydroxy (thio)esters and amides	Chiral *N*-salicylidene oxidovanadium *tert*-leucinate complex anchored on polystyrene resin	CuAAC	[89]
20	CuAAC reactions	Tris(triazole)-CuCl complex immobilized on Merrifield resin	CuAAC	[90]
21	CuAAC reactions	Tris-(benzyltriazolylmethyl) amine (TBTA)-Cu(I) complex immobilized on a TentaGel resin	CuAAC	[91]
22	CuAAC reactions	Tris-(benzyltriazolylmethyl) amine (TBTA)-CuBr complex incorporated on polystyrene copolymers	CuAAC	[92]
23	C–H bond oxidation	Fe(II)-BPMEN complex anchored on polystyrene resin	CuAAC	[93]
24	Epoxidation	Mo(II) acetylacetonate attached on cross-linked polystyrene	Amine—ketone reaction	[94]
25	Asymmetric epoxidation of unfunctionalized olefins	Chiral salen Mn(III) catalyst immobilized on ZnPS-PVPA	CuAAC	[95]
26	Synthesis of RNA oligonucleotide conjugates	Cu(I)-tris(triazolylmethyl)-amine complex immobilized on polystyrene	CuAAC	[96]

(continued)

Table 1 (continued)

Entry	Process	Metal tethering strategy	Click chemistry	Reference
27	Suzuki–Miyaura coupling	NHC-functionalized Pd-PEPPSI complexes tethered with different linker chemistries to polymer and silica	Inverse electron-demand Diels–Alder (IEDDA)	[97]
28	Amination of phenols	Ni particles immobilized on triazole-modified chitosan	CuAAC	[98]
29	CuAAC click reactions	Cu(I)-tris[(1,2,3-triazolyl) methyl]amine immobilized on poly(2-hydroxyethyl methacrylate) polymer	CuAAC	[99]
30*	CuAAC click reactions	Polystyrene-supported tris [(1-benzyl-1H-1,2,3-triazol-4-yl)methyl]amine (TBTA)-Cu(I) catalyst (*highly cross-linked resins were tested in batch and in flow*)	CuAAC	[100]
31	CuAAC click reactions	Polystyrene-supported 1,5,7-triazabicyclo-[4.4.0]dec-5-ene (TBD) as polyvalent ligand for Cu(I)	CuAAC	[101]
32	Isomerization of *E*-cinnamyl acetate	Methacrylate-based cross-linked polymer gel bearing an Ir photocatalyst (*Pumping recycling setup for the Ir catalysts immobilized on the cross-linked polymer gel support*)	CuAAC	[102]
Silica-supported tethering				
33	Catalytic epoxidation	Immobilization of ferrocene, pyrene, tris(pyridylmethyl) amine, and iron porphyrin on mesoporous silica SBA-15	CuAAC	[55]
34	Oxidation of cyclohexane	SBA-supported L-x/Ni catalyst (L = bis(2-pyridylmethyl)(1*H*-1,2,3-triazol-4-ylmethyl)amine)	CuAAC	[57]
35	Oxidation of cyclohexane	Silica-immobilized Co (II) complex catalyst with bis (pyridylmethyl) (triazolylmethyl)amine ligand	CuAAC	[58]
36	Oxidation of alcohols with molecular oxygen	Co(II) Schiff base complex attached on SBA-15 mesoporous silica	CuAAC	[103]
37	Oxidation of sulfides	Oxo–vanadium(IV) tridentate Schiff base complex attached to mesoporous silica	CuAAC	[104]

(continued)

Table 1 (continued)

Entry	Process	Metal tethering strategy	Click chemistry	Reference
38	Tsuji–Trost allylation of ethyl acetoacetate	Mesoporous silica nanoparticles (MSNs) bearing Pd(II) complex	CuAAC	[105]
39	Double Michael addition, and allylic stannylation and alkylation	Pincer Pd and Pt complexes tethered to silica	CuAAC	[106]
40	Epoxidation of alkenes	Mn porphyrins attached to silica	CuAAC	[107]
41*	Ring-closing metathesis	Ru-based metathesis catalyst immobilized on nanoporous silica (*continuous process using a circulating flow reactor*)	CuAAC	[108]
42	Aerobic oxidation of alcohols	SBA-15-supported Pd (II) catalyst	CuAAC	[109]
43	Hashmi phenol synthesis	Au(III)-immobilized silica nanoparticles	CuAAC	[110]
44	Peroxidase mimic	Fe(III) biuret-amide complex immobilized on mesoporous silica NPs	CuAAC	[111]
45	Oxidation of unactivated cyclic hydrocarbons	Fe(II)-coordinated cavitand tethered to the SBA-15 silica	CuAAC	[112]
46	Asymmetric transfer hydrogenation of aromatic ketones	Ir(I)-9-amino *epi*-cinchonine complex immobilized on mesoporous SBA-15	Thiol–ene click reaction	[113]
47	Asymmetric epoxidation of alkenes	Mn(salen) complex on organosiloxane-functionalized mesoporous silica	Thiol–ene click reaction	[114]
48	Asymmetric trimethylsilylcyanation of benzaldehydes	VO(salen) complex on thio-functionalized amorphous silica, delaminated zeolite, and MCM-41	Thiol–ene click reaction	[115]
49	Asymmetric cyclopropanation of alkenes	Ru(salen)-based complex on thio-functionalized amorphous silica	Thiol–ene click reaction	[116]
50	Aerobic oxidation of benzyl alcohol	Silica-supported Cu(I) catalyst	CuAAC	[117]
51	Aerobic oxidation of benzyl alcohol	(bpy)Cu(I)/TEMPO/NMI catalytic system grafted on silica	CuAAC	[118]
52	Aerobic oxidation of alcohols	Co(II) Schiff base immobilized on mesoporous silica	CuAAC	[103]
53	Superoxide dismutase (SOD) mimic	Mn(II) complex grafted to mesoporous silica	CuAAC	[119]
54	Superoxide dismutase (SOD) mimic	Dicationic Cu(II)-N_4-Schiff base complex encapsulated in mesoporous silica	CuAAC	[120]

(continued)

Table 1 (continued)

Entry	Process	Metal tethering strategy	Click chemistry	Reference
55	Asymmetric cyclopropanation	Chiral dirhodium catalyst immobilized on SBA-15	CuAAC	[121]
56	Reduction of methylene blue	Silver hybrids with triazole immobilized on SBA-15 silica	CuAAC	[122]
57	Suzuki coupling	Pd(II) complex immobilized on functionalized mesoporous SBA-15 silica	CuAAC	[123]
58	C–C bond coupling and selective formyl group hydrogenation	Pd(II)- and Rh(III)-impregnated SBA-15 silica	CuAAC	[124]
59	C–C bond formation	Silica-supported Pd NPs	CuAAC	[125]
60	Suzuki–Miyaura C–C coupling in aqueous media	Pd(II) complex supported on triazole-functionalized silica NPs	CuAAC	[126]
61	1,2-Addition of phenylboroxine to *N*-tosylimine	Rh immobilized on selectively functionalized mesoporous SBA-15 silica	CuAAC	[127]
62	Reduction of 4-nitrophenol	Pd nanocatalyst on triazole/triazine-functionalized mesoporous SBA-15 silica support	CuAAC	[128]
63	Asymmetric 1,2-addition of triphenylboroxine to *N*-tosylimines	Mono- and bis-functionalized chiral Rh-diene complexes attached to mesoporous silica inverse opal	CuAAC	[129]
64	Suzuki cross-coupling in water under MW	Pd-triazolium salts supported on halloysite nanotubes	CuAAC	[130]
Carbon, graphene, and graphene oxide (GO)-supported tethering				
65	Oxygen reduction reaction	Cu(I) multilayer catalyst films on carbon electrode surfaces	CuAAC	[131]
66	Selective epoxidation of olefins	Clicked graphene oxide-supported venturello catalyst	Thiol-ene click reaction.	[132]
67	Electron donor–acceptor interactions in the ground and in the excited state	Alkylsulfonyl Zn (II) phthalocyanines bonded to graphene	CuAAC	[133]
68	Catalyst for the oxygen reduction reaction (ORR)	Triazole-pyridine-coordinated Cu(II) complex covalently grafted onto multiwalled carbon nanotubes (CNTs) and reduced GO	CuAAC	[134]
69	Catalyst for the oxygen reduction reaction (ORR)	Triazole-dipyridine multinuclear Cu(II) complex immobilized onto reduced GO	CuAAC	[135]

(continued)

Table 1 (continued)

Entry	Process	Metal tethering strategy	Click chemistry	Reference
70	Oxidation of *o*-phenylenediamine	CuPt nanorod grafted onto GO	CuAAC	[136]
71	Epoxidation of olefins	Peroxophosphotungstate immobilized on GO	CuAAC	[137]
72	Heck and Sonogashira C–C coupling in deep eutectic solvents	Pd supported on cellulose-modified magnetite-GO nanocomposite	CuAAC	[138]
73	Huisgen [3+2] cycloaddition reaction	*N*-heterocyclic carbene (NHC)-Cu(I) complex grafted on G and CNTs	CuAAC	[139]
74	Reduction of 4-nitrophenol	Chitosan-stabilized Au NPs immobilized on magnetite (Fe_3O_4) nanoparticles tethered over CNTs	Thiol-yne "click" chemistry	[140]
75	Suzuki–Miyaura cross-coupling	Ni nanoparticles supported on multiwalled CNTs	CuAAC	[141]
76	Electroenzymatic synthesis	(2,2′-Bipyridyl)(pentamethylcyclopentadienyl)-Rh complex immobilized on porous carbon electrode	CuAAC	[142]
77	Control of the surface coverage	Au NPs deposited on glassy carbon surface	Electro-assisted CuAAC	[143]
78	Electrocatalytic alcohol oxidation	Tris(2-pyridylmethyl)-amine Ru complex immobilized on glassy carbon electrode	CuAAC	[144]
79	Bulk electrolysis and water reduction	Co catalyst immobilized with fullerene (C_{60}) on a carbon cloth electrode	CuAAC	[145]
80	Electrocatalytic reduction of D-fructose to D-sorbitol	Molecular and biological catalyst co-immobilization on carbon electrode	Sequential CuAAC and alkene-thiol coupling	[146]
Nanoparticle (NP)-supported tethering				
81	Electrochemical reduction of CO_2 to CO	Cobalt-porphyrin complex tethered to boron-doped, p-type conductive diamond	CuAAC	[147]
82	Alkene epoxidation	Ni(II) complex immobilized on metal oxide (TiO_2, SiO_2, CeO_2, WO_3)	Maleimide—hydroxy photoclick reaction	[148]
83	Suzuki–Miyaura coupling	Ni(II) catalyst immobilized on MOFs	CuAAC	[149]
84	Artificial heme-peroxidase	"Clickable" biocatalyst, FeMC6*a-PEG4@DBCO immobilized on nanomaterials to afford functional bioconjugates	Strain-promoted azide-alkyne cycloaddition (SPAAC)	[150]

(continued)

Table 1 (continued)

Entry	Process	Metal tethering strategy	Click chemistry	Reference
85	Intracellular CuAAC	Artificial organelles with covalently and non-covalently incorporated tris (triazolylmethyl)amine-Cu(I) complexes in their membrane	CuAAC	[151]
Magnetic nanoparticle (MNP)-supported tethering				
86	Suzuki–Miyaura cross-coupling	(MNP)-supported di(2-pyridyl) methanol palladium dichloride complex	CuAAC	[152]
87	Desymmetrization of racemic 1,2-diols through asymmetric benzoylation	Cu(II)-azabis(oxazoline) complex grafted to magnetite@silica NPs	CuAAC	[153]
88*	Kinetic resolution of racemic 1,2-diphenylethane-1,2-diol	Cu(II)−azabis(oxazoline) complex immobilized on magnetic Co/C MNPs (*Kinetic resolution of 1,2-diphenylethane-1,2-diol under batch and continuous flow conditions*)	CuAAC	[154]
89	Asymmetric Henry reactions between various aldehydes and nitromethane	Chiral bis(oxazoline) Cu (II) complex immobilized on magnetic mesocellular mesoporous silica	CuAAC	[155]
90	Oxidation of alkenes and sulfides	Mn(III) porphyrin anchored onto functionalized silica-coated MNPs	CuAAC	[156]
91	Suzuki cross-coupling	Di(2-pyridyl)methanol palladium dichloride complex immobilized on MNP	CuAAC	[152]
92	Model CuAAC	Cu(I) immobilized onto a triazole-functionalized MNP	CuAAC	[157]
93	Hiyama cross-coupling	MNPs-supported nano-Ni catalyst	CuAAC	[158]
94	New materials for optical, photonic, and biological applications	Zn-complexed metalloporphyrin-bound silica-coated iron oxide particles	CuAAC	[159]
Coordination polymer-supported tethering				
95	Enantioselective cyclopropanation of alkenes	Polytopic chiral azabis (oxazoline) copper (II) complexes	CuAAC	[160]
96	CuAAC click reactions	Cu(I) coordination polymer with hybridized 1,2,3-triazoles	CuAAC	[161]
Miscellaneous-supported tethering				
97	Electrochemical reduction of CO_2	Immobilization of a ligand (post-functionalization by complexation of pyridyl ligands with metals)	CuAAC	[162]

(continued)

Table 1 (continued)

Entry	Process	Metal tethering strategy	Click chemistry	Reference
98	Isomerization of isoxazole to 3H-azirine	*microfluidic "click chip"* Cu-free alkynone-amine click reaction for the immobilization of chiral norbornadiene-based ligands	Amino–yne click reaction	[163]
99	Reduction of Au(III) and in situ generation of Au NPs	Fischer dialkynyl-alcoxy carbene complex anchored to a silanized glass surface	CuAAC	[164]
100	Redox chemistry	Ferrocene redox functionality immobilization onto self-assembled monolayers on gold surface	DNAzyme-catalyzed CuAAC	[165]
101	Redox chemistry	Ferrocene derivative on gold electrode surface	Electro-assisted CuAAC	[166]
102	Hydrolytic kinetic resolution of epoxides	Benzyloxycalix[8]arene-supported Co-salen catalyst	CuAAC	[167]

* Tested in CFC systems.

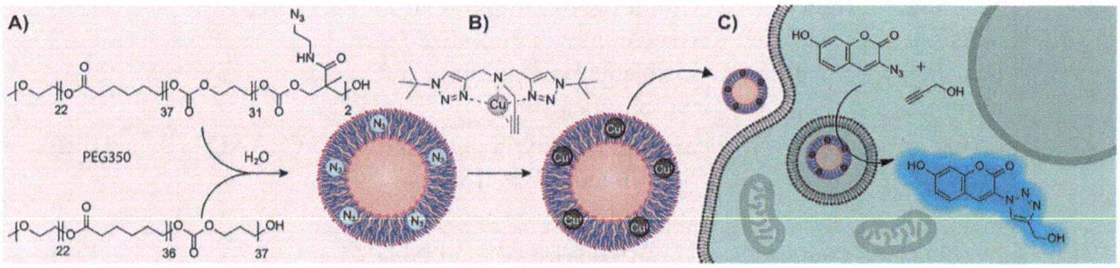

Fig. 4 Schematic overview of the production and intracellular activity of polymersomes with a covalently attached Cu catalyst. (**a**) A mixture of two copolymers self-assembles into polymersomes with azides in the hydrophobic domain. (**b**) The polymersomes are covalently loaded with a copper catalyst, whereby CuAAC supports the chelation. (**c**) After cellular uptake of the nanoreactors, a fluorescent product is formed by intracellular CuAAC, serving as an efficient indicator. (Figure taken from Oerlemans et al. [151])

promoting various transformations, gained prominence when Benjamin List and David MacMillan received the 2021 Nobel Prize in Chemistry for their work in developing asymmetric organocatalysis [169]. Organocatalysts are particularly attractive for continuous flow applications due to their tunable reactivity, robustness, operational simplicity, and metal-free nature. However, their use as homogeneous catalysts presents high catalyst loading, slow kinetics, difficult catalyst separation and recycling, and contamination of the product. Covalent immobilization of organocatalysts is an effective solution to these issues, for batch as well as CFC systems [47].

Among the various immobilization techniques (see Sect. 1, Fig. 2), CuAAC has become prominent because of its mild conditions, high efficiency, modularity, and broad functional group tolerance [17]. Many published studies report the application of CuAAC to attach organocatalysts to a wide range of solid supports including polymer resins, silica particles, dendrimers, and reactor surfaces, enabling effective heterogenization of homogeneous catalysts for use in batch and flow systems. This approach for the immobilization of organocatalysts enables the incorporation of stable triazole linkers, which not only serve as robust covalent bridges, but can also modulate the electronic properties and spatial orientation of the catalyst in the microenvironment of the support. Moreover, in some cases, the triazole motif can act as a functional group that enhances substrate-catalyst interactions or cooperatively contributes to catalytic activity [109]. However, in certain cases, particularly with catalysts containing H-bond donors or Brønsted acids, the triazole can be detrimental [170]. With specific catalysts or substrates, copper residues from the CuAAC reaction can also lead to complications. Additional issues may arise in specific applications, e.g., drug synthesis, where metal contamination is particularly problematic. It is therefore vital to consider the properties of the catalyst, the specific reaction conditions, and the intended application when developing an immobilization strategy.

When integrated into continuous flow systems, click-immobilized organocatalysts have enabled a variety of stereoselective and chemoselective transformations with increased throughput, minimized waste, and improved reaction control. Continuous flow reactors with immobilized catalysts have demonstrated reduced residence times, extended catalyst lifetimes, and scalability for the production of fine chemicals and pharmaceuticals [171–173].

Among the earliest examples of immobilizing organocatalysts on solid supports using click chemistry, two notable studies from 2006 stand out. In Prof. Oliver Reiser's group, an immobilized TEMPO derivative was synthesized on a functionalized Merrifield polystyrene support by CuAAC (Scheme 1a) [174]. The

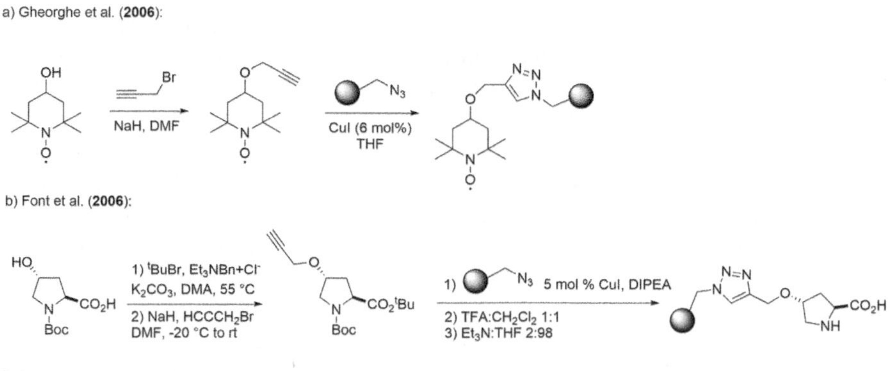

Scheme 1 Immobilization of: (**a**) TEMPO, and (**b**) a 4-hydroxyproline organocatalyst onto a polystyrene support via CuAAC (schemes adapted from Gheorghe et al. [174] and Font et al. [175])

resulting catalyst enabled the efficient oxidation of alcohols to aldehydes using bleach or molecular oxygen as the terminal oxidizing agent. This conversion proceeded with high yields and selectivity over multiple catalytic cycles without significant loss of activity.

At the same time, Miquel A. Pericàs group immobilized 4-hydroxyproline on a similar polystyrene support using the same CuAAC strategy (Scheme 1b) [175]. The resulting supported catalyst demonstrated excellent performance and recyclability in direct aldol reactions in aqueous media. A modified version of the catalyst showed characteristics targeted for an artificial aldolase [176] and was later adapted for continuous flow Mannich reactions [177], α-aminoxylation of aldehydes [178], and aldol reactions [179].

Numerous examples have been reported of click chemistry used to immobilize organocatalysts, including several adapted for use in flow systems [17]. The group of Professor Pericàs has made significant contributions in this area of research [170].

Table 2 summarizes studies featuring the immobilization of organocatalysts via click chemistry and their application in batch and flow systems. Comprehensive reviews of the studies listed below, including comparative analyses between each other and between batch and flow processes, are available in other review articles and book chapters [17, 168, 170, 171, 180].

3.1 Application of Click-Supported Catalysts in Continuous Flow Kinetic Resolution

The enantioselective separation of chiral molecules under continuous flow conditions is an emerging, increasingly important area in modern chemical synthesis, offering significant advantages in efficiency and scalability, especially when combined with supported catalysts that allow reusability. A very promising and interesting application of CuAAC-supported organocatalysts is kinetic resolution under continuous flow conditions. One of the main challenges in developing this strategy is the selectivity of the catalyst and its broad applicability [47]. Recent work in this area has explored various immobilized catalysts, particularly isothiourea-based Lewis bases, on supports such as polystyrene, using CuAAC for efficient anchoring of active sites. In this context, a series of studies from Prof. Pericàs' research group has demonstrated the fundamentals and potential of this approach. Neyyappadath et al. [202] first applied a polystyrene-supported isothiourea ('HyperBTM') catalyst for the acylative kinetic resolution of secondary alcohols in both batch and continuous flow processes. They achieved high selectivity (selectivity factor up to 600), broad substrate scope, and robust reusability of the catalyst over 15 cycles with minimal loss of performance (Fig. 5a).

Guha et al. [203] extended this strategy to tertiary heterocyclic alcohols under batch and flow conditions (Fig. 5b). Importantly, this study also introduced the use of industrially preferred solvents (e.g., dimethyl carbonate) and evaluated various

Table 2 Examples of organocatalysts immobilized using click chemistry in batch and CFC systems

Entry	Process	Organocatalyst/support	Click chemistry	Type of reactor	Reference
1	Stereoselective Mannich reaction of aldehydes and ketones with the *N*-(*p* methoxyphenyl) ethyl glyoxylate imine	(2*S*,4*R*)-hydroxyproline immobilized onto polystyrene	CuAAC	Omnifit glass chromatography column (packed-bed reactor)	[177]
2	Enantioselective domino Michael–Knoevenagel reactions of dimethyl 3-oxoglutarate and 3-substituted acrolein derivatives to form highly functionalized cyclohexane derivatives	Diarylprolinol silyl ethers immobilized onto polystyrene	CuAAC	Omnifit glass chromatography column (packed-bed reactor)	[181]
3	Enantioselective α-aminoxylation of aldehydes	Polystyrene-supported hydroxyproline	CuAAC	Omnifit glass chromatography column (packed-bed reactor)	[178]
4	Asymmetric aldol reactions	Polystyrene-supported proline derivative	CuAAC	Omnifit glass chromatography column (packed-bed reactor)	[179]
5	Enantioselective α-amination of aldehydes	Polystyrene-supported diphenylprolinol silyl ethers	CuAAC	Glass chromatography column (packed-bed reactor)	[182]
6	Asymmetric Michael reactions	Polystyrene-supported 9-amino (9-deoxy)-*epi* quinine derivative	CuAAC	Teflon® tube (packed-bed reactor)	[183]
7	Enantioselective Michael addition of 2-hydroxy-1,4-naphthoquinone to nitroalkenes	Polystyrene-supported bifunctional squaramide	CuAAC	Glass chromatography column (packed-bed reactor)	[184]
8	Asymmetric anti-Mannich reaction of aldehydes and ketones	Polystyrene-supported pyrrolidine-based catalyst	CuAAC	Omnifit glass chromatography column (packed-bed reactor)	[185]
9	Stereoselective Diels–Alder reactions between cyclopentadiene and α,-β-unsaturated aldehydes	Silica nanoparticle-supported enantiomerically pure imidazolidinone (MacMillan catalyst)	CuAAC	Stainless-steel column (packed-bed reactor)	[186]
10	Intermolecular enantioselective alkylation of aldehydes	Polystyrene-supported chiral imidazolidinone (MacMillan catalyst)	CuAAC	Stainless steel HPLC column (packed-bed reactor)	[187]

(continued)

Table 2 (continued)

Entry	Process	Organocatalyst/support	Click chemistry	Type of reactor	Reference
11	Stereoselective Diels–Alder reactions, 1,3-dipolar cycloadditions and Friedel–Crafts alkylations	Chiral imidazolidinone immobilized in a monolithic reactor	CuAAC	Stainless steel column monolithic reactor	[188]
12	Enantioselective three-component anti-Mannich reactions	Polystyrene-supported threonine derivative	CuAAC	Glass chromatography column (packed-bed reactor)	[189]
13	Enantioselective cyclopropanation of enals with dimethyl bromomalonate	Polystyrene-supported diarylprolinol (Jørgensen–Hayashi) catalysts	CuAAC	Omnifit glass chromatography column (packed-bed reactor)	[190]
14	Asymmetric Michael addition of cyclohexanone to nitrostyrenes in aqueous phase	Chiral pyrrolidine-functionalized polyacrylonitrile fiber catalysts	CuAAC	Silicone column (packed-bed reactor)	[191]
15	Enantioselective domino Michael addition/cyclization reactions of in situ activated arylacetic acids with chalcone-type tosylimines to dihydropyridinones	Polystyrene-supported, enantiopure benzotetramisole (BTM) analogue	CuAAC	Omnifit glass chromatography column (packed-bed reactor)	[192]
16	[4+2] Cycloaddition of unsaturated heterocycles with in situ-activated arylacetic acids	Polystyrene-supported isothiourea	CuAAC	Omnifit glass chromatography column (packed-bed reactor)	[193]
17	Catalytic reduction of imines with trichlorosilane	Polystyrene-supported chiral imidazolidinones	CuAAC	Stainless steel HPLC column (packed-bed reactor)	[194]
18	Enantioselective domino reaction between α,β-unsaturated aldehydes and N-protected hydroxylamines	Polystyrene-supported cis-4-hydroxydiphenylprolinol silyl ethers	CuAAC	Omnifit glass chromatography column (packed-bed reactor)	[195]
19	Solvent-free 4-fluorocinnamaldehyde-dimethyl malonate asymmetric conjugate addition	Polystyrene-supported cis-4 hydroxydiphenylprolinol TBS ether	CuAAC	Omnifit glass chromatography column (packed-bed reactor)	[196]

20	Conjugate additions of nitromethane to α,β-unsaturated aldehydes for multigram synthesis of chiral γ-nitrobutyric acids, key intermediates of GABA analogues baclofen, phenibut, and fluorophenibut	Polystyrene-supported cis-4-hydroxydiphenylprolinol TBS ether	CuAAC	Omnifit glass chromatography column (packed-bed reactor)	[197]
21	Oxidation of alcohols into their corresponding aldehydes	TEMPO immobilized onto inverse-opal structured membranes	CuAAC	Membranes assembled into home-made filtration system equipped with a pressurized filtration syringe holder	[198]
22	Esterification of 1-phenylethanol to phenylethylacetate	DMAP immobilized onto porous silica particles and monoliths	CuAAC	Monolithic reactor and packed-bed columns	[199]
23	Addition reaction of *n*-butanal and ß-nitrostyrene to form γ-nitroaldehyde	Peptide-based organocatalyst immobilized onto mesoporous silica	IEDDA ligation	Batch and flow reactor (either a stainless steel column filled with Sil-P10/30-Cat as packed-bed reactor or a functionalized silica monolith)	[200]
24	Asymmetric Diels–Alder reaction	Soluble polymer-bound MacMillan catalysts	Sulfur(VI) fluoride exchange (SuFEx)-click approach	Batch reaction, with solid-liquid biphasic separation for recycling of MacMillan catalyst	[201]

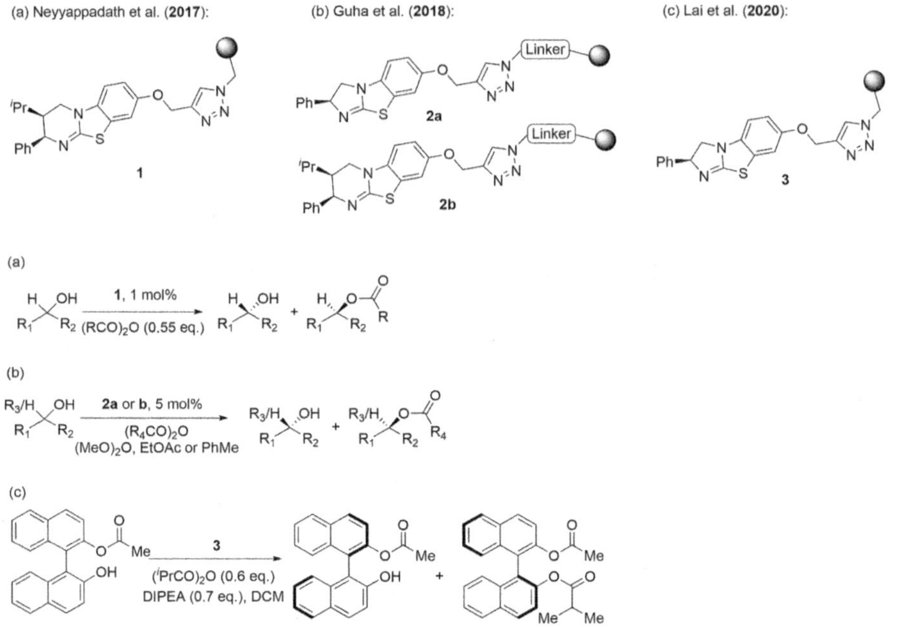

Fig. 5 Isothiourea-based catalysts **1**, **2a** and **b**, and **3**, featuring in studies reported by: (**a**) Neyyappadath et al. [202], (**b**) Guha et al. [203], and (**c**) Lai et al. [204]

polymer supports to optimize swelling and catalyst performance, demonstrating that high selectivity and durability could be maintained outside of traditionally used solvents like chloroform.

Lai et al. (2020) further advanced the field by successfully using immobilized isothiourea for the kinetic resolution of 1,1'-bi-2-naphthol (BINOL) derivatives, sterically complex and industrially relevant materials. The separation was demonstrated in continuous flow operation, achieving scalability and consistent catalyst performance (Fig. 5c) [204].

In addition, click chemistry has also been used to immobilize organic molecules on solid supports, which can then serve as stationary phases in various applications, such as affinity chromatography [205], μ-HPLC separation [206], and chiral HPLC enantiomer separation [207].

Overall, these advances highlight the increasing utility of click-immobilized catalysts for continuous flow kinetic resolution and emphasize the critical role of modular catalyst design, support selection, and process conditions in achieving scalable, enantioselective transformations. Despite the promise of modularity, the field remains complex due to the wide variety of catalytic systems, support materials, and flow reactor configurations, all of which require systematic evaluation to optimize overall performance.

4 Click Chemistry for Immobilization of Biocatalysts

Enzymes are essential to many industrial processes because they serve as "biocatalysts," enabling the use of milder conditions in highly efficient and selective chemical reactions such as asymmetric synthesis [208], while also promoting sustainability [209]. Examples of industries that benefit from enzymes include the fine and bulk chemical, pharmaceutical, food, cosmetic, textile, and pulp and paper industries [210]. To improve their stability and in some cases, their activity and selectivity [211], enzymes can be immobilized. This also facilitates their recovery and reuse in subsequent reactions. Several approaches to enzyme immobilization exist; we recommend excellent reviews by Maghraby et al. [45], Bolivar et al. [212], and Mirsalami et al. [213]

Enzymes have been immobilized using click chemistry onto solid carriers or particles, within hydrogels ("encapsulation"), and onto surfaces. Table 3 summarizes of select enzymes immobilized through a these three approaches.

As far as we know, there is *only one* report of enzymes immobilized via click chemistry in continuous flow systems (Table 3, entry 6) [218]. In this case, α-chymotrypsin (CT) was covalently attached to a monolith through a "click" triazole ring formation. The monolith **6** was prepared by the thermal copolymerization of glycidyl methacrylate (GMA) **4** and ethylene glycol dimethacrylate (EGDMA) **5** in a fused silica capillary using *n*-propanol/1,4-butanediol as a porogen (i.e., to create pores as the molded structure forms) (Scheme 2a). Product **6** was then treated with (3-chloropropyl)trimethoxysilane **7** to form monolith **8** (Scheme 2b). **8** was treated with sodium azide (Scheme 2c) before azide-functionalized **9** reacted with the enzyme (CT) bearing alkyne functionalities **10**, attached by water-soluble *N*-(3-dimethylaminopropyl)-*N*-ethylcarbodiimide hydrochloride (EDCl)-activation to form an amide bond in **10** (Scheme 2d). Finally, functionalized enzyme **10** was linked to monolith **9** through Cu-catalyzed triazole-ring formation, forming CT-encapsulated gel **11** (Scheme 2e).

The derivatization of the monolith was monitored using FTIR, and the final monolith (with covalently attached enzyme) was characterized using scanning electron microscopy (SEM) to verify its attachment to the capillary and to determine the size of any fixed aggregates in the structure. The monolith was also analyzed by mercury porosimetry to assess the pore size distribution.

The activity of the immobilized enzyme reactor (IMER) was studied via microliquid chromatography by observing the conversion of BTEE to BT (with ethanol as a by-product) (Scheme 2f) under different mobile-phase flow rates and substrate feed concentrations. For the monolithic IMER, the final substrate conversion increased with higher feed flow rates and substrate feed concentrations, attributed to convective diffusion in the macropores of the monolith.

The researchers compared the monolithic IMER with particle-based IMER (formed by attaching CT onto 7.5 μm monodisperse and porous poly(GMA-*co*-EGDMA) particles via click chemistry; the particles were prepared by seeded copolymerization of GMA and EGDMA [229]) and found that the monolithic

Table 3 Select examples of enzymes immobilized using click chemistry in batch and CFC systems

Entry	Process	Enzyme, immobilization strategy	Click chemistry	Reference
Carrier- or particle-bound immobilization				
1	Dimerization of phenylacetylene	Laccase on magnetic NPs	CuAAC	[214]
2	Breakdown of pectin in apple juice	Pectinase on polyethyleneimine based support	Amino-yne click reaction	[215]
3	Grafted ATP derivative used in the synthesis of phosphopantetheine	Adenosine derivatives (precursor to many cofactors) on cellulose nanocrystals	CuAAC	[216]
4	Fluorometric lipase activity assay: cleavage of nonfluorescent ester, 5-O-palmitoyl indole to a highly fluorescent 5-OH-indole product (λmax 333 nm)	*Thermomyces lanuginosus* lipase has been attached to azide-functionalized water-soluble gold nanoparticles	CuAAC	[217]
5	Conversion of Nα-p-tosyl-l-arginine methyl ester hydrochloride to Nα-p-tosyl-l-arginine	Trypsin attached to the mesoporous system of large pore SBA-15	CuAAC	[26]
6*	Hydrolysis of BTEE to BTT (EtOH as side product) (Scheme 2f)	α-Chymotrypsin was attached to monodisperse-porous poly(GMA-co-EDMA) particles and used in a flow microreactor	CuAAC	[218]
Encapsulation				
7	Biocatalytic oxidation of glucose to gluconolactone (gluconate), and small electron-transferring molecules deliver an electron from glucose to the electrode	Glucose dehydrogenase immobilization on an electrode	SPAAC (sortase A-mediated click ligation)	[219]
8	Hydrolysis of BTEE to BTT (EtOH as side product) (Scheme 2f)	α-Chymotrypsin was attached to a porous poly (GMA-*co*-EDMA) monolith and used in a flow microreactor	CuAAC	[218]
Surface immobilization				
9	Electrocatalytic reduction of H_2O_2	Horseradish peroxidase on an electrode	CuAAC	[220]
10	Electrocatalytic reduction of H_2O_2	Horseradish peroxidase on electrografted 4-ethynylphenyl film	CuAAC	[221–223]

(continued)

Table 3 (continued)

Entry	Process	Enzyme, immobilization strategy	Click chemistry	Reference
11	Oxidation reactions coupled to reduction of O_2 to H_2O	Laccase on a multiwalled carbon nanotube electrode, with linkers of different lengths, enabling modulation of the enzyme-electrode distance	SPAAC	[224]
12	Electroenzymatic reduction of D-fructose to D-sorbitol	Co-immobilization of NAD-dependent dehydrogenase and molecular electrocatalyst [Cp*Rh (bpy)Cl]$^+$	CuAAC Huisgen reaction	[146]
13	Lactase converts lactose to glucose; Gox converts the glucose to H_2O_2, then used to form a red chinonimine dye (also from phenol and 4-aminoantipyrine)	Glucose oxidase and lactase on alkene-terminated self-assembled monolayers on glass	Photochemical thiol-ene click chemistry	[225]
14	Amylase breaks down starch into simple sugars	Amylase on PEG-maleimide film without any linker	Thiol-ene click reaction	[226]
15	Reactivity of carbonic anhydrase (CA), releases *p*-nitrophenol	CA immobilized on silicon wafers/modified microparticles	IEDDA	[227]
16	Immobilized β-1,4-galactosyltransferase	Immobilized β-1,4-galactosyltransferase	Supramolecular/ non-covalent "click-like" pairs (streptavidin–biotin, host–guest)	[228]

* Tested in CFC system.

IMER had advantages over particle-based conventional high-performance LC–IMERs: smoother fluid flow, lower pressure drop, and efficient mass transfer between phases [230].

5 Summary and Future Implications

Click chemistry offers high efficiency, selectivity, and mild conditions for the immobilization of catalysts (metal complexes, organocatalysts, and biocatalysts) on various supports such as dendrimers, polymers, nanoparticles, and silica. Covalent bonding via CuAAC leads to chemically and thermally stable triazole bonds that can limit leaching and even improve catalytic performance and is particularly attractive for CFC systems. Successful applications include the many examples

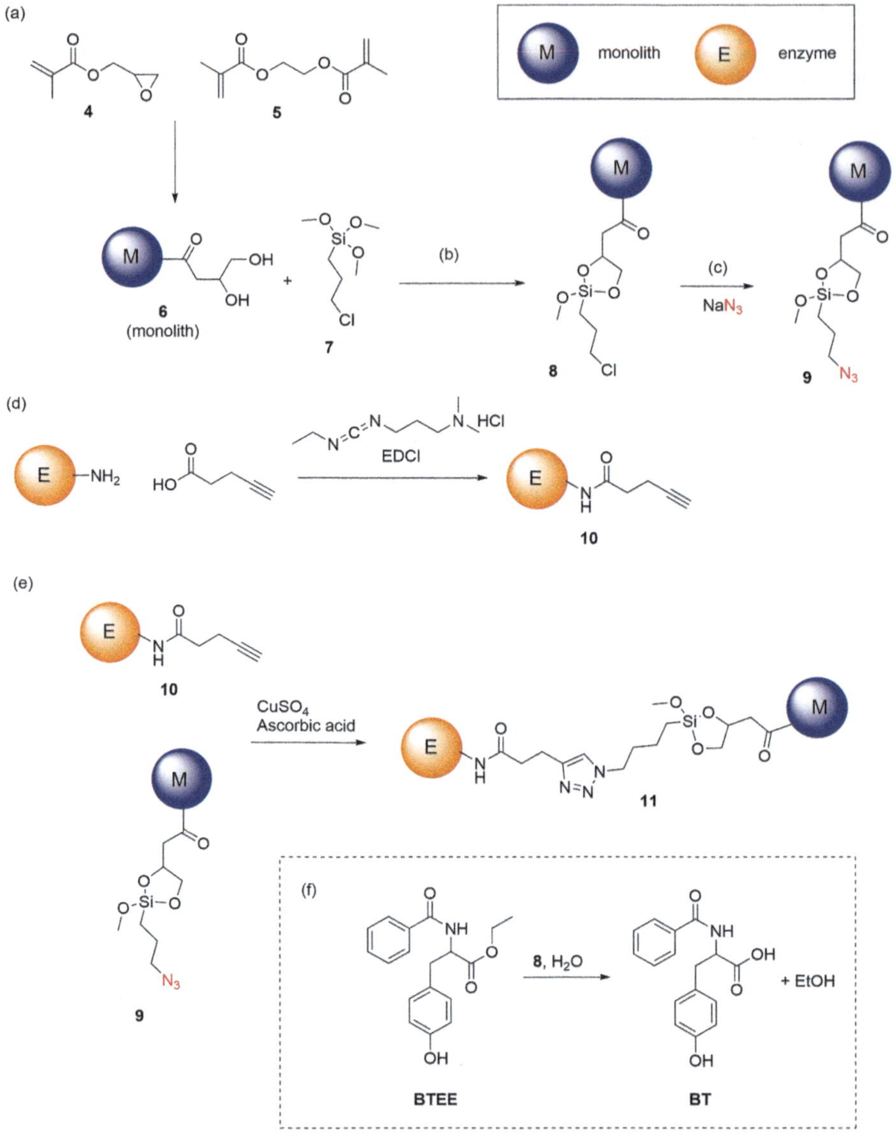

Scheme 2 (**a**) Thermal copolymerization of **4** and **5** to create the initial monolith **6**; (**b**) the reaction of **6** with (3-chloropropyl)trimethoxysilane **7** to form **8**; (**c**) azide formation, resulting in **9**; (**d**) carbodiimide-activated amide formation during the functionalization of enzyme amines with alkyne groups, to form **10**; (**e**) covalent attachment of CT **10** to monolith **9** through triazole ring formation via click chemistry, to form **11**; (**f**) the behavior of the microreactor was evaluated by monitoring the conversion of *N*-benzoyl-L-tyrosine ethyl ester (BTEE) to *N*-benzoyl-L-tyrosine BT, with ethanol as by-product (Scheme adapted from Çelebi et al. [218])

shown in Tables 1, 2, and 3. These advances highlight the versatility of click chemistry for the development of robust, modular catalytic platforms while pointing to several promising directions for the further development of click chemistry-based immobilization strategies in continuous flow systems.

There is a need for the further development of copper-free or low-copper click alternatives (e.g., strain-promoted or photochemical cycloadditions) to avoid residual metal contamination and toxicity issues, especially for pharmaceutical and biological production. Alternative metal-free click reactions are also gaining momentum due to their potential advantages in terms of biocompatibility and environmental impact [65, 231–233]. Reported alternative approaches used for (bio)catalyst immobilization within CFC systems include SPAAC (Table 1, entry 85, and Table 3, entries 7 and 11), thiol-ene click (Table 1, entry 67, Table 3, entries 13 and 14), thiol-yne click (Table 1, entry 75), amino-yne click (Table 3, entry 1), photoclick (Table 1, entry 83), IEDDA (Table 2, entry 15; Table 3, entry 15), (SuFEx)-click (Table 2, entry 24), and supramolecular streptavidin–biotin, host–guest chemistry (Table 3, entry 16). For further information concerning these alternative approaches, please refer to comprehensive reviews by Upadhyay et al. [234] (SPAAC), Knall and Slugovc [235] and Tomarchio et al. [236] (IEDDA), Fairbanks et al. [237] (photoclick), Hoyle and Bowman [238] (thiol–ene click)/Lowe et al. [239] (thiol–yne click), Dong et al. [240] (SuFEx-click), and McGuire et al. [241] (supramolecular streptavidin–biotin, host–guest chemistry). Other alternative approaches not yet studied for catalyst immobilization but worth further exploration to expand the toolkit in this area include the following:

- Strain-promoted alkyne-nitrone cycloaddition (SPANC) [242]
- Strain-promoted sydnone-alkyne cycloaddition (SPSAC) [243, 244]
- Oxime/hydrazone ligation (and aniline-catalyzed variants) [245]
- Nitrile-imine-mediated tetrazole "photo-click" (NITEC) [246]
- Diazaborine formation/boron-based bioorthogonal reactions [247]

These significant literature gaps are worthy of exploration, since each might offer different advantages over the other for certain catalyst/substrate systems.

Other potentially high-impact research directions include the following:

- Development of stimulatory or cleavable linkers that enable reversible immobilization or controlled release of catalysts in flow reactors to support on-demand regeneration and dynamic process control [248–250].
- Development of triazole-based linkers with tailored steric and electronic properties to reduce interference with adjacent chiral ligands and improve stereoselectivity in asymmetric catalytic transformations.
- Optimization and diversification of support materials, including hierarchical porous frameworks, 3D printed structures, MOFs/COFs, and biodegradable or biocompatible substrates, e.g., using biopolymers as catalyst supports, to promote circular chemistry [251], with pre-functionalized click handles for efficient and sustainable immobilization.

- Integration of modular, orthogonal click strategies that enable site-specific, sequential immobilization of multiple catalysts on a single support to facilitate advanced cascade and cooperative reactions in flow systems.
- Despite there being a plethora of alternative approaches to carrier-free enzyme immobilization [252], self-cross-linking of enzymes (i.e., carrier-free immobilization) via click chemistry could broaden the field and open new opportunities.
- With only one example of an immobilized enzyme in flow chemistry, there are ample opportunities for investigating such systems. Since click chemistry has been used to immobilize enzymes in batch systems (Table 3), translating these to continuous flow systems would be the next step. Such an integration could enhance the efficiency and stability of biocatalytic processes in flow applications.

We note a recent review by Fernandes et al. [253] published at the same time as our book chapter, and point readers to this useful resource.

Acknowledgements We acknowledge financial support from the Slovenian Research and Innovation Agency (ARIS) (Research Core Funding Grant No. P1-0230 and Project #J7-50041).

References

1. Bergbreiter DE, Tian J, Hongfa C (2009) Chem Rev 109:530–582. https://doi.org/10.1021/cr8004235
2. Dickerson TJ, Reed NN, Janda KD (2002) Chem Rev 102:3325–3344. https://doi.org/10.1021/cr010335e
3. van de Coevering R, Klein Gebbink RJM, van Koten G (2005) Prog Poly Sci 30:474–490. https://doi.org/10.1016/j.progpolymsci.2005.01.002
4. Jedlovčnik L, Höfferle J, Gaber A, Pashaj F, Kladnik J, Košmrlj J, Derdau V, Jansen-van Vuuren RD (2025) Green Synt Catal 6:1–35. https://doi.org/10.1016/j.gresc.2024.09.008
5. Piermatti O, Abu-Reziq R, Vaccaro L (2020) Strategies to immobilized catalysts: a key tool for modern chemistry. In: Benaglia M, Puglisi A (eds) Catalyst immobilization: methods and applications. Wiley-VCH, Verlag GmbH & Co. KGaA. https://doi.org/10.1002/9783527817290
6. Adeyeye NS, Ajeebiv AM, Alghamdi HS, Aziz A, Shaikh MN (2023) Asian J Org Chem 12:e202300051. https://doi.org/10.1002/ajoc.202300051
7. Lambert S, Job N, Souza LD, Fernando M, Pereira R, Pirard R, Heinrichs B, Figueiredo JL, Pirard J-P, Regalbuto JR (2009) J Catal 261:23–33. https://doi.org/10.1016/j.jcat.2008.10.014
8. Jiao L, Regalbuto JR (2008) J Catal 260:329–341. https://doi.org/10.1016/j.jcat.2008.09.022
9. Shao M, Jiang X, Wang A, Tan G, Wang L (2025) ChemSusChem 18:e202401903. https://doi.org/10.1002/cssc.202401903
10. Schwab E, Mecking S (2001) Organometallics 20:5504–5506. https://doi.org/10.1021/om0107542
11. Liu H-K, Nakama T, Funami S, Ebihara R, Mitsuhashi T, Fujita M (2025) Angew Chem Int Ed Early View:e202517027. https://doi.org/10.1002/anie.202517027
12. Gao C, Lyu F, Yin Y (2021) Chem Rev 121:834–881. https://doi.org/10.1021/acs.chemrev.0c00237
13. Imam HT, Marr PC, Marr AC (2021) Green Chem 23:4980–5005. https://doi.org/10.1039/D1GC01852C

14. Moore BL, Moatsou D, Lua A, O'Reilly RK (2014) Polym Chem 5:3487–3494. https://doi.org/10.1039/C3PY01734F
15. Prabhakar T, Giaretta J, Zulli R, Rath RJ, Farajikhah S, Talebian S, Dehghani F (2025) Chem Eng J 503:158054. https://doi.org/10.1016/j.cej.2024.158054
16. Narimani AZ, Landarani-Isfahani A, Bahadori M, Moghadam M, Tangestaninejad S, Mohammadpoor-Baltork I, Mirkhani V (2025) ACS Appl Bio Mater 8:5067–5077. https://doi.org/10.1021/acsabm.5c00321
17. Fernandes AE, Jonas AM, Riant O (2014) Tetrahedron 70:1709–1731. https://doi.org/10.1016/j.tet.2013.12.034
18. Jesionowski T, Zdarta J, Krajewska B (2014) Adsorption 20:801–821. https://doi.org/10.1007/s10450-014-9623-y
19. Cantillo D, Kappe CO (2014) ChemCatChem 6:3286–3305. https://doi.org/10.1002/cctc.201402483
20. Chen X, Fu M, He J, Lu J, Wang C, Jiang D, Li Y, Yu C (2025) Coord Chem Rev 538:216723. https://doi.org/10.1016/j.ccr.2025.216723
21. Lin G, Qiu H (2022) Chem Eur J 28:e202200069. https://doi.org/10.1002/chem.202200069
22. Kánya N, Zsigmond TS, Hergert T, Lövei K, Dormán G, Kálmán F, Darvas F (2024) Org Process Res Dev 28:1288–1307. https://doi.org/10.1021/acs.oprd.3c00364
23. Ötvös SB, Fülöp F (2015) Cat Sci Technol 5:4926–4941. https://doi.org/10.1039/C5CY00523J
24. de Sá DS, Bustamante RA, Rocha CER, da Silva VD, Rodrigues EJR, Müller CDB, Ghavami K, Massi A, Pandoli OG (2019) ACS Sustain Chem Eng 7:3267–3273. https://doi.org/10.1021/acssuschemeng.8b05273
25. Wu M, Zhang H, Wang Z, Shen Z, Le XC, Li X-F (2013) Chem Commun 49:1407–1409. https://doi.org/10.1039/C2CC37974K
26. Schlossbauer A, Schaffert D, Kecht J, Wagner E, Bein T (2008) J Am Chem Soc 130:12558–12559. https://doi.org/10.1021/ja803018w
27. Guo Z, Lei A, Liang X, Xub Q (2006) Chem Commun 43:4512–4514. https://doi.org/10.1039/B610733H
28. Slater MD, Fréchet JMJ, Svec F (2009) J Sep Sci 32:21–28. https://doi.org/10.1002/jssc.200800493
29. Wiles C, Watts P (2008) Eur J Org Chem 2008:1655–1671. https://doi.org/10.1002/ejoc.200701041
30. Dallinger D, Kappe CO (2017) Curr Opinion Green Sus Chem 7:6–12. https://doi.org/10.1016/j.cogsc.2017.06.003
31. Hessel V, Mukherjee S, Mitra S, Goswami A, Tran NN, Ferlin F, Vaccaro L, Galogahi FB, Nguyen NT, Escribà-Gelonch M (2024) Green Chem 26:953–9528. https://doi.org/10.1039/D4GC01882F
32. Baumann M, Baxendale IR (2015) Beilstein J Org Chem 11:1194–1219. https://doi.org/10.3762/bjoc.11.134
33. Wilms D, Klos J, Frey H (2008) Macromol Chem Phys 209:343–356. https://doi.org/10.1002/macp.200700588
34. Martins GM, Braga FC, de Castro PP, Brocksom TJ, de Oliveira KT (2024) Chem Commun 60:3226–3239. https://doi.org/10.1039/D4CC00418C
35. Porta R, Benaglia M, Puglisi A (2016) Org Process Res Dev 20:2–25. https://doi.org/10.1021/acs.oprd.5b00325
36. Kamio S, Okamoto K, Yamagishi T, Nagaki A (2024) ChemPlusChem 89:e202300744. https://doi.org/10.1002/cplu.202300744
37. Minamihara H, Kusada K, Yamamoto T, Toriyama T, Murakami Y, Matsumura S, Kumara LSR, Sakata O, Kawaguchi S, Kubota Y, Seo O, Yasuno S, Kitagawa H (2023) J Am Chem Soc 145:17136–17142. https://doi.org/10.1021/jacs.3c03713

38. Kulkarni AA, Jundale RB (2019) Chapter 9: Continuous flow synthesis of nanomaterials. In: Luis SV, Garcia-Verdugo EE (eds) flow chemistry: integrated approaches for practical applications. Royal Society of Chemistry. https://doi.org/10.1039/9781788016094-00316
39. Lukas TJ, Prystowsky MB, Erickson BW (1981) Proc Nati Acad Sci USA 78:2791–2795. https://doi.org/10.1073/pnas.78.5.2791
40. Hartrampf N, Saebi A, Poskus M, Gates ZP, Callahan AJ, Cowfer AE, Hanna S, Antilla S, Schissel CK, Quartararo AJ, Ye X, Mijalis AJ, Simon MD, Loas A, Liu S, Jessen C, Nielsen TE, Pentelute BL (2020) Science 368:980–987. https://doi.org/10.1126/science.abb2491
41. Hone CA, Kappe CO (2021) Chem Methods 1:454–467. https://doi.org/10.1002/cmtd.202100059
42. Tu NP, Sarris K, Djuric SW (2015) RSC Adv 5:4754–4757. https://doi.org/10.1039/C4RA13931C
43. Gürsel IV, Noël T, Wanga Q, Hessel V (2015) Green Chem 17:212–226. https://doi.org/10.1039/C4GC02160F
44. Kirschning A, Jas G (2004) Top Curr Chem 242:209–239. https://doi.org/10.1007/b96877
45. Maghraby YR, El-Shabasy RM, Ibrahim AH, Azzazy HMES (2023) ACS Omega 8:5184–5196. https://doi.org/10.1021/acsomega.2c07560
46. Coloma J, Guiavarc'h Y, Hagedoorn P-L, Hanefeld U (2021) Chem Commun 57:11416–11428. https://doi.org/10.1039/D1CC04315C
47. Colella M, Carlucci C, Luisi R (2018) Top Curr Chem 376:6. https://doi.org/10.1007/s41061-018-0225-0
48. Cozzi F (2006) Adv Synth Catal 348:1367–1390. https://doi.org/10.1002/adsc.200606096
49. Hong J, Zaera F (2012) J Am Chem Soc 134:13056–13065. https://doi.org/10.1021/ja304181q
50. Fernandes A, Hensenne P, Mathy B, Guo W, Nysten B, Jonas AM, Riant O (2012) Chem Eur J 18:788–792. https://doi.org/10.1002/chem.201102245
51. Lim J, Lee SS, Ying JY (2008) Chem Commun 2008:4312–4314. https://doi.org/10.1039/B803663B
52. Pugin B, Blaser H-U (2010) Top Catal 53:953–962. https://doi.org/10.1007/s11244-010-9514-8
53. Fernandes AE, Dirani A, d'Haese C, Deumer G, Guo W, Hensenne P, Nahra F, Laloyaux X, Haufroid V, Nysten B, Riant O, Jonas AM (2012) Chem Eur J 18:16226–16233. https://doi.org/10.1002/chem.201202531
54. Brunelli NA, Didas SA, Venkatasubbaiah K, Jones CW (2012) J Am Chem Soc 134:13950–13953. https://doi.org/10.1021/ja305601g
55. Nakazawa J, Smith BJ, Stack TDP (2012) J Am Chem Soc 134:2750–2759. https://doi.org/10.1021/ja210400u
56. Fernandes AE, Devillez S, d'Haese C, Deumer G, Haufroid V, Nysten B, Riant O, Jonas AM (2012) Langmuir 28:14822–14828. https://doi.org/10.1021/la3025292
57. Nakazawa J, Hori T, Stack TDP, Hikichi S (2013) Chem Asian J 8:1191–1199. https://doi.org/10.1002/asia.201300165
58. Nakazawa J, Yata A, Hori T, Stack TDP, Naruta Y, Hikichi S (2013) Chem Lett 42:1197–1199. https://doi.org/10.1246/cl.130464
59. Zaera F (2022) Chem Rev 122:8594–8757. https://doi.org/10.1021/acs.chemrev.1c00905
60. Liang L, Astruc D (2011) Coord Chem Rev 255:2933–2945. https://doi.org/10.1016/j.ccr.2011.06.028
61. Lu J, Toy PH (2009) Chem Rev 109:815–838. https://doi.org/10.1021/cr8004444
62. Itsuno S, Hassan MM (2014) RSC Adv 4:52023–52043. https://doi.org/10.1039/C4RA09561H
63. Cheng T, Zhao Q, Zhanga D, Liu G (2015) Green Chem 17:2100–2122. https://doi.org/10.1039/C4GC02204A
64. Tăbăcaru A, Furdui B, Ghinea IO, Cârâc G, Dinică RM (2017) Inorg Chim Acta 455:329–349. https://doi.org/10.1016/j.ica.2016.07.029

65. Escorihuela J, Marcelis ATM, Zuilhof H (2015) Adv Mater Interfaces 2:1500135. https://doi.org/10.1002/admi.201500135
66. Sadjadi S, Heravi MM (2017) RSC Adv 7:3815–3838. https://doi.org/10.1039/C7RA04833E
67. Navalón S, Herance JR, Álvaro M, García H (2017) Chem Eur J 23:15244–15275. https://doi.org/10.1002/chem.201701028
68. Campisciano V, Gruttadauria M, Giacalone F (2019) ChemCatChem 11:90–133. https://doi.org/10.1002/cctc.201801414
69. Ivanytsya MO, Ryabukhin SV, Volochnyuk DM, Kolotilov SV (2020) Theor Exp Chem 56: 283–308. https://doi.org/10.1007/s11237-020-09660-4
70. Ding S, Motokura K (2023) Curr Opin Green Sustainable Chem 40:100753. https://doi.org/10.1016/j.cogsc.2023.100753
71. Gissibl A, Padié C, Hager M, Jaroschik F, Rasappan R, Cuevas-Yañez E, Turrin C-O, Caminade A-M, Majoral J-P, Reiser O (2007) Org Lett 9:2895–2898. https://doi.org/10.1021/ol071128p
72. Meise M, Haag R (2008) ChemSusChem 1:637–642. https://doi.org/10.1002/cssc.200800042
73. Janssen M, Müller C, Vogt D (2009) Adv Synth Catal 351:313–318. https://doi.org/10.1002/adsc.200900058
74. Wang D, Denux D, Ruiz J, Astruc D (2013) Adv Synth Catal 355:129–142. https://doi.org/10.1002/ADSC.201200619
75. Sheet D, Bera A, Fu Y, Desmecht A, Riant O, Hermans S (2019) Chem Eur J 25:9191–9196. https://doi.org/10.1002/chem.201901735
76. Ornelas C, Aranzaes JR, Salmon L, Astruc D (2008) Chem Eur J 14:50–64. https://doi.org/10.1002/chem.200701410
77. Deraedt C, Salmon L, Astruc D (2014) Adv Synth Catal 356:2525–2538. https://doi.org/10.1002/adsc.201400153
78. Wang Q, Fu F, Yang S, Moro MM, Ramirez MLA, Moya S, Salmon L, Ruiz J, Astruc D (2019) ACS Catal 9:1110–1119. https://doi.org/10.1021/acscatal.8b04498
79. Lukowiak MC, Meise M, Haag R (2014) Synlett 25:2161–2165. https://doi.org/10.1055/s-0034-1378533
80. Gissibl A, Finn MG, Reiser O (2005) Org Lett 7:2325–2328. https://doi.org/10.1021/ol0505252
81. Rasappan R, Olbrich T, Reiser O (2009) Adv Synth Catal 351:1961–1967. https://doi.org/10.1002/adsc.200900129
82. Khatri PK, Jain SL, Sivakumar KLN, Sain B (2011) Org Biomol Chem 9:3370–3374. https://doi.org/10.1039/C0OB01163K
83. Chinnusamy T, Rodionov V, Kühn FE, Reiser O (2012) Adv Synth Catal 354:1827–1831. https://doi.org/10.1002/adsc.201100289
84. Jain S, Reiser O (2008) ChemSusChem 1:534–541. https://doi.org/10.1002/cssc.200800025
85. Detz RJ, Arévalo Heras S, de Gelder R, van Leeuwen PWNM, Hiemstra H, Reek JNH, van Maarseveen JH (2006) Org Lett 8:3227–3230. https://doi.org/10.1021/ol061015q
86. Tilliet M, Lundgren S, Moberg C, Levacher V (2007) Adv Synth Catal 349:2079–2084. https://doi.org/10.1002/adsc.200700112
87. Popa D, Marcos R, Sayalero S, Vidal-Ferran A, Pericàs MA (2009) Adv Synth Catal 351: 1539–1556. https://doi.org/10.1002/adsc.200900163
88. He Y, Cai C (2011) Chem Commun 47:12319–12321. https://doi.org/10.1039/C1CC14898B
89. Salunke SB, Babu NS, Chen C-T (2011) Adv Synth Catal 353:1234–1240. https://doi.org/10.1002/adsc.201100062
90. Ozkal E, Özçubukçu S, Jimeno C, Pericàs MA (2012) Cat Sci Technol 2:195–200. https://doi.org/10.1039/C1CY00297J
91. Chan TR, Fokin VV (2007) QSAR Comb Sci 26:1274–1279. https://doi.org/10.1002/qsar.200740131
92. Lammens M, Skey J, Wallyn S, O'Reilly R, Du Prez F (2010) Chem Commun 46:8719–8721. https://doi.org/10.1039/C0CC01451F

93. Feng Y, Moschetta EG, Jones CW (2014) Chem Asian J 9:3142–3152. https://doi.org/10.1002/asia.201402461
94. Fan W, Shi D, Feng B (2016) Catal Commun 74:1–4. https://doi.org/10.1016/j.catcom.2015.10.022
95. Huang J, Liu S, Ma Y, Cai J (2019) J Organomet Chem 886:27–33. https://doi.org/10.1016/j.jorganchem.2019.02.008
96. Eltepu L, Jayaraman M, Rajeev KG, Manoharan M (2013) Chem Commun 49:184–186. https://doi.org/10.1039/C2CC36811K
97. Marquard AN, Slaymaker LE, Hamers RJ, Goldsmith RH (2017) Mol Catal 429:10–17. https://doi.org/10.1016/j.molcata.2016.12.001
98. Hajipour AR, Abolfathi P (2018) Appl Organomet Chem 32:e4273. https://doi.org/10.1002/aoc.4273
99. Fernandes AE, Ye Q, Collard L, Le Duff C, D'Haese C, Deumer G, Haufroid V, Nysten B, Riant O, Jonas AM (2015) ChemCatChem 7:856–864. https://doi.org/10.1002/cctc.201402913
100. Pucci A, Albano G, Pollastrini M, Lucci A, Colalillo M, Oliva F, Evangelisti C, Marelli M, Santalucia D, Mandoli A (2020) Catalysts 10:434. https://doi.org/10.3390/catal10040434
101. Coelho A, Diz P, Caamaño O, Sotelo E (2010) Adv Synth Catal 352:1179–1192. https://doi.org/10.1002/adsc.200900680
102. Abramov A, Maiti B, Reiser O, Díaz Díaz D (2021) Chem Commun 57:7762–7765. https://doi.org/10.1039/D1CC03020E
103. Rana BS, Jain SL, Singh B, Bhaumik A, Sain B, Sinha AK (2010) Dalton Trans 39:7760–7767. https://doi.org/10.1039/C0DT00208A
104. Jain SL, Rana BS, Singh B, Sinha AK, Bhaumik A, Nandi M, Sain B (2010) Green Chem 12:374–377. https://doi.org/10.1039/B917382J
105. Dickschat AT, Behrends F, Surmiak S, Weiß M, Eckert H, Studer A (2013) Chem Commun 49:2195–2197. https://doi.org/10.1039/C3CC00235G
106. McDonald AR, Dijkstra HP, Suijkerbuijk BMJM, van Klink GPM, van Koten G (2009) Organometallics 28:4689–4699. https://doi.org/10.1021/om900237g
107. McDonald AR, Franssen N, van Klink GPM, van Koten G (2009) J Organomet Chem 694:2153–2162. https://doi.org/10.1016/j.jorganchem.2009.02.020
108. Lim J, Seong Lee S, Ying JY (2010) Chem Commun 46:806–808. https://doi.org/10.1039/B917986K
109. Zhang G, Wang Y, Wen X, Ding C, Li Y (2012) Chem Commun 48:2979–2981. https://doi.org/10.1039/C2CC18023E
110. Ganai AK, Bhardwaj R, Hotha S, Gupta SS, Prasad BLV (2010) New J Chem 34:2662–2670. https://doi.org/10.1039/C0NJ00292E
111. Malvi B, Panda C, Dhar BB, Gupta SS (2012) Chem Commun 48:5289–5291. https://doi.org/10.1039/C2CC30970J
112. Hong J, Djernes KE, Lee I, Hooley RJ, Zaera F (2013) ACS Catal 3:2154–2157. https://doi.org/10.1021/cs400471b
113. Shen Y, Chen Q, Lou L-L, Yu K, Ding F, Liu S (2010) Catal Lett 137:104–109. https://doi.org/10.1007/s10562-010-0341-0
114. Park D-W, Choi S-D, Choi S-J, Lee C-Y, Kim G-J (2002) Catal Lett 78:145–151. https://doi.org/10.1023/A:1014945926583
115. Baleizão C, Gigante B, García H, Corma A (2003) J Catal 215:199–207. https://doi.org/10.1016/S0021-9517(03)00007-1
116. Gill CS, Venkatasubbaiah K, Jones CW (2009) Adv Synth Catal 351:1344–1354. https://doi.org/10.1002/adsc.200800779
117. Fernandes AE, Riant O, Jensen KF, Jonas AM (2016) Angew Chem Int Ed 55:11044–11048. https://doi.org/10.1002/anie.201603673
118. Chandra P, Jonas AM, Fernandes AE (2018) ACS Catal 8:6006–6011. https://doi.org/10.1021/acscatal.8b01399

119. Richezzi M, Signorella S, Palopoli C, Pellegri N, Hureau C, Signorella SR (2023) Inorganics 11:359. https://doi.org/10.3390/inorganics11090359
120. Richezzi M, Ferreyra J, Signorella S, Palopoli C, Terrestre G, Pellegri N, Hureau C, Signorella SR (2023) Inorganics 11:425. https://doi.org/10.3390/inorganics11110425
121. Li Z, Rösler L, Wissel T, Breitzke H, Gutmann T, Buntkowsky G (2021) J CO_2 Util 52: 101682. https://doi.org/10.1016/j.jcou.2021.101682
122. Saad A, Snoussi Y, Abderrabba M, Chehimi MM (2016) RSC Adv 6:57672–57682. https://doi.org/10.1039/C6RA12061J
123. Pathak A, Singh AP (2017) J Porous Mater 24:327–340. https://doi.org/10.1007/s10934-016-0266-0
124. Han YR, Kim JS, Park W-J, Lee C-H, Cheon J, Jun C-H (2021) Chem Asian J 16:197–201. https://doi.org/10.1002/asia.202001152
125. Hajipour AR, Abolfathi P, Mohammadsaleh F (2016) RSC Adv 6:78080–78089. https://doi.org/10.1039/C6RA11734A
126. Hajipour AR, Mohammadsaleh F (2018) Catal Lett 148:1035–1046. https://doi.org/10.1007/s10562-018-2316-5
127. Beurer A-K, Kirchhof M, Bruckner JR, Frey W, Baro A, Dyballa M, Giesselmann F, Laschat S, Traa Y (2021) ChemCatChem 13:2407–2419. https://doi.org/10.1002/cctc.202100229
128. Saad A, Vard C, Abderrabba M, Chehimi MM (2017) Langmuir 33:7137–7146. https://doi.org/10.1021/acs.langmuir.7b01247
129. Deimling M, Kousik SR, Abitaev K, Frey W, Sottmann T, Koynov K, Laschat S, Atanasova P (2021) ChemCatChem 13:2242–2252. https://doi.org/10.1002/cctc.202001997
130. Massaro M, Riela S, Cavallaro G, Colletti CG, Milioto S, Noto R, Parisi F, Lazzara G (2015) J Mol Catal A-Chem 408:12–19. https://doi.org/10.1016/j.molcata.2015.07.008
131. Kallick JK, Feng W-J, McCrory CCL (2020) ACS Appl Energy Mater 3:6222–6231. https://doi.org/10.1021/acsaem.0c00332
132. Masteri-Farahani M, Modarres M (2017) Mater Chem Phys 199:522–527. https://doi.org/10.1016/j.matchemphys.2017.07.028
133. Ragoussi M-E, Katsukis G, Roth A, Malig J, de la Torre G, Guldi DM, Torres T (2014) J Am Chem Soc 136:4593–4598. https://doi.org/10.1021/ja411830x
134. Wang R-C, Yin T-L, Wei P-J, Liu J-G (2015) RSC Adv 5:66487–66493. https://doi.org/10.1039/C5RA12972A
135. Xi Y-T, Wei P-J, Wang R-C, Liu J-G (2015) Chem Commun 51:7455–7458. https://doi.org/10.1039/C5CC00963D
136. Yang H, Kwon Y, Kwon Y, Kwoon T, Lee H, Kim BJ (2012) Small 8:3161–3168. https://doi.org/10.1002/smll.201201002
137. Masteri-Farahani M, Modarres M (2017) Colloids Surf A Physicochem Eng Asp 529:886–892. https://doi.org/10.1016/j.colsurfa.2017.06.073
138. Niakan M, Masteri-Farahani M, Shekaari H, Karimi S (2021) Carbohydr Polym 251:117109. https://doi.org/10.1016/j.carbpol.2020.117109
139. Nia AS, Rana S, Döhler D, Jirsa F, Meister A, Guadagno L, Koslowski E, Bron M, Binder WH (2015) Chem Eur J 21:10763–10770. https://doi.org/10.1002/chem.201501217
140. Mondal I, Samanta D, Shaik MAS, Basu R, Bhattacharya A, Shaw M, Pathak A (2023) ChemSelect 8:e202303707. https://doi.org/10.1002/slct.202303707
141. Hajipour AR, Abolfathi P (2016) RSC Adv 6:110622–110628. https://doi.org/10.1039/C6RA23004K
142. Zhang L, Vilà N, Kohring G-W, Walcarius A, Etienne M (2017) ACS Catal 7:4386–4394. https://doi.org/10.1021/acscatal.7b00128
143. Lesniewski A, Matyjewicz J, Palys B, Niedziolka-Jonsson J (2015) Electrochem Commun 53: 20–23. https://doi.org/10.1016/j.elecom.2015.01.019
144. Cook TC, Stenehjem ED, Ainsworth J, Stack TDP (2022) J Inorg Biochem 231:111784. https://doi.org/10.1016/j.jinorgbio.2022.111784

145. Chen Y, Chen H, Tian H (2015) Chem Commun 51:11508–11511. https://doi.org/10.1039/C5CC03856A
146. Zhang L, Vilà N, Walcarius A, Etienne M (2018) ChemElectroChem 5:2208–2217. https://doi.org/10.1002/celc.201800258
147. Yao SA, Ruther RE, Zhang L, Franking RA, Hamers RJ, Berry JF (2012) J Am Chem Soc 134:15632–15635. https://doi.org/10.1021/ja304783j
148. Ghosh D, Febriansyah B, Gupta D, Ng LK-S, Xi S, Du Y, Baikie T, Dong Z, Soo HS (2018) ACS Nano 12:5903–5912. https://doi.org/10.1021/acsnano.8b02118
149. Elumalai P, Mamlouk H, Yiming W, Feng L, Yuan S, Zhou H-C, Madrahimov ST (2018) ACS Appl Mater Interfaces 10:41431–41438. https://doi.org/10.1021/acsami.8b16136
150. Zambrano G, Chino M, Renzi E, Di Girolamo R, Maglio O, Pavone V, Lombardi A, Nastri F (2020) Biotechnol Appl Biochem 67:549–562. https://doi.org/10.1002/bab.1969
151. Oerlemans RAJF, Shao J, Huisman SGAM, Li Y, Abdelmohsen LKEA, van Hest JCM (2023) Macromol Rapid Commun 44:2200904. https://doi.org/10.1002/marc.202200904
152. Lv G, Mai W, Jin R, Gao L (2008) Synlett 9:1418–1422. https://doi.org/10.1055/s-2008-1072597
153. Schätz A, Hager M, Reiser O (2009) Adv Funct Mater 19:2109–2115. https://doi.org/10.1002/adfm.200801861
154. Schätz A, Grass RN, Kainz Q, Stark WJ, Reiser O (2010) Chem Mater 22:305–310. https://doi.org/10.1021/cm9019099
155. Lee J-M, Kim J, Shin Y, Yeom C-E, Lee JE, Hyeon T, Kim BM (2010) Tetrahedron Asymmetry 21:285–291. https://doi.org/10.1016/j.tetasy.2010.01.024
156. Bagherzadeh M, Hosseini M, Mortazavi-Manesh A (2019) Inorg Chem Commun 107:107495. https://doi.org/10.1016/j.inoche.2019.107495
157. Moghaddam FM, Ayati SE (2015) RSC Adv 5:3894–3902. https://doi.org/10.1039/C4RA13330G
158. Hajipour AR, Abolfathi P (2018) Catal Commun 103:92–95. https://doi.org/10.1016/j.catcom.2017.09.021
159. Hollingsworth JV, Bhupathiraju NVSDK, Sun J, Lochner E, Vicente MGH, Russo PS (2016) ACS Appl Mater Interfaces 8:792–801. https://doi.org/10.1021/acsami.5b10034
160. García JI, Herrerías CI, López-Sánchez B, Mayoral JA, Reiser O (2011) Adv Synth Catal 353:2691–2700. https://doi.org/10.1002/adsc.201100213
161. Jiang L, Wang Z, Bai S-Q, Hor ATS (2013) Dalton Trans 42:9437–9443. https://doi.org/10.1039/C3DT50987G
162. Li H, Whittenberg JJ, Zhou H, Ranganathan D, Desai AV, Koziol J, Zeng D, Kenisb PJA, Reichert DE (2014) RSC Adv 5:6142–6615. https://scholar.umw.edu/student_research/622/ (a student's thesis)
163. Manuel Kirchhof, Sanja Krause, Liam Rautenbach, Jakoba Wacker, Ann-Katrin Beurer, Hoang-Huy Nguyen, Wolfgang Frey, Yvonne Traa, Deven P. Estes, Anna Zens, Sabine Laschat, Advanced Synthesis & Catalysis 2025, 0, e70213. https://doi.org/10.1002/adsc.70213
164. Bertolino MC, Granados AM (2016) Appl Surf Sci 383:375–381. https://doi.org/10.1016/j.apsusc.2016.03.214
165. Gan N, Liu K, Qi L, Lat P, Yu H-Z, Sen D (2020) J Phys Chem C 124:19083–19090. https://doi.org/10.1021/acs.jpcc.0c05150
166. Ripert M, Farre C, Chaix C (2013) Electrochim Acta 91:82–89. https://doi.org/10.1016/j.electacta.2012.12.108
167. Abdellah I, Martini C, Dos Santos A, Schulz E, Huc V (2018) ChemCatChem 10:4761–4767. https://doi.org/10.1002/cctc.201801164
168. De Risi C, Martín-Matute B, Pérez-Temprano MC (2020) React Chem Eng 5:117–152. https://doi.org/10.1039/D0RE00076K
169. Ooi T, Crudden CM (2021) ACS Catal 11:15234. https://doi.org/10.1021/acscatal.1c05384

170. Rodríguez-Escrich C, Pericàs MA (2019) Chem Rec 19:1872–1890. https://doi.org/10.1002/tcr.201800097171
171. Rodríguez-Escrich C, Pericàs MA (2015) Eur J Org Chem 2015:1173–1185. https://doi.org/10.1002/ejoc.201403042
172. Amenós L, Alza E, Pericàs MA (2021) Organocatalysis in continuous flow for drug discovery, in flow chemistry in drug discovery. In: Alcazar J, de la Hoz A, Díaz-Ortiz A (eds) Topics in medicinal chemistry. Springer, Cham, pp 241–274. https://doi.org/10.1007/7355_2021_119
173. Barbaro P, Liguori F (2010) Heterogenized homogeneous catalysts for fine chemicals production: materials and processes.1st edn. Springer, Dordrecht. https://doi.org/10.1007/978-90-481-3696-4
174. Gheorghe A, Matsuno A, Reiser O (2006) Adv Synth Catal 348:1016–1021. https://doi.org/10.1002/adsc.200606043
175. Font D, Jimeno C, Pericàs MA (2006) Org Lett 8:4653–4655. https://doi.org/10.1021/ol061964j
176. Font D (2008) Org Lett 10:337–340. https://doi.org/10.1021/ol702901z
177. Alza E, Rodríguez-Escrich C, Sayalero S, Bastero A, Pericàs MA (2009) Chem Eur J 15:10167–10172. https://doi.org/10.1002/chem.200901310
178. Cambeiro XC, Martín-Rapún R, Miranda PO, Sayalero S, Alza E, Llanes P, Pericàs MA (2011) Beilstein J Org Chem 7:1486–1493. https://doi.org/10.3762/bjoc.7.172
179. Ayats C, Henseler AH, Pericàs MA (2012) ChemSusChem 5:320–325. https://doi.org/10.1002/cssc.201100570
180. Rodríguez-Escrich C, Pericàs MA (2023) Immobilized organocatalysts for enantioselective continuous flow processes. In: Albrecht Ł, Albrecht A, Dell'Amico L (eds) Asymmetric organocatalysis. WILEY-VCH GmbH, pp 349–391. https://doi.org/10.1002/9783527832217.ch11
181. Alza E, Sayalero S, Cambeiro XC, Martín-Rapún R, Miranda PO, Pericàs MA (2011) Synlett 2011:464–468. https://doi.org/10.1055/s-0030-1259528
182. Fan X, Sayalero S, Pericàs MA (2012) Adv Synth Catal 354:2971–2976. https://doi.org/10.1002/adsc.201200887
183. Izquierdo J, Ayats C, Henseler AH, Pericàs MA (2015) Org Biomol Chem 13:4204–4209. https://doi.org/10.1039/C5OB00325C
184. Kasaplar P, Rodríguez-Escrich C, Pericàs MA (2013) Org Lett 15:3498–3501. https://doi.org/10.1021/ol400974z
185. Martín-Rapún R, Sayalero S, Pericàs MA (2013) Green Chem 15:3295–3301. https://doi.org/10.1039/C3GC41444B
186. Porta R, Benaglia M, Chiroli V, Coccia F, Puglisi A (2014) Isr J Chem 54:381–394. https://doi.org/10.1002/ijch.201300106
187. Porta R, Benaglia M, Puglisi A, Mandoli A, Gualandi A, Cozzi PG (2014) ChemSusChem 7:3534–3540. https://doi.org/10.1002/cssc.201402610
188. Chiroli V, Benaglia M, Puglisi A, Mandoli A, Jumde RP, Mandoli A (2014) Green Chem 16:2798–2806. https://doi.org/10.1039/C4GC00031E
189. Ayats C, Henseler AH, Dibello E, Pericàs MA (2014) ACS Catal 4:3027–3033. https://doi.org/10.1021/cs5006037
190. Llanes P, Rodríguez-Escrich C, Sayalero S, Pericàs MA (2016) Org Lett 18:6292–6295. https://doi.org/10.1021/acs.orglett.6b03156
191. Du J, Shuai B, Tao M, Wang G, Zhang W (2016) Green Chem 18:2625–2631. https://doi.org/10.1039/C5GC02622A
192. Izquierdo J, Pericàs MA (2016) ACS Catal 6:348–356. https://doi.org/10.1021/acscatal.5b02121
193. Wang S, Izquierdo J, Rodríguez-Escrich C, Pericàs MA (2017) ACS Catal 7:2780–2785. https://doi.org/10.1021/acscatal.7b00360
194. Porta R, Benaglia M, Puglisi A (2017) Adv Synth Catal 359:2106–2112. https://doi.org/10.1002/adsc.201700376

195. Lai J, Sayalero S, Ferrali A, Osorio-Planes L, Bravo F, Rodríguez-Escrich C, Pericàs MA (2018) Adv Synth Catal 360:2914–2924. https://doi.org/10.1002/adsc.201800572
196. Ötvös SB, Pericàs MA, Kappe CO (2019) Chem Sci 10:11141–11146. https://doi.org/10.1039/C9SC04752B
197. Ötvös SB, Llanes P, Pericàs MA, Kappe CO (2020) Org Lett 22:8122–8126. https://doi.org/10.1021/acs.orglett.0c03100
198. Na H, Choi GH, Eom T, Bang J, Yoo PJ (2020) Sep Purif Technol 240:116621. https://doi.org/10.1016/j.seppur.2020.116621
199. Schulze JS, Wesselbaum S, Timoshkin AV, Baumann W, Smarsly B (2022) ChemCatChem 14:e202101845. https://doi.org/10.1002/cctc.202101845
200. Brand RD, Busche SA, Börner HG, Smarsly BM (2023) ChemCatChem 15:e202300778. https://doi.org/10.1002/cctc.202300778
201. Lee W-S, Li L, Kim BM (2021) Catalysts 11:1044. https://doi.org/10.3390/catal11091044
202. Neyyappadath RM, Chisholm R, Greenhalgh MD, Rodríguez-Escrich C, Pericàs MA, Hähner G, Smith AD (2018) ACS Catal 8:1067–1075. https://doi.org/10.1021/acscatal.7b04001
203. Guha NR, Neyyappadath RM, Greenhalgh MD, Chisholm R, Smith SM, McEvoy ML, Rodríguez-Escrich C, Young CM, Pericàs MA, Hähner G, Smith AD (2018) Green Chem 20:4537–4546. https://doi.org/10.1039/C8GC02020E
204. Lai J, Neyyappadath RM, Smith AD, Pericàs MA (2020) Adv Synth Catal 362:1370–1377. https://doi.org/10.1002/adsc.201901420
205. Punna S, Kaltgrad E, Finn MG (2005) Bioconjug Chem 16:1536–1541. https://doi.org/10.1021/bc0501496
206. Slater M, Snauko M, Svec F, Fréchet JMJ (2006) Anal Chem 78:4969–4975. https://doi.org/10.1021/ac060006s
207. Kacprzak KM, Maier N, Lindner W (2006) Tetrahedron Lett 47:8721–8726. https://doi.org/10.1016/j.tetlet.2006.10.018
208. Hall M (2021) RSC Chem Biol 2:958–989. https://doi.org/10.1039/D1CB00080B
209. Sheldon RA, Woodley JM (2018) Chem Rev 118:801–838. https://doi.org/10.1021/acs.chemrev.7b00203
210. Choi JM, Han SS, Kim HS (2015) Biotechnol Adv 33:1443–1454. https://doi.org/10.1016/j.biotechadv.2015.02.014
211. Rodrigues RC, Ortiz C, Berenguer-Murcia Á, Torres R, Fernández-Lafuente R (2013) Chem Soc Rev 42:6290–6307. https://doi.org/10.1039/C2CS35231A
212. Bolívar JM, Woodley JM, Fernandez-Lafuente R (2022) Chem Soc Rev 51:6251–6290. https://doi.org/10.1039/D2CS00083K
213. Mirsalami SM, Mirsalami M, Ghodousian A (2024) Results Chem 7:101486. https://doi.org/10.1016/j.rechem.2024.101486
214. Mogharabi-Manzari M, Heydari M, Sadeghian-Abadi S, Faramarzi MA (2019) Biocatal Biotransform 37:455–465. https://doi.org/10.1080/10242422.2019.1611788
215. Oktay B, Demir S, Kayaman-Apohan N (2020) Food Bioprod Process 122:159–168. https://doi.org/10.1016/j.fbp.2020.04.010
216. Bourgery C, Mendoza DJ, Garnier G, Mouterde LMM, Allais F (2024) ACS Appl Mater Interfaces 16:11315–11323. https://doi.org/10.1021/acsami.3c19025
217. Brennan JL, Hatzakis G, Dordick JS (2006) Bioconjug Chem 17:1379–1385. https://doi.org/10.1021/bc0601018
218. Çelebi B, Bayraktar A, Tuncel A (2012) Anal Bioanal Chem 403:2655–2663. https://doi.org/10.1007/s00216-012-6075-7
219. Matsumoto T, Isogawa Y, Tanaka T, Kondo A (2018) Biosens Bioelectron 99:56–61. https://doi.org/10.1016/j.bios.2017.07.044
220. Ran Q, Peng R, Liang C, Ye S, Xian Y, Zhang W, Jin L (2011) Talanta 83:1381–1385. https://doi.org/10.1016/j.talanta.2010.11.024

221. Ran Q, Peng R, Liang C, Ye S, Xian Y, Zhang W, Jin L (2011) Anal Chim Acta 697:27–31. https://doi.org/10.1016/j.aca.2011.04.035
222. Hayat A, Marty JL, Radi AE (2012) Electroanalysis 24:1469–1475. https://doi.org/10.1002/elan.201200053
223. Hayat A, Sassolas A, Rhouati A, Marty JL (2013) Immobilization of enzymes on Ethynyl-modified electrodes via click chemistry. In: Guisan JM (ed) Immobilization of enzymes and cells. Methods in molecular biology, vol 1051. Humana Press, Totowa, NJ, pp 209–216. https://doi.org/10.1007/978-1-62703-550-7_13
224. Guan D, Kurra Y, Liu W, Chen Z (2015) Chem Commun 51:2522–2525. https://doi.org/10.1039/C4CC09179E
225. Buhl M, Vonhören B, Ravoo BJ (2015) Bioconjug Chem 26:1017–1023. https://doi.org/10.1021/acs.bioconjchem.5b00282
226. Oktay B, Demir S, Kayaman-Apohan N (2019) ChemSelect 4:6055–6059. https://doi.org/10.1002/slct.201900296
227. Chaparro Sosa AF, Bednar RM, Mehl RA, Schwartz DK, Kaar JL (2021) J Am Chem Soc 143:7154–7163. https://doi.org/10.1021/jacs.1c02375
228. Makrydaki E, Donini R, Krueger A, Royle K, Moya Ramirez I et al (2024) Nat Chem Biol 20:732–741. https://doi.org/10.1038/s41589-023-01539-4
229. Yurtsever A, Saraçoğlu B, Tuncel A (2009) Electrophoresis 30:589–598. https://doi.org/10.1002/elps.200800417
230. Roy S, Bauer T, Al-Dahhan MH, Lehner P, Turek T (2004) AIChE J 50:2918–2938. https://doi.org/10.1002/aic.10268
231. Lallana E, Riguera R, Fernández-Megia E (2011) Angew Chem Int Ed 50:8794–8804. https://doi.org/10.1002/anie.201101019
232. Lutz J-F (2008) Angew Chem Int Ed 47:2182–2184. https://doi.org/10.1002/anie.200705365
233. Becer CR, Hoogenboom R, Schubert US (2009) Angew Chem Int Ed 48:4900–4908. https://doi.org/10.1002/anie.200900755
234. Upadhyay R, Rastogi S, Mishra AK, Yadav S, Yadav AK, Maurya SK (2025) Asian J Org Chem 14:e00505. https://doi.org/10.1002/ajoc.202500505
235. Knall A-C, Slugovc C (2013) Chem Soc Rev 42:5131–5142. https://doi.org/10.1039/C3CS60049A
236. Tomarchio EG, Turnaturi R, Saccullo E, Patamia V, Floresta G, Zagni C, Rescifina A (2024) Bioorg Chem 150:107573. https://doi.org/10.1016/j.bioorg.2024.107573
237. Fairbanks BD, Macdougall LJ, Mavila S, Sinha J, Kirkpatrick BE, Anseth KS, Bowman CN (2021) Chem Rev 121:6915–6990. https://doi.org/10.1021/acs.chemrev.0c01212
238. Hoyle CE, Bowman CN (2010) Angew Chem Int Ed 49:1540–1573. https://doi.org/10.1002/anie.200903924
239. Lowe AB, Hoyle CE, Bowman CN (2010) J Mater Chem 20:4745–4750. https://doi.org/10.1039/B917102A
240. Dong J, Krasnova L, Finn MG, Sharpless KB (2014) Angew Chem Int Ed 53:9430–9448. https://doi.org/10.1002/anie.201309399
241. McGuire K, He S, Gracie J, Bryson C, Zheng D, Clark AW, Koehnke J, France DJ, Nau WM, Lee T-C, Peveler WJ (2023) ACS Nano 17:21585–21594. https://doi.org/10.1021/acsnano.3c0660
242. MacKenzie DA (2014) Curr Opin Chem Biol 21:81–88. https://doi.org/10.1016/j.cbpa.2014.05.023
243. Wallace S, Chin JW (2014) Chem Sci 5:1742–1744. https://doi.org/10.1039/C3SC53332H
244. Krell K, Pfeuffer B, Rönicke F, Chinoy ZS, Favre C, Friscourt F, Wagenknecht H-A (2021) Chem Eur J 27:16093–16097. https://doi.org/10.1002/chem.202103026
245. Rashidian M, Mahmoodi M, Dozier JK, Wagner CR, Distefano MD (2013) Bioconjug Chem 24:534–542. https://doi.org/10.1021/bc3004167
246. Blasco E, Sugawara Y, Lederhose P, Blinco JP, Kelterer A-M, Barner-Kowollik C (2017) ChemPhotoChem 1:159–163. https://doi.org/10.1002/cptc.201600042

247. Chowdhury A, Chatterjee S, Kushwaha A, Nanda S, Dhilip Kumar TJ, Bandyopadhyay A (2023) Chem Eur J 29:e202300393. https://doi.org/10.1002/chem.202300393
248. Fulton DA (2016) Nat Chem 8:899–900. https://doi.org/10.1038/nchem.2626
249. Liu B, Ianosi-Irimie M, Thayumanavan S (2019) ACS Nano 13:9408–9420. https://doi.org/10.1021/acsnano.9b04198
250. Alge DL, Azagarsamy MA, Donohue DF, Anseth KS (2013) Biomacromolecules 14:949–953. https://doi.org/10.1021/bm4000508
251. Mohan MK, Silenko O, Krasnou I, Volobujeva O, Kulp M, Ošeka M, Lukk T, Karpichev Y (2024) ChemSusChem 17:e202301588. https://doi.org/10.1002/cssc.202301588
252. Ölçücü G, Krauss U, Jaeger K-E (2023) Chem Ing Tech 95:531–542. https://doi.org/10.1002/cite.202200167
253. Fernandes AE, Riant O, Jonas AM, (2026) Recent advances in the application of CuAAC for the covalent immobilization of homogeneous catalysts, a decade of progress, Tetrahedron, 192, 135108, https://doi.org/10.1016/j.tet.2025.135108

GPSR Compliance

The European Union's (EU) General Product Safety Regulation (GPSR) is a set of rules that requires consumer products to be safe and our obligations to ensure this.

If you have any concerns about our products, you can contact us on ProductSafety@springernature.com

In case Publisher is established outside the EU, the EU authorized representative is:

Springer Nature Customer Service Center GmbH
Europaplatz 3
69115 Heidelberg, Germany

Batch number: 10918163

Printed by Printforce, the Netherlands